# Design of Reinforced Concrete

# Design of Reinforced Concrete

**Samuel E. French**
*University of Tennessee*

*Prentice Hall, Englewood Cliffs, New Jersey 07632*

*Library of Congress Cataloging-in-Publication Data*

French, Samuel E., (date)
Design of reinforced concrete.
Bibliography: p.
Includes index.
1. Reinforced concrete construction. I. Title.
TA683.F826 1988 624.1'8341 87-2485
ISBN 0-13-201203-0

Editorial/production supervision and
interior design: Carolyn Fellows and Elaine Lynch
Cover design: Ben Santora
Manufacturing buyer: John Hall

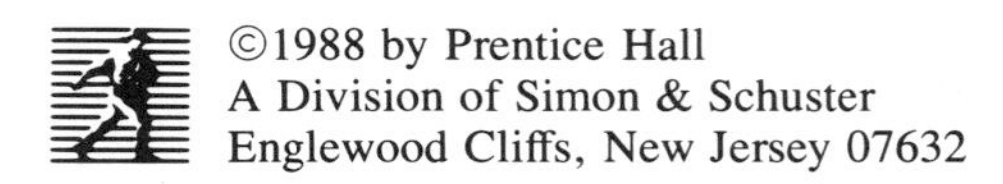


A Division of Simon & Schuster
Englewood Cliffs, New Jersey 07632



Printed in the United States of America

10 9 8 7 6 5 4 3 2 1

ISBN 0-13-201203-0 025

Prentice-Hall International (UK) Limited, *London*
Prentice-Hall of Australia Pty. Limited, *Sydney*
Prentice-Hall Canada Inc., *Toronto*
Prentice-Hall Hispanoamericana, S. A., *Mexico*
Prentice-Hall of India Private Limited, *New Delhi*
Prentice-Hall of Japan, Inc., *Tokyo*
Prentice-Hall of Southeast Asia Pte. Ltd., *Singapore*
Editora Prentice-Hall do Brasil, Ltda., *Rio de Janeiro*

# Contents

PREFACE ix

1 PROPERTIES OF CONCRETE 1

General, *1*
Coarse Aggregate, *1*
Fine Aggregate, *2*
Portland Cement, *2*
Water, *3*
Strength, *3*
Density and Weight, *6*
Mixing, Placing, and Curing, *6*
Water/Cement Ratio, *6*
Elasticity, *7*
Thermal Coefficient, *9*
Shrinkage and Creep, *10*
Related Properties, *11*
Additives, *11*
Reinforcement, *12*
Bond Strength, *14*

2 LIMITATIONS, DESIGN METHODS, AND CODES 16

General, *16*
Inelastic Deformations, *16*
Strength Method, *17*
Working Stress Method, *17*
Comparison of Methods, *18*
Deflection Analysis, *18*
References to ACI 318-83, *19*

3 DESIGN AND DETAILING PRACTICES 20

General, *20*
Accuracy of Computations, *21*
Nominal Loads, *22*
Minimum Dimensions, *22*
Minimum and Maximum Reinforcement, *22*
Minimum Cover, *24*
Limits on Deflections, *25*
Control of Deflections, *26*
Rules of Thumb, *27*
Temperature and Shrinkage Joints, *29*
Construction Joints, *30*
Differential Settlements, *31*
Strength Reduction Factors, *32*
Load Factors, *32*
Strength of Steel, *33*
Strength of Concrete, *36*

4 FLEXURE IN CONCRETE BEAMS 38

General, *38*
Flexure Formula, *38*
Elastic Flexure, *39*
Elastic Flexure plus Axial Force, *46*
Investigation of Elastic Flexure, *53*
Ultimate Flexure, *62*
Ultimate Flexure plus Axial Force, *69*
Investigation of Ultimate Flexure, *79*
Comparison of Strength Method to Working Stress Method, *83*
Floor Slabs and Roof Slabs, *85*

## 5 SHEAR IN BEAMS 91

General, *91*
Shear as a Measure of Diagonal Tension, *92*
Types of Effective Shear Reinforcement, *94*
Code Requirements for Strength Method, *97*
Examples in the Strength Method, *100*
Code Requirements for Working Stress Method, *104*

## 6 TEE BEAMS AND JOISTS 108

General, *108*
Tee Beams Subject to Negative Moment, *109*
Code Requirements for Tee Shapes, *112*
Tee Beams Subject to Positive Moment, *114*
Neutral Axis below the Flange, *120*
Approximate Analysis of Tee Shapes, *121*
One-Way Joist Construction, *123*

## 7 BOND AND ANCHORAGE 133

General, *133*
Cutoff Points, *134*
Development Lengths of Straight Bars, *136*
Splice Lengths of Straight Bars, *137*
Development Lengths with Standard Hooks, *138*
Abbreviated Criteria for Cutoff Points, *140*
Development of Stirrups and Ties, *144*
Special Provisions and Limitations, *146*
Elastic Analysis, *147*

## 8 REINFORCED CONCRETE COLUMNS 153

General, *153*
Configurations and Practices, *154*
Behavior under Load, *156*
ACI Column Formula, *158*
Buckling Criteria, *159*
Ultimate Strength Analysis, *160*
Examples in the Strength Method, *166*
Elastic Analysis, *174*

Examples in the Elastic Range, *176*
Discussion of the Strength Analysis, *179*
Discussion of the Elastic Analysis, *180*

9 CONTINUITY 183

General, *183*
Shear and Moment Envelopes, *184*
Exact Maximum Values of Shear and Moment, *189*
Summary of Design Values for Shear and Moment in Beams, *192*
Loads and Moments on Columns, *193*
Summary of Design Loads on Columns, *194*
Examples, *195*

10 FOUNDATIONS AND SLABS ON GRADE 218

General, *218*
Types of Shallow Foundations, *218*
Foundation Loading and Failure Modes, *220*
Allowable Soil Pressures, *220*
Footing Rotations, *221*
Spread Footings, *222*
Strip Footings, *235*
Grade Beams, *239*
Other Types of Footings, *243*
Slabs on Grade, *245*

SELECTED REFERENCES 250

APPENDIX 252

INDEX 265

# Preface

This book is about selecting the size and reinforcement for structural members of reinforced concrete. The larger problem of choosing an overall structural system is left to more advanced study. The same word, "design," is used to describe both activities.

The very rudimentary approach used in this book is directed at those architects, technicians, and construction pushers who are not frequently called upon to design concrete members and who, between times, are inclined to forget the specialized approach to concrete that is used by engineers. While using conventional concepts, the simplified practices herein use the ultimate strength method for design of concrete members, the same as that used in the more comprehensive approach.

The ultimate strength method does not require a check of stresses at service levels, but there are still those who believe that the day-to-day service stresses in a building are worth noting. The elastic analysis of concrete is therefore developed concurrently with the ultimate strength analysis in this text. Results from the two methods are periodically compared and contrasted; each serves to reinforce the other.

The treatment in this introductory text is necessarily limited in scope. The following general limitations apply:

1. Only the most common strengths of concrete and common types of steel are considered; design tables are limited to these materials.
2. Only regular-weight concrete is considered; design tables do not extend to lightweight concretes.
3. Only vertical stirrups for shear reinforcement are considered; bent bars and inclined stirrups are not included.

4. Only rectangular tied columns are considered; design tables for circular columns and spiral reinforcement are not included.
5. Moment magnification and critical loading are mentioned only briefly.
6. Only those structural systems are considered where lateral loads are taken by bracing or shearwalls; moment-resistant frames are not considered.
7. A detailed analysis of deflections is not presented; only the ''blanket'' method for control of deflections is used.

Although condensed and simplified, the approach used in this textbook is not an approximate method. It is an accurate theoretical solution to concrete design, applicable to the overwhelming majority of smaller structures. The textbook would of course be entirely suitable for use by practicing engineers who are engaged in such designs, although it is judged to be too limited in scope to be used as a general reference.

The simplified approach used in this book has been found to be useful in the preliminary design of buildings up to four or five stories, where lateral loads due to wind or earthquake are not of major concern. The approach has also been found to be useful in plan review where a quick independent check on a beam design or a column size is frequently necessary. One of its greater values, however, has been in allowing a quick check to be made on stresses and deformations at service levels, both in plan review and in active design.

Although brief, this book is also intended for use in a basic course of design in concrete for architecture students or technology students. It is presumed that the students have taken the usual preparatory courses in statics and strength of materials and are familiar with concepts of stress and strain. The flexure formula and the shear equation are the starting points for the example problems; it is assumed that the students have had some drill in their use.

The initial edition of the text was prepared in SI units. For the United States and Canada, this edition of the text has been prepared in Imperial units. There are occasional references to SI units in this edition where significant features or comparisons between the two systems merit special attention.

It is intended that this book be used without additional references, although every concrete designer should have his own marked-up copy of ACI 318-83, *Building Code Requirements for Reinforced Concrete*. Design tables for beams and columns are included in the back of the book. Although the text uses only Imperial units, the design tables for beams and columns are dimensionless and are equally applicable either in SI or in Imperial units. If a project is encountered where more advanced tables or methods are necessary, the project is beyond the scope of this book and a more comprehensive approach should be used.

The approach used in this textbook was developed by the author over a period of twenty years in design practice. The actual preparation of the text, however, was undertaken to fill the off-duty hours during the five years the author was employed as a structural engineer on the Jubail Project in Saudi Arabia. The author is indebted to his many friends and colleagues on the project who shared freely from their extensive and authoritative knowledge of field practices. In particular, the author expresses his thanks to Kumar Varma, Manah Bannout and Safa'a Bayati for their encouragement and advice during the many rewritings of the text and reformatting of the design tables.

The author owes particular thanks to his wife Sherry, who typed the manuscript. And who, with quiet patience, retyped the manuscript. And retyped the manuscript. And retyped . . . .

# Design of Reinforced Concrete

# 1

# Properties of Concrete

## GENERAL

To be a good structural material, a material should be homogeneous, isotropic, and elastic. Portland cement concrete is none of these. Nonetheless, it is very popular.

Portland cement concrete is manufactured using a coarse aggregate of stone, a fine aggregate of sand, and a cementing paste of portland cement and water. The end result is a man-made stone that can be shaped while in its plastic state and allowed to harden into its final configuration. Compared to other construction materials, such as steel, aluminum, or timber, portland cement concrete is weak and heavy, but its properties of durability, adaptability , and availability have made it a popular material of construction.

Portland cement concrete is quite weak in tension. In most structural applications, its tensile strength is assumed to be zero and steel is provided at all points where the structure will experience tension. The result is concrete reinforced for tension, or more simply, reinforced concrete.

## COARSE AGGREGATE

The coarse aggregate used in making concrete may be crushed stone or a natural gravel. The size of an individual particle may range from the size of a pea to that of a boulder, depending on whether the concrete is being cast into a thin decorative panel or a large dam. Since the remaining constituents are proportioned to fill the void spaces between the particles of coarse aggregate, the coarse aggregate will obviously form the bulk of the volume of the concrete. The quality of the finished concrete is heavily dependent on the quality of the coarse aggregate. A hard, durable concrete can only be made from a hard, durable coarse aggregate.

## FINE AGGREGATE

The fine aggregate used in making concrete is usually a sharp silica sand. Calcareous sands (derived from limestone) may be somewhat weaker and, although frequently used, can impart a limit on the ultimate strength of the concrete. The mixture of sand, cement, and water is called the *paste* or the *matrix* of the mix.

## PORTLAND CEMENT

Portland cement is a hydraulic cement, developed in 1824 by an English bricklayer, Joseph Aspdin. The name ''portland'' was applied because the color of the finished concrete resembled a building stone quarried on the Isle of Portland, off the coast of Dorset, England. The cement is manufactured from a mixture of about 4 parts limestone

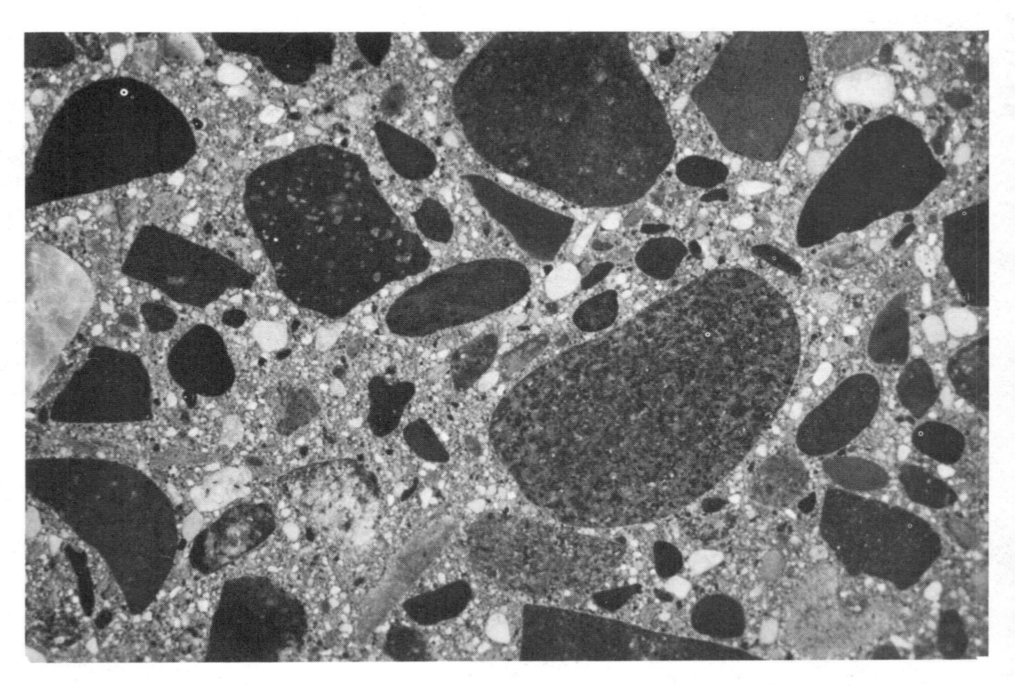

Cut and polished section of concrete. The void spaces between the particles of coarse aggregate are filled by the paste of sand, cement and water (Courtesy of Portland Cement Association.)

and 1 part clay, heated to incipient melting (about 2700°F) and then ground to a fine powder. A small amount of gypsum is added to control set.

Portland cement works by hydration, that is, it forms a chemical bond with the water in the mix and, in doing so, forms a bond to other cement particles, to the aggregate, and to any reinforcement that it contacts. It does not need air to harden or "set"; it will harden as well under water as when exposed to air.

Five standard types of portland cement are manufactured:

*Type I:* regular portland cement
*Type II:* increased resistance to sulfate attack
*Type III:* high early strength
*Type IV:* lowered heat of hydration
*Type V:* high resistance to sulfate attack

By itself, type I portland cement concrete is reasonably resistant to most forms of chemical attack that might occur naturally, but high concentrations of waterborne sulfates can have very deleterious effects on it. Commonly, such concentrations can occur in sewage, in groundwater carrying dissolved gypsum, and in some seawater exposures (from sulfur-producing marine organisms). For castings in a high-sulfate environment, special cements are manufactured, types II and V, which have an increased resistance to sulfate attack. Type III cement achieves its specified strength in 7 days rather than in 28 days. It does so at a penalty in increased heat of hydration as well as in higher cost.

Heat is generated by all five types of cement during hydration. For the more common sizes of structural members this heat is readily dissipated and presents no problems. For extremely large castings such as heavy equipment foundations, the trapped heat can pose serious problems and must be provided for. For such large castings a special cement is manufactured, type IV, which has a lower heat of hydration than other types and helps to alleviate the problem.

## WATER

As a general rule, any water suitable for drinking is suitable for making concrete. It should be remembered, however, that the hydration of portland cement is a very complex chemical reaction and that undesirable results can occur from even small amounts of certain compounds in solution. Nonetheless, if drinking water of standard quality is used as the mixing water, the chances of introducing a deleterious compound through the water is minimal.

## STRENGTH

As with other structural materials, the strength of a concrete and the configuration of its stress–strain curve form the basis of all structural calculations. The strength of a concrete

Concrete cylinder test. The compressive stress in the concrete cylinder at failure load is called the ''ultimate strength'' of the concrete. (Courtesy of Portland Cement Association).

is gauged at 28 days by its compressive stress at failure. To measure the stress at failure, a concrete cylinder 12 in. long and 6 in. in diameter is cast from a sample of the concrete. The cylinder is allowed to cure under controlled conditions for 28 days and is then placed in a compression test machine and loaded to failure. The compressive stress computed for the highest load is the ''ultimate'' stress, or 28-day strength, of that concrete.

A typical strength–time curve is shown in Fig. 1-1. It should be noted that the

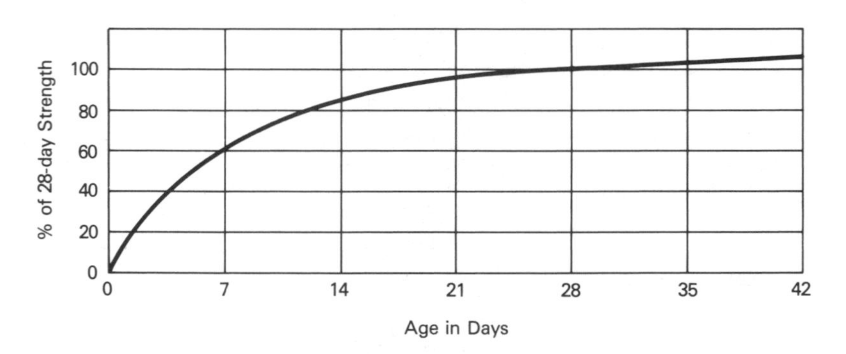


**Figure 1-1** Strength Gain

gain in strength after 28 days is not acknowledged, but that concrete continues to gain strength for months or even years after manufacture. It should also be noted that the concrete attains about 60% of its design strength at about 7 days, a useful bit of information to know when precast elements are to be lifted or forms are to be removed.

There is no immediate test that can be performed at the time the concrete is being cast to assure that the concrete will reach a certain strength 28 days later. In practice, concrete is sampled periodically and cylinders are made as the concrete is being cast into the forms. These cylinders are tested 28 days later, affording a verification that the concrete had been manufactured properly last month. It is thus apparent that the cylinder test is a "report card," revealing how well a job was done 28 days ago.

To assure that the concrete is of proper strength and quality at the time it is cast, rigorous measures of quality control must be instituted and maintained during its production. Over the years, simple but adequate measures have been developed which, with proper enforcement, will assure the production of good-quality concrete. These measures are unique in the industry; no other structural material is sampled for its strength and quality at the time it is being placed in the structure.

All of the desirable properties of the finished concrete, such as durability, hardness, abrasion resistance, and so on, improve as the strength increases. By controlling the strength of the concrete, its other properties may also be controlled. As a consequence, it is common practice to identify a concrete by its ultimate strength, greatly simplifying the matter of identification.

Reinforced concrete building. Small buildings built of reinforced concrete can be readily designed to be functional, durable and economical. (Courtesy of Portland Cement Association).

## DENSITY AND WEIGHT

The weight of concrete varies with the density of its aggregates. An average value for unreinforced concrete is 140 pounds per cubic foot (pcf) and for reinforced concrete is 150 pcf. A median weight of 145 pcf is used throughout this text.

The weight of concrete can be reduced significantly by the use of lightweight aggregates. Concretes that weigh as low as 90 pcf are produced this way and are in common use. In such concretes, the decrease in dead load can often offset the higher materials costs, even though the strength of the lighter concrete is also reduced.

It is almost axiomatic that concrete strength is reduced as density is reduced. Whatever causes a decrease in density, such as entrapped air or lighter aggregates, will almost always produce a reduction in strength.

There are many specialty concretes of all types: gas-formed concrete, gap-graded (popcorn) concrete, nailable concrete, fiberboard concrete, insulating concrete, and others. Specialty concretes are rarely used in structural applications and are not included here. All references to concrete in this book will mean portland cement concrete of standard weight and density.

Reinforced concrete bridge. Low maintenance costs and long life have made reinforced concrete a popular material for small bridges. (Courtesy Portland Cement Association).

## MIXING, PLACING, AND CURING

Volumes have been written on field practices concerning the mixing, placing, and curing of concrete. One of the most authoritative references on such matters is the *Concrete Manual* (U.S. Bureau of Reclamation, 1975). Field practices are outside the scope of this book. It is noted, however, that the tolerances, inaccuracies, and approximation used in standard field practices will have a profound effect on the final dimensions. The cumulative effects of all such deviations is accounted for in the design calculations by the strength reduction factor $\phi$, discussed at more length in Chapter 3.

## WATER/CEMENT RATIO

The strength of portland cement concrete is dependent on the number of pounds of water used per pound of cement. This ratio of water to cement, shown as the water/cement W/C ratio in Fig. 1-2, is the single most important parameter used to control the strength of concrete. As indicated in the figure, there is a minimum W/C ratio required for hydration of all the cement molecules. Any water in the mix in excess of that amount will reduce the final strength of the concrete. It is assumed, of course, that the aggregate is at least as strong as the cement paste.

In introductory comments such as these, it is difficult to overstate the importance of the W/C ratio. An excess of water reduces not only strength but hardness, durability, resistance to chemical attack, resistance to freeze–thaw, and all other desirable properties of the concrete. Thus the construction worker who adds water to the mix to make it more workable is, by this single act, significantly reducing all the desirable characteristics of the finished concrete.

At the higher W/C ratios, the "yield" or volume of usable concrete per pound of cement is higher, affording a more economical mix. Unfortunately, as water content increases, the concrete has a higher tendency to shrink as it hardens. This tendency to

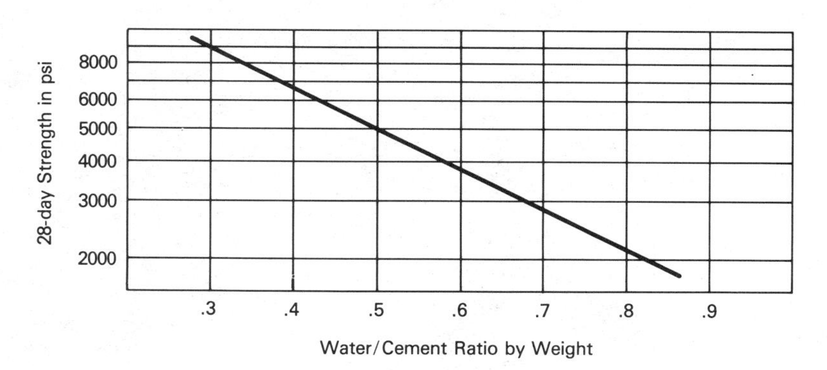


**Figure 1-2** Concrete Strength

shrink, combined with the reduced strength, durability, and hardness, imposes a practical upper limit on the W / C ratio of about 0.80. At the lower W / C ratios, the shrinkage of the concrete can be minimized, but the mix is expensive and workability during placement is reduced. Thus the lower limit of the W / C ratio is limited by practical considerations, usually to about 0.40. Almost all reinforced concrete used in today's practice will have W / C ratios between these general limits of 0.40 to 0.80, although some prestressed concrete units may use lower values. In this book, consideration is limited to concretes having W / C ratios within these limits.

## *ELASTICITY*

The stress–strain curve for concrete can be plotted readily from the load–deflection data of a standard cylinder test. Typical stress–strain curves for various strengths of concrete are shown in Fig. 1-3, where ultimate stress is denoted $f'_c$. It should be noted that there are no distinct "breaks" in these stress–strain curves; they are continuous smooth curves.

An elastic material under load will deform along a straight line up to its yield point. Within this elastic range, it will resume its exact original configuration upon release of the load. From the curves of Fig. 1-3 it is seen that concrete is not a truly elastic material under this definition, although it is reasonably elastic within the lower half of its ultimate strength. It is noted, however, that the normal range of working stresses for concrete is also within the lower half of its ultimate stress. Consequently, with a normal factor of safety and under its day-to-day service loads, concrete can be expected to work at less than half of its ultimate stress and will therefore behave as an elastic material under short-term service loads.

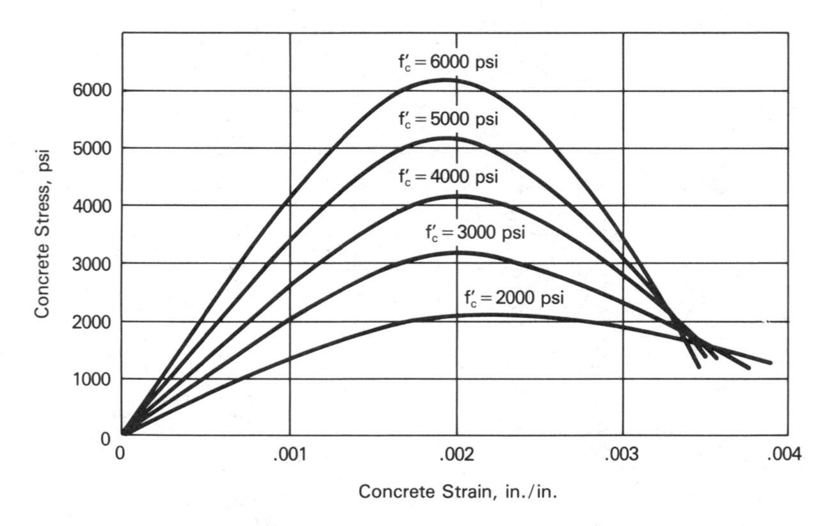


**Figure 1-3** Typical Stress Strain Curves

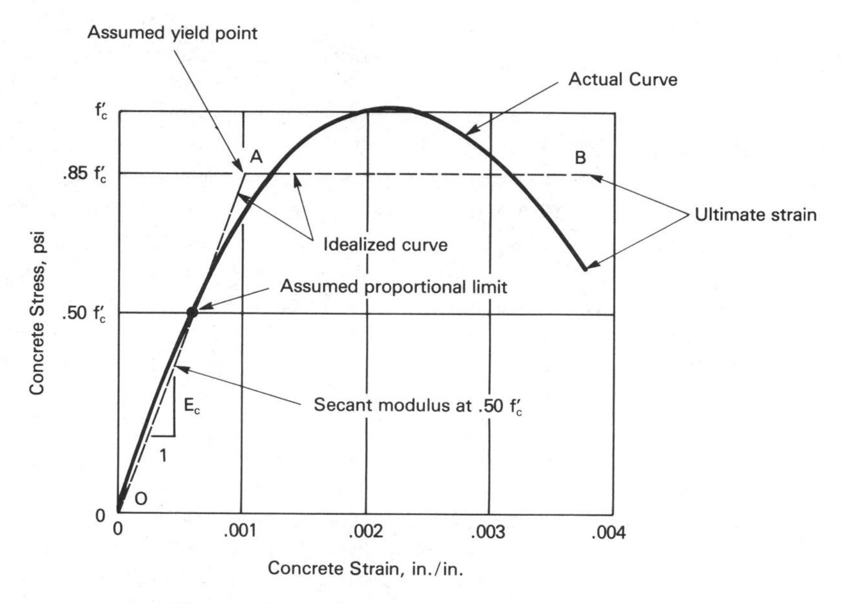


**Figure 1-4** Idealized Stress-strain Curve for Concrete.

An exaggerated stress–strain curve is shown in Fig. 1-4. An approximation of the slope of the initial portion of this curve is usually defined as the modulus of elasticity; its value is the stress divided by the strain at any point. The value of the modulus of elasticity increases as the strength of the concrete increases. The modulus of elasticity $E_c$ in pounds per square inch is computed empirically by the relationship

$$E_c = 57{,}000 \sqrt{f'_c} \tag{1-1}$$

where $f'_c$ is the ultimate strength of the concrete in pounds per square inch.

Since the actual stress–strain curve of Fig. 1-4 has no well-defined yield point, an idealized curve is substituted for the top portion of the actual curve. The resultant idealized curve $0AB$ forms the basis of the derivations and calculations in the following chapters. The position of line $AB$ has been found to provide the best correlation to comprehensive test data when it is taken at $0.85 f'_c$. The distance $AB$ is called the *plastic range* of the material.

## THERMAL COEFFICIENT

The coefficient of thermal expansion and contraction for concrete varies somewhat with the aggregates. An average value is 0.0000055 in./in. per degree Fahrenheit. Since the corresponding coefficient for reinforcing steel is quite close, 0.0000065 in./in. per degree Fahrenheit, the effects of differential thermal growth between the concrete and its reinforcement is negligible at common atmospheric temperature ranges.

## SHRINKAGE AND CREEP

Only a relatively small amount of water is required to hydrate all the cement in the concrete mix. Any excess water that is free to migrate can evaporate as the concrete hardens, causing a slight reduction in the size of the member. This reduction in size, called *shrinkage*, can produce undesirable internal stresses and cracking.

As a very general rule of thumb, the drying shrinkage of a typical structural concrete will produce a dimensional change roughly equal to that of a temperature drop of 40°F. Unlike a temperature drop, however, which affects both the concrete and its reinforcement, shrinkage affects only the concrete. The reinforcement tries to stay at its original dimensions, while the concrete shrinks around it, creating internal stresses in the member.

Concrete also undergoes another type of reduction in size if it is subjected to sustained stresses over a long period. This reduction, called *creep* or *plastic flow*, is a permanent deformation and is only partially recovered when the load is released. Creep varies both with time and with intensity of load, as shown in the typical curves of Fig. 1-5. Roughly, creep strain is directly proportional to stress. For verification, it may be seen in Fig. 1-5 that after the first few months, the strain corresponding to a stress of 600 psi is roughly double that of 300 psi.

It is also evident from Fig. 1-5 that creep will occur even under very low levels of stress where the load is sustained over long periods. However, for both low levels

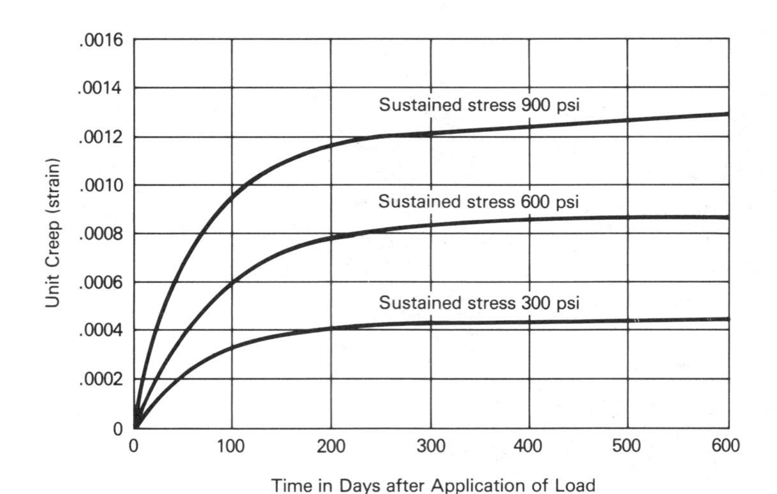


**Figure 1-5** Rate of Creep in Concrete.

and high levels of stress, it can be seen that the rate of creep diminishes sharply after about 3 months and is negligible after about 2 years. For comparison, the ultimate strength of a typical grade of structural concrete is 3000 psi; in Fig. 1-5 the stress of 300 psi is therefore 10% of ultimate and 900 psi is 30% of ultimate.

These two permanent dimensional changes, shrinkage and creep, can cause serious problems in the structure unless they are properly accounted for in the design. A means to account for them is discussed in Chapter 2.

## RELATED PROPERTIES

There are many other properties of concrete that may be highly desirable in the design of the project but which do not affect the load-carrying capabilities of the concrete. Properties such as impermeability, hardness, resistance to abrasion, resistance to freeze–thaw, resistance to chemical attack, and other desired properties could have a pronounced influence on choice of aggregates and choice of ultimate strength. Once the ultimate strength is established, however, the succeeding structural calculations are based only on that ultimate strength and the corresponding deformation properties. Other properties would have only a secondary influence on the structural design.

## ADDITIVES

Additives are chemicals added to the mix while it is plastic to impart a particular property to the concrete. The decision to use an additive should not be taken lightly; the additive cannot be withdrawn should the concrete fail at its 28-day tests.

*Water-reducing agents* (WRA) are one type of additive. They act as a lubricant on the plastic mix, allowing the W/C ratio to be reduced while keeping the workability. As the W/C ratio is reduced, all the desirable properties of concrete improve, including strength. Water-reducing agents are also used to permit a reduction in the amount of cement being used; strength and workability can thus be kept constant while shrinkage is reduced.

*Air-entraining agents* are another type of additive. They impart billions of microscopic bubbles of air, distributed uniformly throughout the mix. The presence of this air (3 to 7% by volume) makes the hardened concrete much more resistant to freeze–thaw and to chemical attack. It also reduces overall density by the same percentage and produces a corresponding slight loss of strength. The improvement in properties is reliable and dependable, but maintaining the required amount of air is sometimes quite difficult. The chemical reaction that produces air entrainment is a separate reaction; the chemical hydration of the cement is not involved.

*Accelerators*, another type of additive, serve to accelerate the setting time of the mix. In doing so, additional heat of hydration is generated. Some batching plants use accelerators indiscriminately during cold weather to keep the mix from getting cold in transit; such practices are very questionable and may often do more harm than good, particularly where the accelerator contains chlorides.

*Retardants* are additives that are used to slow the setting rate of the mix. They are used where long finishing times are required. Retardation up to 8 hours or longer is commonly produced, but the results are sometimes erratic and unpredictable.

*Plasticizers* and *superplasticizers* are additives that are used to lubricate a ''stiff'' mix, allowing it to be placed in intricate formwork while maintaining a low W/C ratio. The effects are temporary, usually about 30 minutes, after which time the concrete supposedly returns to its unplasticized state. Long-term performance records on concrete placed while in a ''plasticized'' state are scattered and incomplete.

There are other additives being sold to those who are willing to take the risk. Except for air entrainment, few additives can be recommended. Any additive that interferes with the hydration process should be used with extreme caution.

## REINFORCEMENT

Reinforcement is manufactured under several international specifications for a variety of yield stresses, or grades. In this text, consideration is limited to reinforcement grades 40, 50 and 60. For these grades, the yield stress is 40,000, 50,000 and 60,000 psi, respectively.

Reinforcement is rolled in round bars, having deformed surfaces designed to improve the connection to the adjacent concrete (see Fig. 1-6). In the United States, the

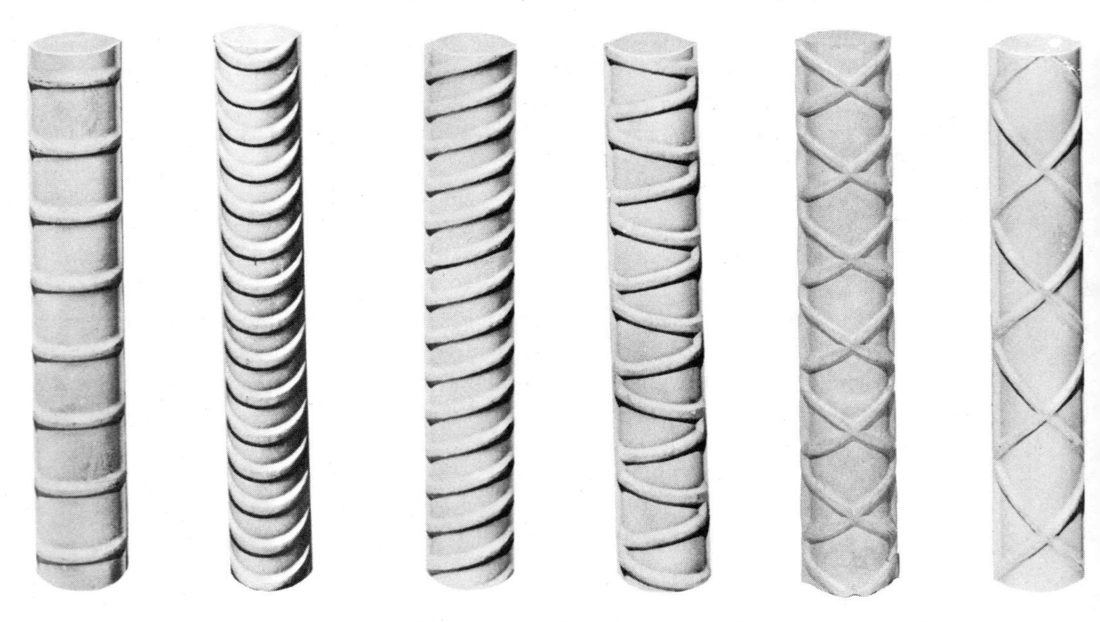

**Figure 1-6** Deformed Reinforcing Bars.

identifying number of a reinforcing bar, up to a No. 9 bar, indicates its diameter in eighths of an inch; thus a No. 5 bar has a nominal diameter of $\frac{5}{8}$ in. Bar sizes No. 9, No. 10, and No. 11 are also round, but their diameters are set to provide the same areas as older square bars 1 in., $1\frac{1}{8}$ in., and $1\frac{1}{4}$ in. respectively. A No. 9 bar has a diameter 1.128 in., a No. 10 bar has a diameter 1.27 in., and a No. 11 bar has a diameter 1.41 in. Bar sizes larger than No. 11 are not included in this text.

Although deformed bars are used almost without exception in the United States, smooth bars or drawn wire are frequently used throughout much of the rest of the world. In small diameters (6, 8, 10 and 12 mm), smooth wire can be shipped in rolls occupying a low volume at reduced shipping costs. It should be noted that labor costs are considerably higher to place a large number of small smooth wires in the forms compared to fewer larger deformed bars. Where labor is cheap and shipping is expensive, however, smooth wire will probably remain popular.

Stress–strain curves for steel are treated in detail in elementary strength-of-materials texts. For its applications in reinforced concrete, an idealized curve similar to that of concrete (Fig. 1-4) is required. Such idealized curves for steel are shown in Fig. 1-7 with the complete stress–strain curve shown alongside for reference. Since steel is such an ideal structural material, the idealized curves of Fig. 1-7 are but little different from the same portion of the actual curve.

From Table A-1 of the Appendix it is noted that the elastic modulus for steel is $29 \times 10^6$ psi. It is also noted that the strain at the yield point is 0.00138 for grade 40 steel, 0.00172 for grade 50 steel, and 0.00207 for grade 60 steel.

In addition to individual smooth wires and deformed bars, reinforcement is also fabricated into a mesh of smooth wires, called *welded wire fabric* (WWF) or into a mat of deformed bars called *structural fabric*. These fabrics are made up as a convenience to construction; the structural design is the same regardless whether the reinforcement is placed as individual bars or as a mesh.

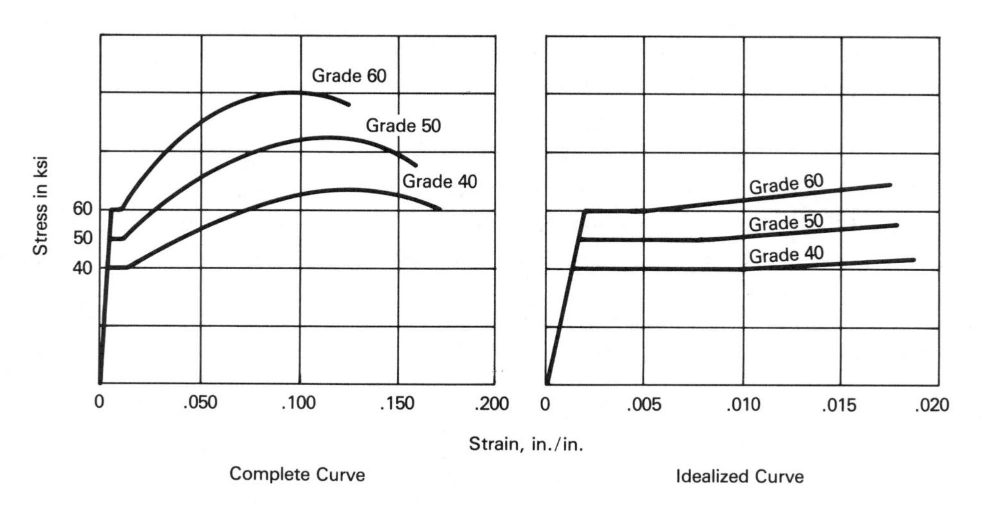


**Figure 1-7** Idealized Stress-strain Curve for Steel.

## BOND STRENGTH

Bond strength between concrete and its reinforcement is probably the most highly variable property of concrete that must be used in design. Accordingly, bond strength is assigned a high factor of safety to account for the many uncertainties and variables. The resulting conservative values for bond strength requires that reinforcement be deeply embedded in concrete to assure full development of the steel strength.

Bond strength increases as the ultimate strength of the concrete increases. It can be adversely affected by excess water, by air bubbles, by excessive rust on the reinforcement, by improper consolidation of the concrete, by accidental movement or vibration of the reinforcement after the concrete has set, and by the type and number of deformations on the reinforcement. Even with identical placement conditions and practices, two embedded bars can have strikingly different pullout strengths.

The ACI Code no longer permits smooth bars or wire to be used as reinforcement. A provision of older codes, such as ACI 318-63, allowed half as much bond strength for smooth bars as for deformed bars, and many buildings throughout the world are still being built to that provision. The success of the many older concrete buildings still in use today attests to the validity of that provision.

## REVIEW QUESTIONS

1. List the four basic ingredients used in making portland cement concrete.
2. What does the term ''calcareous'' mean when applied to aggregates?
3. What is the ''matrix'' of a concrete mix?
4. Why is it necessary to reinforce concrete?
5. List the five types of portland cement and the distinguishing characteristic of each.
6. If a city water supply is ''hard'' water, could that water be safely used to make concrete?
7. What is the ''ultimate strength'' of concrete?
8. What are the units of ultimate strength?
9. At what age is the concrete when its ultimate strength is usually determined?
10. How long does the strength of concrete continue to increase?
11. What is the weight of a cubic foot of concrete? Of a cubic yard?
12. A concrete has a specified ultimate strength of 4000 psi at 28 days. Approximately what is its water/cement ratio?
13. Is water/cement ratio a ratio of volumes or weights?
14. What is the primary detrimental effect of adding water to a mix, with all other ingredients unchanged?
15. What is the primary detrimental effect of adding cement to a mix, with all other incredients unchanged?
16. A concrete has a specified ultimate strength of 3000 psi. What is its modulus of elasticity?

17. A reinforced concrete bridge spans 60 ft. From a low atmospheric temperature of $-5°F$ to a high temperature of $+115°F$, what is the overall dimensional change in its length?
18. How does creep differ from shrinkage?
19. How is creep similar to shrinkage?
20. In general terms, what is the primary cause of shrinkage?
21. At what levels of stress does creep occur?
22. What is the major objection to the use of additives?
23. What is air entrainment?
24. What is grade 50 steel?
25. What percent elongation does a typical grade 60 steel undergo in its plastic range (Fig. 1-7)?
26. In older codes, what is the difference in bond strength between smooth bars and deformed bars?

# 2

# Limitations, Design Methods, and Codes

## GENERAL

The design of structures in reinforced concrete is rigidly governed by codes. One may spend years learning the fine points of these codes, but the major features of design are contained in a surprisingly few paragraphs. It is these major features that are the subject of this text; the fine points are usually learned through exposure, experience, and embarrassment.

In the United States, these sections of building codes that deal with concrete design are usually based on the *Building Code Requirements for Reinforced Concrete* (ACI, 1983). Any reference to a code in the following chapters will mean this code. It is commonly identified as ACI 318-77 or ACI 318-83, where the number following the dash is the year of the edition.

## INELASTIC DEFORMATIONS

It was noted in Chapter 1 that concrete is reasonably elastic under short-term service loads. Where loads are sustained over long periods, however, they can produce inelastic deformations due to creep; the magnitude of these deformations at any given time is unpredictable.

Shrinkage is yet another inelastic deformation in concrete whose magnitude at any time is unpredictable. Shrinkage is attributed to the drying out of excess water in the concrete. The concrete will regain much of the shrinkage loss if it is later submerged, but in dry buildings the shrinkage loss can be considered to be permanent.

Inelastic deformations introduce an unpredictable source of error into any elastic analysis of concrete. ACI has devoted considerable effort to finding analytical methods that produce reasonable agreement with test data, to include the effects of the inelastic deformations. The two most common methods, the strength method and the working stress method, are discussed in the following sections.

The response of reinforced concrete to loads is heavily dependent on the stress–strain relationships of the concrete and its reinforcement. The meaning of the idealized stress–strain curves of Chapter 1 must be fully appreciated if the following analytical methods are to be understood.

## STRENGTH METHOD

The *strength method* of design of individual concrete members is the method preferred by ACI. In this method, a member is designed to an ultimate loading, then allowed to work at some lower service level.

In the strength method of design, the moments, shears, and axial forces acting on a member are found by ordinary analytical methods. They are then multiplied by appropriate "load factors" and combined together to produce peak "ultimate" loads. The size of the member and its reinforcement are then proportioned to sustain these ultimate loads at the ultimate strength of the section.

At its ultimate load, a member is assumed to have reached its ultimate strength, or highest computed strength. All materials are allowed to be in full yield. In establishing the ultimate load, the load factors are carefully set such that under day-to-day service loads, the stresses in the member will be down at some acceptable level. ACI does not require any further check on these day-to-day service stresses, nor is a limit placed on the magnitude of these service stresses.

In using the strength method, the moments, shears, and axial forces acting on the member are computed by the usual elastic methods that would be used for any other structural material. It is only when selecting the size of the concrete member and its reinforcement that the ultimate loads are used. The strength method is therefore seen to be simply a means of sizing and reinforcing a member, not a general method of structural analysis and design.

The strength method has evolved largely since the 1956 ACI Code, when it was first recognized. Earlier methods used a conventional elastic approach, but the inelastic deformations due to shrinkage and creep produced serious discrepancies with test results. The elastic method is still permitted under the ACI Code, where it is now called the *alternate design method*. Since the 1956 Code, however, some corrections have been introduced which approximate the effects of shrinkage and creep.

## WORKING STRESS METHOD

The "alternate design method" discussed in the preceding section is commonly called the *working stress method*. In this method, concrete is designed in the same way as other contemporary structural materials.

In the working stress method, a maximum allowable level of stress is established within the elastic range for both the concrete and the steel. The applied loads are combined as usual to produce the highest working load the structure should ever experience. The members are then sized to sustain this peak working load at this maximum allowable working stress.

At all stages of the design, the members are assumed to be elastic. All design theory is based on simple straight-line elastic deformations. It is acknowledged that the structure will not frequently be subjected to its maximum allowable working stresses. Under day-to-day service, it will undoubtedly be loaded to some lower level of service loads and stresses.

It should be recognized that "service levels" of loads and stresses will occur in both the strength method and the working stress method. The term "service stress" is not always interchangeable with "working stress," The distinction is rigorously observed herein.

## COMPARISON OF METHODS

It is essential to recognize that the two methods of design are not equivalent. Members designed under the strength method may not have the same configuration as members designed under the working stress method. In general, the working stress method is the more conservative and will usually yield larger sizes of members and larger quantities of reinforcement.

## DEFLECTION ANALYSIS

Even though a concrete member is designed using the strength method, its deflection and deformations have meaning only under service conditions. Regardless of what method was used for the initial design, it is necessary to rely on an analysis at service levels to determine deformations. An analysis at service levels is necessarily based on elastic strains. As noted earlier, the primary source of error in such an analysis is the inelastic strains due to shrinkage and creep. Some approximation of the inelastic strains must be included in the elastic analysis.

In the working stress method, ACI assumes that the magnitude of the inelastic strains could be as high as that of the elastic strains. The net results of this assumption is that stresses in the reinforcement in compressive areas could be as much as doubled if the inelastic strains ever reach their maximum value. The inelastic strains in the concrete are assumed to have no direct effect on the stress in the concrete nor in the tensile steel; the increase in stress affects only the compressive steel.

In the following chapters, the analysis of concrete members anywhere in their elastic range will include the foregoing assumption that stresses in compressive reinforcement may be doubled. If the actual inelastic strains in the member should ever develop fully, the analysis will be correct. Until that happens, the calculated results may not correlate accurately with the actual performance of the member.

## REFERENCES TO ACI 318-83

Discussions in the following chapters refer frequently to particular sections of the ACI Code. It is recommended that anyone designing concrete members obtain a current copy of the Code and become familiar with it. Where the meaning of a particular provision becomes clear only after repeated study, a marginal note will save some time several weeks or months hence when the provision is encountered again.

Where appropriate in subsequent discussions, the section number of the Code is given in parentheses following a reference to the Code. For example, in the sentence "Code (10.2.3) requires that ultimate strain in concrete be taken at 0.003 inches per inch," the referenced provision is stated in Section 10.2.3 of ACI 318-83.

## REVIEW QUESTIONS

1. In structural design, a margin of strength is maintained above the known loading to allow for unanticipated loads. How is this margin provided in the ultimate strength method? In the working stress method?
2. In the working stress method, distinguish between the "allowable working stress" and the "service stress."
3. In the ultimate strength method, how are deflections computed?
4. Of the two methods, the ultimate strength method and the working stress method, which is usually more conservative?
5. Which method of design is in more general use?

# 3

# Design and Detailing Practices

## GENERAL

A structural member must be sized and reinforced to carry computed loads; the methods commonly used to accomplish this are called *design practices*. Once the sizes of the members are chosen, the members must be connected with reinforcement, notched for equipment mounts, and penetrated by ducts; the methods used to accomplish these and hundreds of other such details are called *detailing practices*.

General design and detailing practices suited to small projects are learned largely by exposure over a long period of time. Every designer soon develops his or her preferred set of practices and will adapt this consistent set of practices to each project. Even so, there are basic practices common to the industry that are used by most designers with but little variation.

Some of these basic practices are presented in the following sections. Several rules of thumb are presented and several construction practices are discussed. Load factors and strength reduction factors are introduced and factors of safety are compared. All these topics have a bearing on design and detailing practices; their effects are illustrated in later chapters in the examples.

A major factor influencing design and detailing practices is deciding whether a member is to be precast or cast in place. Significant differences apply to the two designs, to include the minimum sizes and reinforcement that must be used. The use of precast components, even those cast on the job site, is so commonplace that the following criteria include both precast and cast-in-place construction whenever appropriate.

## ACCURACY OF COMPUTATIONS

With the advent of the electronic calculator, there came a natural tendency to carry design calculations to four-, five-, or even six-figure accuracy. Such practices do no harm unless one begins to believe in this superficial accuracy. It must be borne in mind that the final accuracy is no better than the least accurate input, and the use of an electronic calculator does nothing to improve the accuracy of the input.

An accuracy of three significant figures will yield a final accuracy, after arithmetic manipulations, of about two significant figures or less. There is little point in performing the calculations for reinforced concrete to any higher accuracy since the loads are known only to two-place accuracy at best and the concrete can meet specifications even when it varies as much as 15%. The manual 10-in. slide rule, with its three-place accuracy, is adequate for design of concrete structures.

The lack of numerical precision in the calculations should not be equated to the lack of sensitivity of concrete. For example, concrete is quite sensitive to torsional shears; the fact that the magnitude of the torsion is known only approximately does not decrease that sensitivity. The same degree of attention to detail must be accorded to concrete as to other materials.

Precast concrete panel. The use of modular precast elements is often used as a means of reducing costs in concrete construction. (Courtesy Portland Cement Association).

## NOMINAL LOADS

Loading on concrete buildings varies considerably for any number of reasons. Nonetheless, it is frequently necessary to estimate the total load (dead plus live) on a beam or a column or a slab. For common spans and routine design, the following values offer a first guess:

*Floor load:* 150 psf, dead plus live load
*Roof load:* 125 psf, dead plus live load
*Column load:* 30 tons per story, total load

For routine concrete buildings, the dead load should be expected to be about 60% of the total load and live load to be about 40% of the total load.

For members at the perimeter of a concrete building, the loads should be expected to be only slightly less (about 15% less) than the loads at the interior. The fenestration and exterior walls commonly used in concrete buildings are so heavy that little difference should be expected.

## MINIMUM DIMENSIONS

The minimum size of some concrete members is set by Code and of others by limitations on forming and casting. The difficulties in forming and casting very thin or very small members in the field should not be underestimated. Labor hours and costs required to cast such members can easily be double or triple those required to cast more conventional sizes.

The following list includes an indication of whether the minimum dimension is fixed by Code and may not be decreased, or whether it is limited by practice and may be decreased if one is willing to pay the price.

Bearing walls, cast in place, minimum 10 in. thick (practice)
Bearing walls, precast, minimum 4 in. thick (Code 14.5.3.1)
Columns, cast in place, minimum dimension 10 in. (practice)
Foundation walls, minimum $7\frac{1}{2}$ in. thick (Code 14.5.3.2)
Nonbearing walls, cast in place, minimum 4 in. thick (Code 14.6.1)
Slabs on grade, minimum thickness $3\frac{1}{2}$ in. (practice)

## MINIMUM AND MAXIMUM REINFORCEMENT

For all members subject to flexure, the ratio of tensile steel area $A_s$ to concrete area $bd$ is widely used. Called the *steel ratio* $\rho$, it is defined as

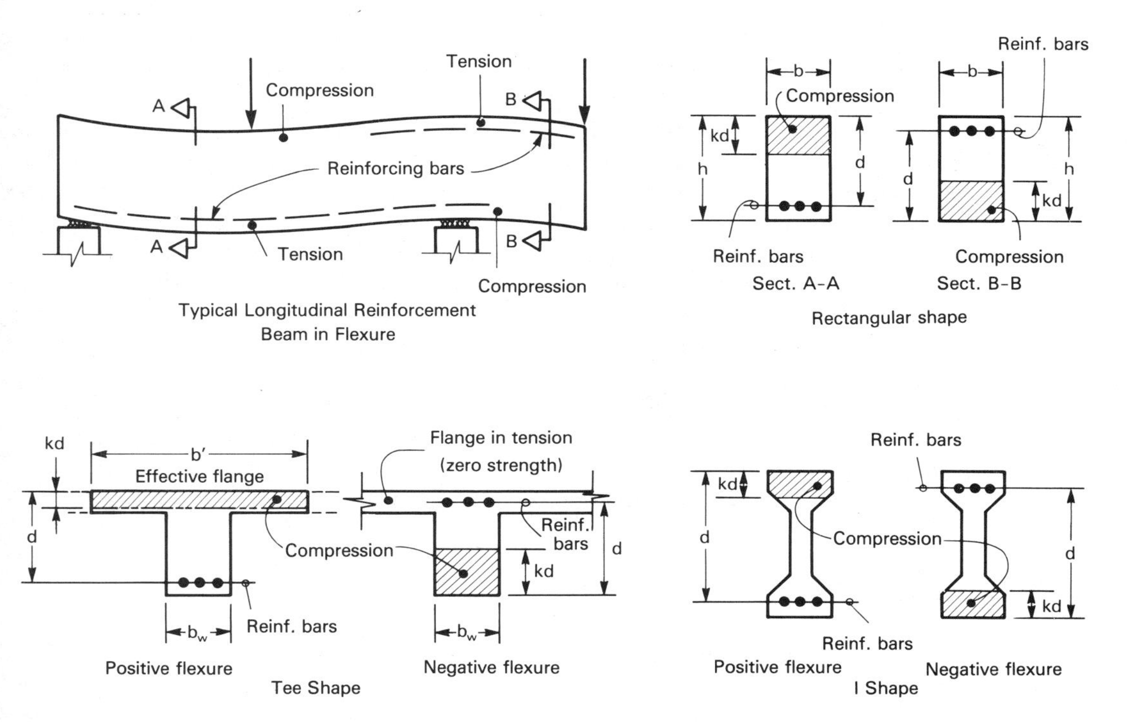


**Figure 3-1** Typical Flexure Sections.

$$\rho = \frac{A_s}{bd}$$

where $A_s$, $b$, and $d$ are shown in Fig. 3-1.

The minimum and maximum area of flexural reinforcement given in the following paragraphs is governed by Code. These limits assure that the member will behave properly under extremes of load and that cracking in the concrete will be within acceptable sizes and will follow acceptable patterns.

For slabs having flexural reinforcement in one direction only, temperature reinforcement in the other direction is required. Minimum area of temperature reinforcement is $0.002bt$ for grades 40 and 50 steel and $0.0018bt$ for grade 60 steel, where $b$ and $t$ are gross dimensions of the section. This requirement applies to all one-way slabs, including those of joist systems.

For flexural members, the steel ratio may not be less than $200/f_y$ nor may it be more than 75% of the balanced steel ratio. The balanced steel ratio is defined as that ratio of steel that will allow the reinforcement to enter yield just as the concrete reaches a strain of 0.003; it is discussed further in Chapter 4.

For columns, the area of longitudinal reinforcement may not be less than $0.01bt$

nor more than 0.08*bt*, where *b* and *t* are gross cross-sectional dimensions. In practice, however, congestion at joints becomes so severe with the higher amounts of reinforcement that the use of more than 4% steel is avoided.

For bearing walls reinforced with grade 60 deformed bars not larger than No. 5, the area of reinforcement may not be less than 0.0012*bt* vertically nor less than 0.0020*bt* horizontally. These same minimums also apply to mesh reinforcement having the same bar sizes regardless of whether the bars are smooth or deformed or whether the bars are grade 40, 50, or 60 steel (smooth bars are permitted in a mesh). For all other reinforcement sizes and grades, the area of reinforcement may not be less than 0.0015*bt* vertically nor less than 0.0025*bt* horizontally. As before, *b* and *t* are gross dimensions. These requirements are for total area of steel, whether placed in the middle of the wall, on one face, or on both faces.

For floor slabs on grade, Code does not require reinforcement. For crack control, however, it is common practice to meet at least Code requirements for temperature reinforcement (in both directions). For slabs more than 5 in. thick, the reinforcement mesh should be placed in two layers, at top and bottom surfaces.

## MINIMUM COVER

The amount of clear concrete cover over the outermost reinforcement is governed by Code (7.7.1). It is this cover that provides fire protection. It also prevents the intrusion of salts or oxygen that would cause corrosion of the reinforcement. Code (7.7.5) requires that the cover be increased above Code minimums where conditions are severe.

The following criteria are taken from the ACI Code (7.7). In this summary, any dimension that is based on bar sizes is the same whether the bars are used singly or made into a mesh. Mesh reinforcement may be made of smooth or deformed bars; bars used singly must be deformed.

Minimum cover for concrete cast in place:

- Concrete cast against earth and permanently exposed to earth — 3 in.
- Concrete exposed to earth or weather:
  - (a) Bars larger than No. 5 — 2 in.
  - (b) Bars No. 5 and smaller — $1\frac{1}{2}$ in.
- Concrete not exposed to earth or weather:
  1. Slabs, walls, and joists
     - (a) Bars larger than No. 11 — $1\frac{1}{2}$ in.
     - (b) Bars No. 11 and smaller — $\frac{3}{4}$ in.
  2. Beams and columns — $1\frac{1}{2}$ in.
  3. Shells and folded plates
     - (a) Bars larger than No. 5 — $\frac{3}{4}$ in.
     - (b) Bars No. 5 and smaller — $\frac{1}{2}$ in.

Minimum cover for precast concrete elements manufactured under plant-controlled conditions:

| | |
|---|---|
| Concrete exposed to earth or weather: | |
| 1. Wall panels | |
| (a) Bars larger than No. 11 | $1\frac{1}{2}$ in. |
| (b) Bars No. 11 and smaller | $\frac{3}{4}$ in. |
| 2. Other members | |
| (a) Bars larger than No. 11 | 2 in. |
| (b) Bars larger than No. 5 but smaller than No. 11 | $1\frac{1}{2}$ in. |
| (c) Bars No. 5 and smaller | $1\frac{1}{4}$ in. |
| Concrete not exposed to earth or weather: | |
| 1. Slabs, walls, and joists | |
| (a) Bars larger than No. 11 | $1\frac{1}{4}$ in. |
| (b) Bars No. 11 and smaller | $\frac{5}{8}$ in. |
| 2. Beams and columns | |
| (a) Primary reinforcement: | |
| one bar diameter, but not less than | $\frac{5}{8}$ in. |
| nor more than | $1\frac{1}{2}$ in. |
| (b) Ties or stirrups | $\frac{3}{8}$ in. |
| 3. Shells and folded plates | |
| (a) Bars larger than No. 5 | $\frac{5}{8}$ in. |
| (b) Bars No. 5 and less | $\frac{3}{8}$ in. |

## LIMITS ON DEFLECTIONS

The following criteria concern the glass, finishes, fillers, or other nonstructural elements of a building that are not themselves part of the load-carrying system but which are attached directly to a load-carrying member. Some of these elements may be damaged if the structural members supporting them undergo excessive deflections. Those nonstructural elements that would be subject to such damage are collectively termed herein the *brittle elements*.

The effects of structural deflections on the brittle elements are no different in concrete structures than in structures made of other materials. The nature and timing of the deflections themselves, however, can be quite different due to the inelastic deformations resulting from shrinkage and creep. In evaluating the effects of deflections in concrete structures, the designer must distinguish between short-term and long-term loadings and the effect that each could have on the deflections at any given time.

The following limits on deflections specified by Code (9.5) apply only to application of the short-term live load:

1. For flat roofs not supporting or attached to brittle elements, deflections are limited to $L/180$, where $L$ is the span.
2. For floors not supporting or attached to brittle elements, deflections are limited to $L/360$, where $L$ is the span.

The following limits on deflections apply only to that part of the total deflection following the placing and attaching of the nonstructural elements; earlier long-term deflections may be excluded but subsequent long-term or short-term deflections may not.

1. For roofs or floors supporting or attached to brittle elements, deflections are limited to $L/480$, where $L$ is the span.
2. For roofs or floors supporting nonstructural elements that do not include any brittle elements, deflections are limited to $L/240$, where $L$ is the span.

The computed deflections due to flexure may not exceed the foregoing limits. The computed deflections must include the elastic deflections due to short-term load, the elastic deflections due to sustained load, and any inelastic deflections due to shrinkage and creep. The elastic deflections are computed by ordinary methods; the Code prescribes the following means to estimate the inelastic deflections due to sustained load.

Deflections due to sustained loads are known to increase with time and decrease with the amount of compressive reinforcement. Code (9.5.2.5) includes these effects in a factor $\lambda$, where

$$\lambda = \frac{\xi}{1 + 50\rho'}$$

where $\xi$ is a time factor and $\rho'$ is the ratio of compressive reinforcement $A'_s/bd$. The time factor is prescribed by Code:

$$\text{For 5 years or more: } \xi = 2.0$$

$$\text{For 12 months: } \xi = 1.4$$

$$\text{For 6 months: } \xi = 1.2$$

$$\text{For 3 months: } \xi = 1.0$$

The inelastic deflection is computed by multiplying the elastic deflection due to sustained loads times the factor $\lambda$.

Code does not place limits on any deflections other than those due to flexure. Where brittle finishes are fixed to concrete columns, such as mosaic tile, mirrors, or marble, the designer must estimate the long-term deformations by his or her own methods. The foregoing time factors offer a useful guide for estimating such long-term inelastic deformations.

## CONTROL OF DEFLECTIONS

The foregoing Code criteria give the limits that must be observed for deflections of flexural members. The foregoing criteria do not, however, stipulate how the deflections are to be determined. In a separate section, Code (9.5.2) prescribes means to compute or to limit deflections when using the strength method of design.

Only the deformations at service levels of loading are of interest to the designer. These are the day-to-day deformations that determine whether the building will perform satisfactorily in service. Such deformations affect cracking of plaster, ''springiness'' of floors, transmission of vibrations, weather seals at joints, and all other effects that arise from the building's deflections.

Code (9.5.2) permits two approaches to the control of deflections. The first approach requires the computation of deflections using special equations developed from experimental data. The constants and section properties, such as elastic modulus and moment of inertia, are prescribed by Code (9.5). The method is sometimes tedious and often introduces complexities into otherwise simple calculations.

The second approach is a ''blanket'' method. In this method, deflections are not computed nor is there any estimate made of their effects. The depths of all members are simply kept above prescribed minimum limits; experience has shown that buildings constructed using such member sizes will perform satisfactorily insofar as deflections are concerned. In this method, therefore, it is never known what the magnitude of the deflections are; it is only known that they will not be a problem. As one may suppose, this blanket approach to control of deflections is somewhat conservative. Nonetheless, the use of this method produces little or no penalty for smaller buildings. As the building size increases, however, the conservatism becomes increasingly expensive.

In this book, only the blanket method for the control of deflections is used. Consequently, when the depth of a member is selected, the check for deflections consists only of a fast check against the minimum allowable depth for the member. The minimum overall depth $h$ of members is adapted from the Code (Table 9.5a) and presented in Table 3-1.

**TABLE 3-1 MINIMUM OVERALL DEPTH OF FLEXURAL MEMBERS[a]**

| | Overall depth of member | |
|---|---|---|
| Support condition | Solid slabs, one-way flexure | Beams or ribbed slabs, one-way flexure |
| Simply supported | $L/20$ | $L/16$ |
| One end continuous | $L/24$ | $L/18.5$ |
| Both ends continuous | $L/28$ | $L/21$ |
| Cantilever | $L/10$ | $L/8$ |

[a] $L$ is the length of span. Values are for grade 60 steel, $f_y = 60{,}000$ psi. For other grades of steel, these values must be multiplied by $(0.4 + f_y/100{,}000)$.

## RULES OF THUMB

Rules of thumb can be useful, particularly when one is trying to make an initial estimate for a beam size or a slab thickness. They can also be useful when one is keeping a running check on a calculation and wishes to know if the results obtained are at least reasonable. The following rules and generalities may be useful as a first guess.

The optimum column module for concrete structures can be expected to be 18 to 20 ft. This module can be extended with little penalty in cost up to 25 ft for joist systems, with many advantages in utilization of space. Spans in reinforced concrete greater than 28 ft are uncommon, although two-way joist systems work reasonably well with spans up to 32 ft or even more.

As a means to distinguish between girders and beams, it is assumed that girders are supported by vertical members and beams are supported by horizontal members. The overall depth of a girder in reinforced concrete can be expected to be about $L/10$ and the overall depth of a beam can be expected to be about $L/12$, where $L$ is the span. For spans greater than about 20 ft, the depths of the members increase rapidly.

Optimum spacing of the stems of tee beams is generally controlled by the floor slab they support. A floor slab 4 in. thick will span about 8 ft and a floor slab 6 in. thick will span about 12 ft. Both these examples apply to the usual range of floor live loads and both examples observe the common 4-ft module for building materials.

The stem of a tee beam is designed to take all the shear on the tee beam. Proportions for width and depth have been found to work well for such shear when the width of the stem $b_w$ is roughly half the effective depth $d$ (see Fig. 3-1).

Rectangular sections occur when a tee beam is placed in negative moment or when a rectangular frame is designed to receive precast floor planks. The proportions for such rectangular sections have been found to be economical when the width $b$ is roughly two-thirds of the effective depth $d$.

For continuous tee beams subject to both positive and negative bending, the width of the stem should be somewhere between half and two-thirds of the effective depth $d$.

Concrete joist systems are floor systems that utilize narrow joists spaced closely together, supporting a thin concrete floor slab. The resulting closely spaced tee joists are significantly shallower than a regular tee-beam floor. The overall depth of a joist system can be estimated at about two-thirds of the depth of a regular tee-beam-and-slab system.

Concrete columns subject to bending (as part of a rigid frame) can vary widely in size depending on the magnitude of the moment. As a first guess the size of a column subject only to axial load may be estimated as 12 in. on a side plus an additional 1 in. for each story above the one being considered. Thus the columns in the third story of a five-story building can be expected to be about 12 in. on a side plus 1 in. times two stories above, for a total of 14-in. square. A rectangular column for this case would have only slightly more cross-sectional area than the square column, say 10 by 20 in.

Cantilevered retaining walls are usually tapered, being considerably thicker at the base than at the top (see Fig. 3-2). The thickness at the top can be expected to be about 8 in. for a wall 10 ft high and as much as 12 in. for a wall 20 ft high. The thickness at the base can be expected to be about one-ninth of the height, with a minimum thickness of 12 in.

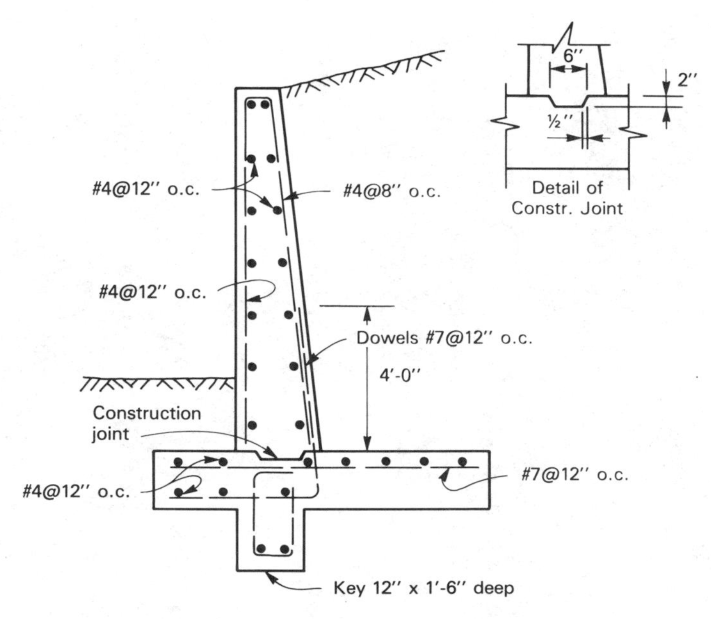


**Figure 3-2** Typical Cantilevered Retaining Wall.

## TEMPERATURE AND SHRINKAGE JOINTS

Due to the phenomenon of shrinkage in concrete, problems of contraction and expansion of structures in concrete become somewhat more severe than in structures of other materials. The contraction due to shrinkage has been likened to that due to a temperature drop of 40°F. With such an addition to overall contraction, contraction joints in concrete structures should be expected to be at closer intervals than in structures of other materials.

Code does not give a maximum distance at which a full separation joint (expansion–contraction joint) is required since such a requirement is necessarily dependent on local temperatures. As a general guide in temperate climates, when routine continuous structures are more than 100 ft long, a full separation joint starts to become desirable. At 150 ft it becomes a pressing consideration and at 200 ft it becomes essential.

Slabs on grade are protected from temperature extremes by the huge heat sink that supports them. Nonetheless, slabs exposed to direct sunshine are subject to considerably higher temperatures than slabs located under shade. A common specification allows slabs located in shaded areas to be cast up to 400 $ft^2$ in area with a maximum dimension of 25 ft between joints or, for exposed locations, up to 250 $ft^2$ with a maximum dimension of 20 ft between joints. The control joints used to meet these requirements are shown in Fig. 10-11.

Heavily reinforced concrete member. The final layout of reinforcement in heavily reinforced sections must always be a prime consideration of the designer. (Courtesy Portland Cement Association).

With the myriads of combinations that could occur due to temperature and shrinkage deformations, it is not common practice to perform calculations for such stresses. Rather, the building is designed and detailed to prevent such stresses from occurring; no special analysis is then required. Any accidental effects due to random thermal stress is then relegated to the factor of safety.

Where thermal stresses are allowed to occur, however, they should be expected to be significant. The associated forces are quite large and can be extremely difficult to handle. Such problems are far beyond the scope of this book.

## CONSTRUCTION JOINTS

Construction joints are used to interrupt a casting; the casting can then be resumed at that joint at a later time. With few exceptions, Code permits construction joints in vertical members to be placed anywhere in the member, to include points of maximum shear and moment. A typical example may be seen at the base of a retaining wall (Fig. 3-2); the construction joint is almost always located at the base of the stem, where both moment and shear on the stem are highest.

Notable exceptions are drawn by the Code (6.4.4) for joints in horizontal members. When joints are to be located within the span, they must be located within the middle third of the span. Additionally, the stem of a tee must be cast monolithically with the slab unless specifically designed and detailed otherwise. Similarly, haunches and drop panels must be cast monolithically unless detailed otherwise. It may be concluded that for horizontal members, the existence or absence of high flexural stress has no significant effect when placing a joint, but placing a joint in an area of high shear stress is to be avoided.

When properly designed and detailed, a construction joint does not define a plane of weakness, neither at working levels nor at ultimate strength. "Properly designed and detailed," however, when applied to joints in monolithic frames can be quite complex. Some of these problems are treated in detail by *Recommendations for Design of Beam–Column Joints in Monolithic Reinforced Concrete Structures* (American Concrete Institute, 1976).

## DIFFERENTIAL SETTLEMENTS

Uniform settlement of a structure may have highly deleterious effects on sewer, water, and power connections to a building, but it has no effect on the structure. Only the differential settlement between adjacent footings or supports will affect the stress levels in the structure. The structural analysis is therefore concerned only with differential settlements between supports, not with total settlements.

If, however, the largest settlement at any footing in a group of footings is limited to 1 in., the differential settlement between any two footings in the group can be expected to be somewhat less than 1 in., say a maximum of $\frac{3}{4}$ in. It is this approach that is commonly used in practice for the design of shallow footings—that the largest settlements will be limited to about 1 in. and that differential settlements, if any occur, can then be expected to be less than about $\frac{3}{4}$ in.

As with thermal stresses, there are myriads of combinations of differential settlements that might occur in a routine structure. If, however, the differential settlements are less than about $\frac{3}{4}$ in., the change in total stress due to any reasonable combination can be expected to be less than about 15% of the total. For this amount, a separate analysis for potential differential settlements is not usually considered to be necessary and is rarely performed in practice. Where circumstances are unusual, however, or where a key foundation is known to be subject to large settlements, a thorough study of the effects of differential settlements is necessary. For the sake of the foregoing discussion, a "routine" structure is one in which the largest column load is no more than four times the smallest column load and the shallow spread footings are between 3 ft and 8 ft on a side.

In some circumstances, a foundation system may be composed of a mixture of isolated spread footings at columns and continuous strip footings at bearing walls. Where both the spread footings and the strip footings have the same contact pressure on the soil, the strip footings can be expected to settle more than the spread footings, up to about 50% more. The use of a lesser allowable pressure under the strip footings will help equalize the settlements.

## STRENGTH REDUCTION FACTORS

Construction and placement tolerances are more numerous and more generous for concrete than for other materials, presumably because concrete members are usually fabricated on-site. All aspects of concrete construction have allowable tolerances, which, under normal circumstances, combine to cancel each other. If only a few of these happen to accumulate, however, a significant loss of strength could result.

Code (9.3) requires that the computed ultimate strength of concrete members be reduced somewhat to account for any accidental buildup in tolerances. The amount of the reduction varies, depending on the type of loading; a higher reduction is required for columns than for beams. The following strength reduction factors, $\phi$, apply to the indicated type of loading at ultimate capacity:

Beam flexure, without axial loads: $\phi = 0.90$

Axial tension, with or without flexure: $\phi = 0.90$

Axial compression, with or without flexure, square or rectangular tied columns: $\phi = 0.70$

Shear or torsion: $\phi = 0.85$

Bearing: $\phi = 0.70$

Strength reduction factors are not used when computing stresses at service levels, nor are they used in the working stress method of design.

## LOAD FACTORS

The strength reduction factor $\phi$ of the preceding section provides for accidental undercapacity of a section. Its purpose is to provide extra capacity in the event the section is not built exactly as intended; it is not related to improper loading. In arriving at the value of $\phi$, it is tacitly assumed that the external loads are properly applied and that no overload occurs.

Overloading can in fact occur, however, and the chances of its occurring are completely independent of construction practices. It has already been noted that small loads may occur due to temperature or settlements that are ignored in the analysis and design of the structure. An allowance for such overload, accidental or planned, is an essential part of the design.

It is not reasonable to design a structure for every maximum load that can be imagined during its life and to assume that all these maximum loads will occur at the same time. Rather, the structure is designed for reasonable loads compatible with the intended service; a reasonable margin is then provided for unpredictable circumstances.

In the working stress method this margin is provided by the *safety factor* and in the strength method by the *load factors*.

Load factors are prescribed by Code (9.2). The following requirements are taken from the code, with $D$ representing the effects for dead load, $L$ for live load, $W$ for wind load, $T$ for settlement or temperature, and $E$ for earthquake. With this notation, the required strength $U$ at ultimate load must be at least equal to

$$U = 1.4D + 1.7L, \text{ or}$$

$$U = 0.75(1.4D + 1.7L + 1.7W), \text{ or}$$

$$U = 0.9D + 1.3W, \text{ or}$$

$$U = 0.75(1.4D + 1.7L + 1.1E), \text{ or}$$

$$U = 0.75(1.4D + 1.7L + 1.7T)$$

Where lateral loads are small, the first equation will usually yield the highest loads.

It is evident that the load factor for dead load is 40% above the specified dead load and for live load it is 70% above the specified live load. Recognizing that for concrete structures the dead load averages about 60% of the total and live load about 40% of total, an overall average load factor is about 1.52 times the total working load. For the strength method, the load factors thus provide an average 52% margin of strength above that nominally required for dead and live loads at working levels.

The safety factor for working stress design is also prescribed by Code (B.3). For working stress design, the maximum allowable compressive stress is $0.45f'_c$, which is 53% of the idealized ''yield stress'' discussed in Chapter 1. There is considerably more margin beyond the yield point which is not considered here, so the overall margin to ultimate is considerably higher in the working stress method than in the strength method.

## STRENGTH OF STEEL (FOR THE TABLES)

The ACI Code is published in two editions: Imperial units (feet, inches, pounds, slugs) and Système International (meters, millimeters, newtons, kilograms). The two editions are not exactly identical. Some differences are due only to round-off error, but other more significant disparities occur due to the use of different standards.

An example of where different standards are applied occurs in the specifications for reinforcement. In the edition containing Imperial units, the yield stress specified by ASTM A-615 for the two grades of reinforcing steel are 40,000 and 60,000 psi, respectively. In the SI edition there are also two steels, but these steels have yield stresses specified by ASTM A-615M as 300 N/mm$^2$ (43,500 psi) and 400 N/mm$^2$ (58,000 psi). These differences in yield strength are enough to affect a design significantly.

Another example of disparities in standards occurs in the selection of ''standard'' sizes for reinforcement. Here the United States and Canada are somewhat at variance with other countries. For metric sizes, the United States and Canada specify the sizes

**TABLE 3-2 U.S.–CANADIAN METRIC STANDARD STEEL SIZES**

| Bar No. | Diameter (mm) | Area (mm$^2$) |
|---|---|---|
| 10 | 11.3 | 100 |
| 15 | 16.0 | 200 |
| 20 | 19.5 | 300 |
| 25 | 25.2 | 500 |
| 30 | 29.9 | 700 |
| 35 | 35.7 | 1000 |
| 45 | 43.7 | 1500 |
| 55 | 56.4 | 2500 |

shown in Table 3-2, which yields even increments for the cross-sectional area of the reinforcing bars rather than even increments for diameters.

The U.S.–Canadian system for metric sizes has not been widely adopted and is not widely used even in North America. Nonetheless, these sizes are sometimes available at discount prices. The sizes of Table 3-2 are provided in two grades of steel, $f_y$ = 300 N/mm$^2$ and $f_y$ = 400 N/mm$^2$.

In countries where the metric system (rather than SI) is in use, there is very little standardization of reinforcement sizes. Perhaps the most widely applicable standard is the UNESCO-recommended standard (UNESCO, 1971), although in any given country some of these "standard" sizes may not be available. The UNESCO standard is shown

**TABLE 3-3 UNESCO-RECOMMENDED REINFORCEMENT SIZES**

| Bar diameter (mm) | Area (mm$^2$) | Mass (kg/m) |
|---|---|---|
| 6 | 0.28 | 0.211 |
| 8 | 0.50 | 0.377 |
| 10 | 0.79 | 0.596 |
| 12 | 1.13 | 0.852 |
| 14 | 1.54 | 1.16 |
| 16 | 2.01 | 1.52 |
| 18 | 2.54 | 1.92 |
| 20 | 3.14 | 2.37 |
| 22 | 3.80 | 2.87 |
| 25 | 4.91 | 3.70 |
| 28 | 6.16 | 4.65 |
| 30 | 7.07 | 5.33 |
| 32 | 8.04 | 6.07 |
| 40 | 12.56 | 9.47 |
| 50 | 19.63 | 14.80 |
| 60 | 28.27 | 21.30 |

in Table 3-3. These sizes are commonly imported into the United States in various strengths of steel.

The older standard Imperial sizes are discussed in Chapter 1. These sizes and their corresponding cross-sectional areas are given in Table 3-4. These are the most widely available sizes in North America and are generally available in steel grades 40, 50, and 60.

All three sets of standard sizes shown in Tables 3-2 through 3-4 are available in the United States. Consequently, it is somewhat difficult to determine design standards that will be widely applicable for any length of time. For the near future, however, it seems probable that the Imperial standards of Table 3-4 will continue to dominate the North American market.

Among engineers, a strong influence causing resistance to change is the data on which the Codes are based. The overwhelming bulk of those data was developed over the past 60 years using the older Imperial sizes of reinforcement, with steel having yield strengths of 40,000, 50,000, and 60,000 psi. An understandable preference to use Code values based on the actual experimental sizes and strengths should continue to be a strong influence.

In this text, the design tables of the Appendix are based on the traditional U.S. sizes of reinforcement shown in Table 3-4, with steel grades 40, 50, and 60. Where sizes other than these are to be used, the conversions should be based on the cross-sectional areas rather than the closest equivalent diameter. The design of reinforcement will be seen later to be based on cross-sectional areas; conversions using approximate diameters (which are then squared) can reduce the accuracy of the conversion.

For the design tables of the Appendix, only the most frequently used bar sizes have been adopted. Since the Code does not permit bar sizes less than No. 3 to be used structurally, those sizes have been dropped from the tables. Similarly, for bar sizes greater than No. 11, the Code introduces numerous restrictions and complications, so for the sake of simplicity those sizes have also been dropped. The remaining sizes, 3 through 11, will apply to the overwhelming majority of routine structures.

**TABLE 3-4 U.S.-CANADIAN IMPERIAL STANDARD STEEL SIZES**

| Size designation | Bar diameter (in.) | Area (in$^2$) |
|---|---|---|
| 3 | 0.375 | 0.11 |
| 4 | 0.500 | 0.20 |
| 5 | 0.625 | 0.31 |
| 6 | 0.750 | 0.44 |
| 7 | 0.875 | 0.60 |
| 8 | 1.000 | 0.79 |
| 9 | 1.128 | 1.00 |
| 10 | 1.270 | 1.27 |
| 11 | 1.410 | 1.56 |

## STRENGTH OF CONCRETE (FOR THE TABLES)

In practice, concrete strengths lower than 3000 psi are rarely used for structural concrete. The design tables have been set up for only the more common concrete strengths: 3000, 4000, and 5000 psi. Beams and girders are commonly cast from concrete having an ultimate strength of 3000 or 4000 psi and columns from 4000 or 5000 psi.

## REVIEW QUESTIONS

1. About how many pounds of structural material is commonly required to carry 2 pounds of live load in an average concrete building?
2. What is the minimum steel ratio for concrete members in flexure using grade 60 steel? Grade 40 steel?
3. What is the minimum required cover for reinforcement in a footing when the footing is cast against the soil (without forms)?
4. At what length of continuous structure does a full expansion/contraction joint become a consideration?
5. What maximum allowable settlement is commonly used for foundation design? How does this help to control differential settlement between any two footings?
6. What is grade 50 steel?
7. What is the controlling factor in concrete design that restricts the accuracy of all calculations?
8. What is the minimum thickness of a reinforced concrete foundation wall? A cast-in-place bearing wall?
9. What is the minimum and the maximum steel ratio for concrete columns?
10. Where soil is in contact with the retaining wall of Fig. 3-2, what is the required concrete cover over the reinforcement?
11. Why isn't a thermal analysis commonly made for routine concrete structures?
12. What is the strength reduction factor $\phi$, and why is it necessary?
13. What range of bar sizes are most commonly used in reinforced concrete construction in American practice?
14. How many significant figures are usually considered to be adequate in design calculations for reinforced concrete?
15. What is the minimum temperature reinforcement in slabs where grade 60 steel is used?
16. What is the required steel ratio for concrete slabs on grade?
17. What is the limit on live load deflection of a typical roof slab when it supports a brittle plastered ceiling? When there is no ceiling finish?
18. When a construction joint must be made in an upper-level floor slab, would the joint be better located at midspan or at the quarter-point? Why?
19. What is the ultimate load factor for dead load? For live load?
20. Where can smooth bars be used for structural reinforcement of concrete in American practice?
21. In a routine modular concrete construction, about how much weight should the designer expect to be supported by an interior footing of a three-story building?

22. How is the ''steel ratio'' computed?
23. What purpose is served by providing concrete cover outside the reinforcement?
24. A certain beam supporting a window frame has a computed deflection (3 months or less) of $\frac{1}{2}$ in. The beam has no compressive reinforcement. What total allowance should be made for deflections at 5 years or more?
25. Why isn't a foundation settlement analysis commonly made for routine concrete structures?
26. Using a ratio of dead load to live load of 3 : 2, derive the average overall load factor of 1.52 for concrete buildings.
27. What range of concrete strengths are most commonly used in American practice?
28. A rectangular concrete floor beam spans 24 ft. About what depth should be expected for the beam?
29. A continuous floor slab is to be supported by beams spaced at 8 ft on center. What thickness of slab should be anticipated?
30. What size should be expected for a concrete column at the bottom floor of a five-story building?

# 4

# Flexure in Concrete Beams

## GENERAL

As discussed in Chapter 2, flexure in concrete beams may be analyzed either in the elastic range of the materials or at ultimate load. The elastic analysis applies to all stresses at service levels regardless of which method of design was used for the original design. At ultimate load, the stress analysis is a completely separate and independent analysis.

The approach developed in the following sections is somewhat at variance with the usual engineering approach to reinforced concrete design. The usual engineering approach is a highly specialized approach, unique to concrete. The approach herein follows the same approach as that used for other contemporary materials. An elastic section modulus and an ultimate section modulus are derived, from which standard design tables are developed.

The following development treats elastic behavior first, using familiar ideas and notations. The development then proceeds into the less familiar plastic range, but retains the same design concepts as in elastic design.

## FLEXURE FORMULA

The flexure formula is well known from elementary strength of materials. With $f_c$ denoting the stress on the compression side of the beam, the flexure formula is

$$f_c = \frac{Mc}{I} \tag{4-1a}$$

where $M$ = moment acting on the section
$I$ = moment of inertia about the neutral axis
$c$ = distance from the neutral axis to the level where $f_c$ is being computed

In an alternative form, the flexure formula is used to compute the required magnitude of the section modulus $S_c$, where $S_c = I/c$:

$$\frac{M}{f_c} = S_c \tag{4-1b}$$

The symbol $S_c$ is used to denote the section modulus, taken to the outermost compression fiber. The section modulus is seen to be the ratio between the moment acting on a section and the stress it produces at the outermost fibers. It is a property of the section; its value depends only on the size and shape of the cross section.

For design in steel, the section modulus of a structural steel member is rarely computed; it is almost always found by looking it up in a table in the steel manual. For American practice, the *Steel Construction Manual* (American Institute of Steel Construction, 1980) lists the section modulus for each of the several hundred steel sections commonly manufactured in the United States. To try to design a steel structure without such a section modulus table would be impractical.

For design in timber, the section modulus may be computed by formula ($S = bh^2/6 = 0.1667bh^2$) or again, it may be looked up in a standard section modulus table for timber sections. It must be remembered that a timber section nominally designated 6 in. × 12 in. is actually $5\frac{1}{2}$ in. × $11\frac{1}{4}$ in. and it is the actual dimensions that must be used in the flexure formula. It is usually more convenient to use a table to find a suitable section modulus in timber than to try to remember the rules for finding the actual dimensions.

For design in concrete, the procedure used herein is no different from that for steel or timber. The section modulus is found from a section modulus table and used in the same way. Admittedly, the existence of the larger number of variables in a concrete section makes the tables somewhat longer than for other materials, but there is no difference in concept. Even with a large number of variables, however, it is a relatively simple matter to develop a table of values for the section modulus of a rectangular concrete section. The derivation of such a table follows.

## ELASTIC FLEXURE

The flexure formula itself can be adapted readily to beams of two materials, such as reinforced concrete beams. To do so, it is only necessary to derive the moment of inertia and the section modulus for such a beam. In the following derivations, the section modulus is always taken to the compression side of the cross section.

Wherever concrete and steel are bonded together in a beam, their strains at any point are equal. As a consequence the relationship between their stresses is fixed:

$$\epsilon_c = \epsilon_s = \frac{f_{ca}}{E_c} = \frac{f_{sa}}{E_s} \tag{4-2}$$

where $\epsilon_c, \epsilon_s$ = strains in adjacent concrete and steel
$f_{ca}, f_{sa}$ = stresses in adjacent concrete and steel
$E_c, E_s$ = moduli of elasticity of concrete and steel

Equation (4-2) is readily solved for stress in steel,

$$f_{sa} = nf_{ca} \tag{4-3}$$

where $n = E_s/E_c$. It was noted in Chapter 2 that the amount of stress in steel in compressive areas (where used) may eventually be as much as twice the expected amount. For long-term loading, therefore,

$$f_{sa} = 2nf_{ca} \quad \text{for steel in compression} \tag{4-4a}$$

$$f_{sa} = nf_{ca} \quad \text{for steel in tension} \tag{4-4b}$$

A typical section of a concrete beam is shown in Fig. 4-1a. The stress diagram is shown in Fig. 4-1b. By ratio, the stress in the concrete adjacent to the compressive steel is found to be

$$f_{c2} = \frac{f_c(k - g)}{k} \tag{4-5a}$$

The stress in the compressive steel is, then, from Eq. (4-4a):

$$f_{sc} = 2nf_{c2} = \frac{2nf_c(k - g)}{k} \tag{4-5b}$$

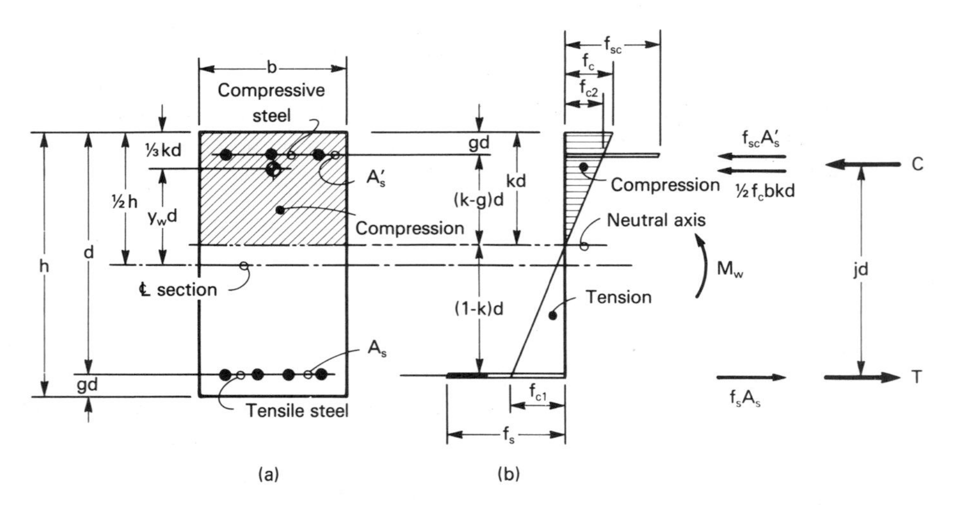


**Figure 4-1** Beam Section.

The stress in the tensile steel is found similarly:

$$f_s = nf_{c1} = \frac{nf_c(1 - k)}{k} \tag{4-5c}$$

It should be noted that the section shown in Fig. 4-1 contains reinforcement on the compression side of the beam. Such compressive reinforcement is commonly used where the size of the beam is so restricted that the concrete alone is not adequate to carry the compressive load. Alternatively, it is far less expensive (but not always possible) to make the concrete section bigger and avoid the use of compression steel.

It should also be noted that the resisting moment of the section can be shown as an internal couple formed by the forces $C$ and $T$ in Fig. 4-1. An average value for the distance between the forces, designated $jd$ in Fig. 4-1, will be shown later to be about $0.93d$. The average distance $jd$ is affected but little by the existence of compressive steel.

Refer again to Fig. 4-1. The symbol $A'_s$ is used to denote compressive steel and $A_s$ to denote tensile steel. The steel ratios are then

$$\rho' = \frac{A'_s}{bd} \quad \text{and} \quad \rho = \frac{A_s}{bd} \tag{4-6a, b}$$

Horizontal forces shown in Fig. 4-1 are now summed, yielding

$$\sum H = 0 = \tfrac{1}{2} f_c bkd - f_{c2}A'_s + f_{sc}A'_s - f_s A_s \tag{4-7}$$

The second term in Eq. (4-7) deducts the concrete displaced by the compressive steel.

Equations (4-5) and (4-6) are substituted into Eq. (4-7), yielding a solution for $k$:

$$k = \sqrt{[n\rho + (2n - 1)\rho']^2 + 2[n\rho + (2n - 1)g\rho']} - [n\rho + (2n - 1)\rho'] \tag{4-8}$$

Refer again to Fig. 4-1. Moments are summed about the neutral axis, yielding

$$M_w = \tfrac{1}{2} f_c bkd \frac{2kd}{3} - f_{c2}A'_s(k - g)d + f_{sc}A'_s(k - g)d + f_s A_s(1 - k)d \tag{4-9}$$

Equations (4-5) and (4-6) are substituted into Eq. (4-9) and the result solved for the elastic section modulus $S_c$,

$$\frac{M_w}{f_c} = S_c = \left[\frac{k^2}{3} + (2n - 1)\rho' \frac{(k - g)^2}{k} + n\rho \frac{(1 - k)^2}{k}\right] bd^2 \tag{4-10}$$

where the value of $k$ is given by Eq. (4-8).

Equations (4-8) and (4-10) would be rather formidable expressions to solve manually in order to find the elastic section modulus $S_c$. Fortunately, it is not necessary to solve them manually. They can be solved readily by a small computer and the results tabulated for ready reference. Such a tabulation is presented in the Appendix in Tables A-5 through A-7.

With the section modulus known, however computed, the moment of inertia is readily computed from the definition of the section modulus:

$$S_c = \frac{I}{c} \qquad \text{hence} \qquad I = S_c kd \tag{4-11}$$

The distance from the centerline of section to the center of force on the concrete will also be needed later:

$$y_w d = \frac{d + gd}{2} - \frac{kd}{3} = \left(\frac{1 + g}{2} - \frac{k}{3}\right) d \tag{4-12}$$

Since the factor $y_w$ in Eq. (4-12) is based on the effective depth $d$ rather than the total depth $h$, the value of $y_w$ may sometimes be larger than 0.5.

It will also be useful to know the ratio of stresses $f_s/f_c$:

$$\frac{f_s}{f_c} = \frac{n(1 - k)}{k} \tag{4-13}$$

A table of the foregoing section properties is presented in Tables A-5 through A-7 of the Appendix. The tables are entered with known values of $f_s/f_c$ and the ultimate strength of the concrete. The columns on the tables with the heading ''At Service Loads'' reflect the tabulated values of Eqs. (4-8) and (4-10) through (4-13). To distinguish the ratio $k$ from that of the ultimate load analysis (presented later), a subscript $w$ has been added to $k$ in the tables.

Tables A-5 through A-7 include all the foregoing limitations and variations. No values are given that are not permitted by Code. The elastic constants of Tables A-5 through A-7 are applicable for any service stress up to and including the maximum allowable working stress.

Only those section constants that are used directly are included in the tables. Other section constants that are needed only occasionally may be computed from the tabled values. For example, the section modulus on the tension side (taken at the tension steel) may be computed:

$$S_t = \frac{S_c}{f_s/f_c} \tag{4-14}$$

Other constants may be derived as necessary.

The use of the foregoing relationships is demonstrated in the subsequent examples using the working stress method. The allowable flexure stresses in the working stress method are prescribed by Code (B.3) and are given in Table A-1. The examples consider only rectangular cross sections for beams; tee sections are discussed in detail in Chapter 6.

**Example 4-1**

*Working stress method*
*Design of a rectangular section in flexure*
*No limitations on dimensions*

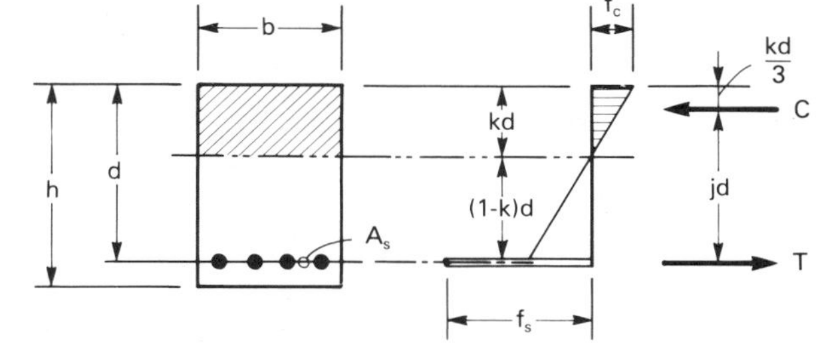


*Given:*
$M_w = 78$ kip-ft
Grade 60 steel
$f'_c = 3000$ psi
Interior exposure

From Table A-1, $f_c = 0.45f'_c = 1350$ psi, $f_s = 24{,}000$ psi.

Compute $f_s/f_c = 24{,}000/1350 = 17.78$ (maximum allowable).

Enter Table A-5: select $\rho = 0.010$, $S_c = 0.154bd^2$.

Compute the section modulus:

$$\frac{M_w}{f_c} = S_c \qquad \frac{78 \times 10^3 \times 12 \text{ lb-in.}}{1350 \text{ lb/in}^2} = 693 \text{ in}^3 = 0.154bd^2$$

Assume that $b = 0.6 \times d$; then

$$693 \text{ in}^3 = 0.154 \times 0.6d \times d^2$$

Solve for $d$: $d = 19.6$ in., $b = 11.7$ in., and $A_s = \rho bd = 0.010 \times 19.6 \times 11.7 = 2.29$ in$^2$.

From Table A-3, select 3 No. 8 bars; hence, the $A_s$ provided is 2.36 in$^2$, slightly more than the 2.29 in$^2$ required. The dimensions are rounded, yielding

$$b = 12 \text{ in.} \qquad d = 20 \text{ in.} \qquad A_s = 3 \text{ No. 8 bars}$$

The final section is shown in the following sketch. Note that the overall dimension $h$ does not enter directly into the calculations, but is set at some even number once $d$ is known.

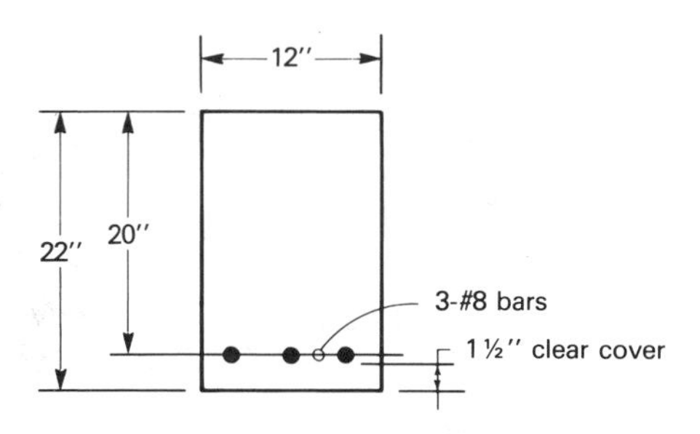


On occasion, it becomes necessary to limit the overall depth of a beam. When there is no corresponding limit on the width $b$, one way to design the member is to make it wider and flatter. Such a case is considered in the next example.

**Example 4-2**

*Working stress method*
*Design of a rectangular section in flexure*
*Depth of section limited*

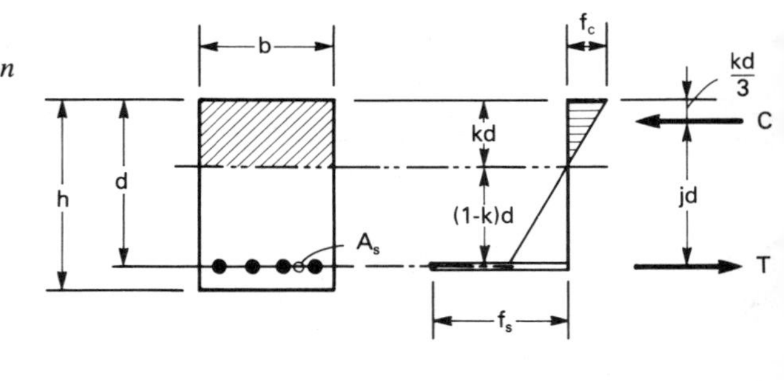


*Given:*
$M_w$ = 68 kip-ft
Grade 50 steel
$f'_c$ = 3000 psi
Overall depth limited to 16 in.
Interior exposure

With overall depth $h$ limited, it is necessary to estimate the maximum allowable value for $d$, as shown in the following sketch.

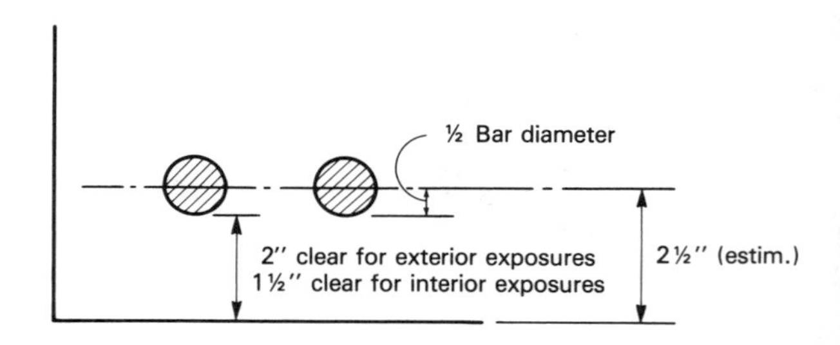


From Table A-1, $f_c = 0.45f'_c = 1350$ psi, $f_s = 20{,}000$ psi.

Compute $f_s/f_c = 20{,}000/1350 = 14.8$ (maximum allowable).

Enter Table A-5 and interpolate: select $\rho = 0.0130$; $S_c = 0.169bd^2$.

Compute the section modulus:

$$\frac{M_w}{f_c} = S_c \qquad \frac{68 \times 10^3 \times 12}{1350} = 604 \text{ in}^3 = 0.169bd^2$$

For $d = h - 2.5$ in. $= 13.5$ in., solve for $b = 19.6$ in., and $A_s = \rho bd = 0.013 \times 19.6 \times 13.5 = 3.44$ in$^2$.

From Table A-3, select 2 No. 9 and 2 No. 8 bars.

Use $b = 20$ in., $h = 16$ in., 2 No. 9 and 2 No. 8 bars.

Note that if this width $b$ had been unacceptable for some reason, it would have been necessary to limit the width also. That case is considered in the next example.

**Example 4-3**

*Working stress method*
*Design of a rectangular section in flexure*
*Both depth d and width b limited*

*Given:*
$M_w$ = 76 kip-ft
Grade 40 steel
$f'_c$ = 3000 psi
Overall depth limited to 20 in. and width to 12 in.
Exterior exposure

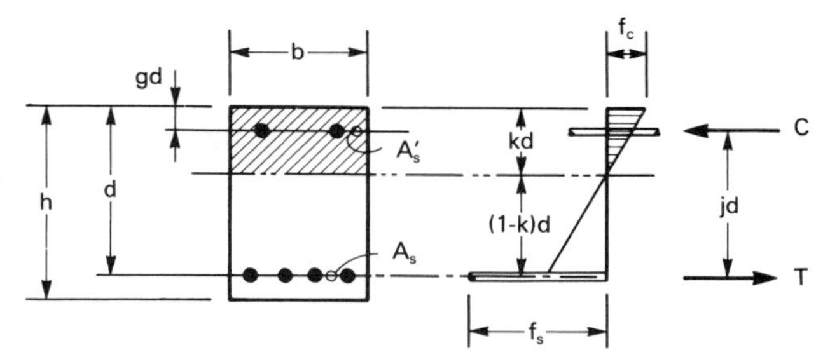


With depth $h$ limited, it is necessary to estimate maximum effective depth $d$, as shown in the following sketch.

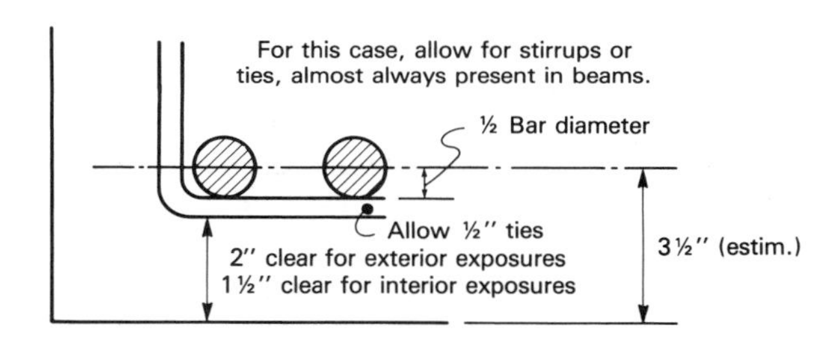


From Table A-1, $f_c = 0.45f'_c = 1350$ psi, $f_s = 20{,}000$ psi.

Compute $f_s/f_c = 20{,}000/1350 = 14.8$ (maximum allowable).

The required $S_c$ from the tables will have the form:

$$S_c = \text{coefficient} \times bd^2, \text{ maximum } d = 20 - 3.5 = 16.5 \text{ in.}$$

Compute:

$$\frac{M_w}{f_c} = S_c \qquad \frac{76 \times 10^3 \times 12}{1350} = \text{coeff.} \times 12 \times 16.5 \times 16.5$$

Solve for coeff.: coeff. = 0.207. This coefficient is seen to be larger than the coefficient of 0.169 interpolated from Table A-5, for $f_s/f_c = 14.8$ and $\rho' = 0$, but considerably less than the coefficient of 0.284, where $f_s/f_c = 14.8$ and $\rho' = 0.5$. Interpolate between the two values of $\rho'$; keep $f_s/f_c$ at 14.8 and the coefficient of $S_c$ at 0.207. Find the steel ratio $\rho$ to be 0.016 and $\rho' = 0.33\rho$. The required steel area is then:

$$A_s = 0.016 \times 12 \times 16.5 = 3.17 \qquad A_s = 2 \text{ No. 9 and 2 No. 7 bars}$$

$$A'_s = 0.33A_s, \text{ use 2 No. 7 bars}$$

Note that if $f_s/f_c$ were greater than 14.8, the tensile steel would be overstressed when the concrete stress is at its allowable value of $0.45f'_c$.

## ELASTIC FLEXURE PLUS AXIAL FORCE

The foregoing discussions and examples considered rectangular beams loaded only in flexure. In addition to flexure, a beam may be subjected occasionally to an axial load, either in compression or tension. Where the axial load is small, the design of such members can be accomplished by a relatively simple innovation; a separate analytical approach is not necessary.

Although not common, the phenomenon of a beam being subject to a small axial load is not a rare occurrence. A typical example is a basement floor being subject to inward axial compression from the soil pressure against the walls. An example of a tension member might be a roof beam over a water tank, where the water pressure exerts an outward load against the walls.

When the term ''small axial load'' is used, it means that the axial load $P_w$ should be no greater than about $0.06 f'_c bh$ (author's estimate—not specified by Code). When the axial load is greater than this, the member ceases being a beam carrying an axial load and becomes a column carrying a flexural load. The design of column members is quite different from flexural members; it is discussed in detail in Chapter 8.

The design of members subject both to a moment $M_w$ and to an axial compressive force $P_w$ (Fig. 4-2) can be taken in three steps:

1. Using regular procedures, find the width $b_1$ and the depth $d$ for a reinforced section subject only to the design moment $M_w$.
2. Expand the triangular compression block laterally (without reinforcement) by a width $b_2$ such that the design load $P_w$ is exactly opposed.

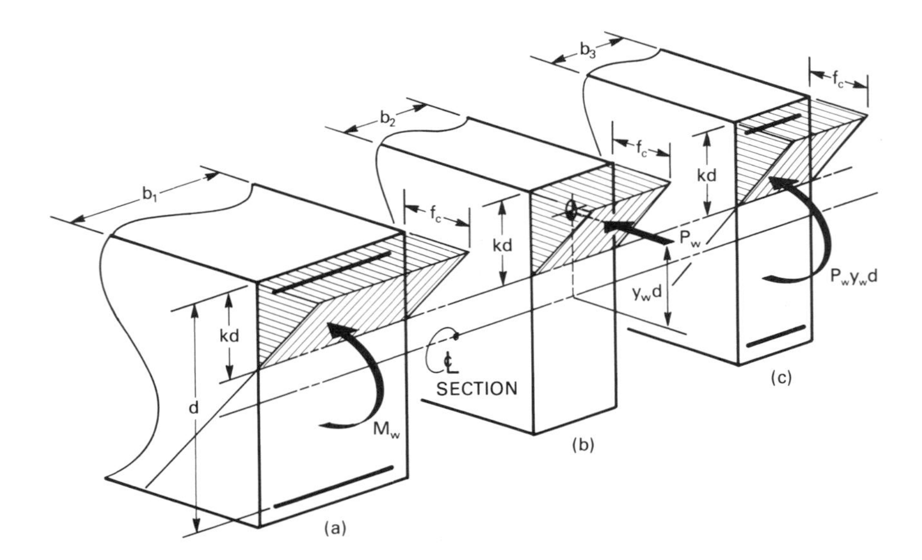


**Figure 4-2** Flexure Plus Small Axial Force.

3. Deduct a width $b_3$ to compensate for the moment induced due to the eccentricity of the load $P_w$.

The final required width is the algebraic sum of these three widths.

When the force $P_w$ is in tension rather than compression, the sign is simply reversed; steps 2 and 3 become "deduct" and "expand," respectively. For either case, the line of action of the axial load is assumed to be at the centerline of the section.

The foregoing steps are illustrated in Fig. 4-2. Figure 4-2a shows a stress diagram for a beam subjected to a design moment $M_w$. When no axial force $P_w$ exists, this stress diagram is the final stress diagram, and is the same as those discussed earlier.

If a compressive load $P_w$ exists, the triangular compression stress block of Fig. 4-2a can be imagined to be expanded laterally without reinforcement to oppose this load as shown in Fig. 4-2b. When expanding the stress block, the rotation of the section is held constant, such that $f_c$ and $kd$ remain unchanged. The force in the extension of the stress block is then equated to the load $P_w$:

$$P_w = \tfrac{1}{2} f_c b_2 kd \tag{4-15}$$

(For the sake of symmetry, it may be imagined that the actual expansion of the stress block takes place in strips between the reinforcing bars.)

As a complication to the procedure, it is noted that the resultant of the stress block is eccentric to the centerline of the member by the amount $y_w d$. The resulting moment $P_w y_w d$ permits a reduction in the flexural width $b_1$. The width $b_3$ is shown in Fig. 4-2c and represents this reduction.

Note that the final steel area $A_s$ is computed for the least flexural width $b_1 - b_3$, not for the total width of $b$. This "least flexural width" is also designated $b_m$ in later discussions. The steel ratio $\rho$, however, is the same for both $b_1$ and $b_3$.

Note also that the existence of the axial force $P_w$ permits a net reduction in $A_s$ due to the "prestressing" effect of $P_w$. If the force $P_w$ is in fact intermittent, the design should be done both with and without the axial force $P_w$ and the more conservative section used as the final design. Some examples will illustrate the procedure. In this first example, the load $P_w$ is assumed to be compressive.

**Example 4-4**

*Working stress method*
*Design of a rectangular section for flexure in combination with a small axial compression*
*No limitations on dimensions*

*Given:*
$M_w$ = 66 kip-ft
$P_w$ = 28 kips
$f'_c$ = 4000 psi
Grade 40 steel

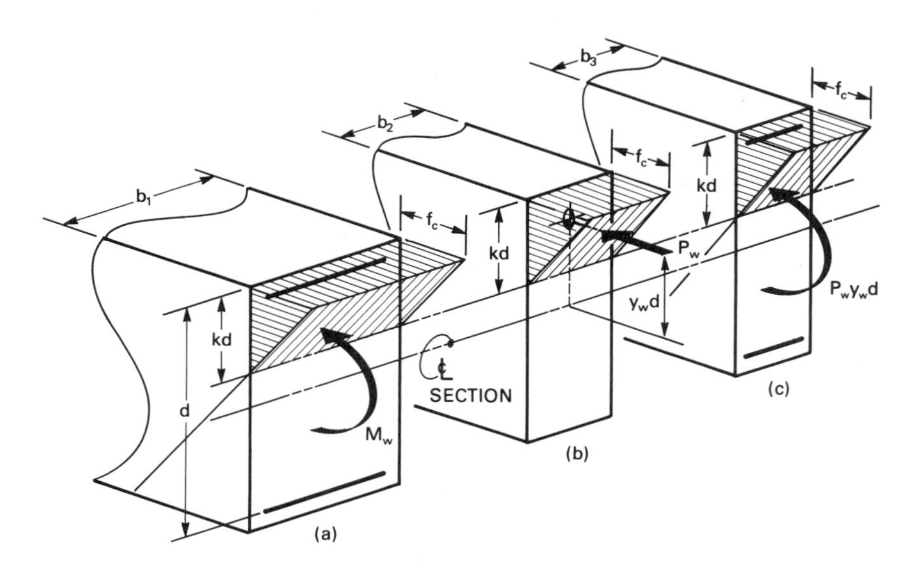


From Table A-1, $f_c = 0.45f'_c = 1800$ psi, $f_s = 20{,}000$ psi.

Compute $f_s/f_c = 20{,}000/1800 = 11.11$ (maximum allowable).

Enter Table A-6: $S_c = 0.181bd^2$, $k_w = 0.422$, $\rho = 0.019$, $y_w = 0.422$.

Solve for $b_1$ for the reinforced section:

$$\frac{M_w}{f_c} = \frac{66 \times 10^3 \times 12}{1800} = S_c = 0.181b_1d^2$$

Assume that $b_1 = 0.5d$; then solve for $d = 17.0$ in., $b_1 = 8.5$ in.

Add $b_2$ due to axial force $P_w$, where $P_w = \frac{1}{2}f_cb_2k_wd$:

$$b_2 = \frac{P_w}{\frac{1}{2}f_ck_wd} = \frac{28{,}000}{\frac{1}{2} \times 1800 \times 0.422 \times 17.0} = 4.3 \text{ in.}$$

Deduct $b_3$ due to eccentricity of $P_w$, where $P_wy_wd = f_cS_c$:

$$b_3 = \frac{P_wy_wd}{0.181 \times f_cd^2} = \frac{28{,}000 \times 0.422 \times 17}{0.181 \times 1800 \times 17 \times 17} = 2.1 \text{ in.}$$

$$\text{final } b = b_1 + b_2 - b_3 = 8.5 \text{ in.} + 4.3 \text{ in.} - 2.1 \text{ in.} = 10.7 \text{ in., say 11 in.}$$

$$\text{least flexural width } b_m = 8.5 - 2.1 = 6.4 \text{ in.,}$$

$$A_s = 0.019 \times 6.4 \times 17 = 2.07 \text{ in}^2$$

Use $b = 11$ in., $d = 17$ in., $A_s = 2$ No. 7 and 2 No. 6 bars.

Check: $P_{max} = 0.06f'_cbh = 0.06 \times 4000 \times 11\,(17 + 3.5)$.

Solve for $P_{max}$, $P_{max} = 54$ kips $> 28$ kips applied (O.K.).

A second example will illustrate the procedure when the axial load is tensile.

**Example 4-5**

*Working stress method*
*Design of a rectangular section for flexure in combination with a small axial tension*
*No limitations on dimensions*

*Given:*
$M_w$ = 60 kip-ft
$T_w$ = 20 kips
Grade 60 steel
$f'_c$ = 3000 psi

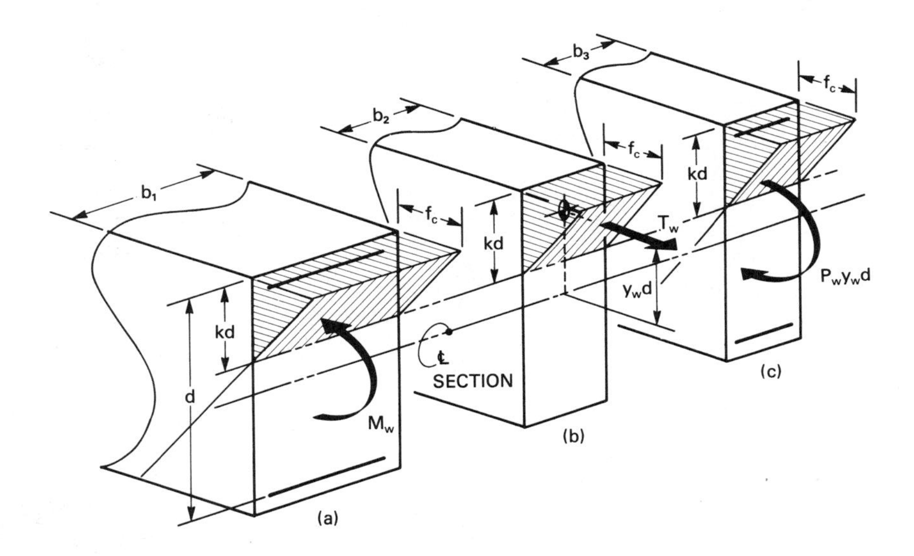


From Table A-1, $f_c = 0.45f'_c = 1350$ psi, $f_s = 24{,}000$ psi.

Compute $f_s/f_c = 24{,}000/1350 = 17.8$.

Enter Table A-5: $\rho = 0.010$, $S_c = 0.154bd^2$, $k_w = 0.349$, $y_w = 0.446$.

Solve for $b_1$ for reinforced section due to moment $M_w$:

$$\frac{M_w}{f_c} = S_c \qquad \frac{60 \times 10^3 \times 12}{1350} = 533 \text{ in}^3 = 0.154b_1d^2$$

For $b_1 = 0.7d$, solve for $d = 17$ in., $b_1 = 11.9$ in.

Deduct $b_2$ due to tensile load $T_w$, where $T_w = \frac{1}{2}f_c b_2 k_w d$,

$$b_2 = \frac{T_w}{\frac{1}{2}k_w d f_c} = \frac{20{,}000}{\frac{1}{2} \times 0.349 \times 17 \times 1350} = 5.0 \text{ in.}$$

Add $b_3$ due to the eccentricity of $T_w$; the moment is then $T_w y_w d$:

$$b_3 = \frac{T_w y_w d}{0.154 f_c d^2} = \frac{20{,}000 \times 0.446 \times 17}{0.154 \times 1350 \times 17 \times 17} = 2.5 \text{ in.}$$

$$\text{final } b = b_1 - b_2 + b_3 = 11.9 \text{ in.} - 5.0 \text{ in.} + 2.5 \text{ in.} = 9.4 \text{ in.}$$

$$\text{least flexural width } b_m = 11.9 \text{ in.} + 2.5 \text{ in.} = 14.4 \text{ in.}$$

$$A_s = 0.010 \times 14.4 \times 17 = 2.45 \text{ in}^2$$

Use $b = 10$ in., $d = 17$ in., $A_s = 2$ No. 10 bars.

When the depth of the section is restricted, the same procedure applies, as shown in the next example.

**Example 4-6**

*Working stress method*
*Design of a rectangular section for flexure in combination with a small axial compression*
*Depth of section limited*

*Given:*
$M_w = 96$ kip-ft
$P_w = 22$ kips
Grade 60 steel
$f'_c = 3000$ psi
Depth $d$ limited to 18 in.

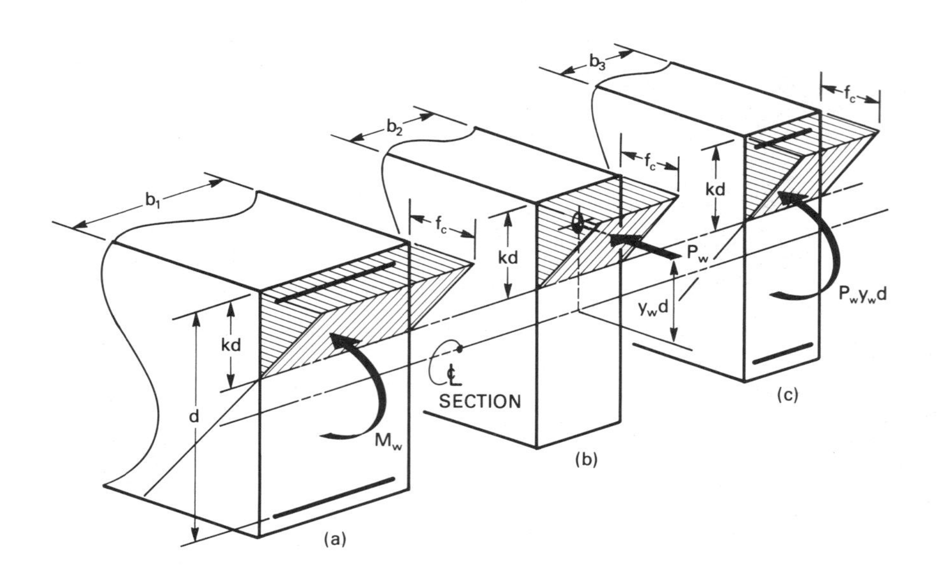

From Table A-1, $f_c = 0.45f'_c = 1350$ psi, $f_s = 24{,}000$ psi.

Compute $f_s/f_c = 24{,}000/1350 = 17.8$.

Enter Table A-5: $\rho = 0.010$, $S_c = 0.154bd^2$, $k_w = 0.349$, $y_w = 0.446$.

Solve for $b_1$ for the reinforced section:

$$\frac{M_w}{f_c} = S_c \qquad \frac{96 \times 10^3 \times 12}{1350} = 853 \text{ in}^3 = 0.154b_1d^2$$

With $d$ fixed at 18 in., solve for $b_1 = 17.1$ in.

Add $b_2$ due to axial load $P_w$, where $P_w = \frac{1}{2}f_cb_2k_wd$.

$$b_2 = \frac{P_w}{\frac{1}{2}f_ck_wd} = \frac{22{,}000}{\frac{1}{2} \times 1350 \times 0.349 \times 18} = 5.2 \text{ in.}$$

Deduct $b_3$ due to eccentricity of $P_w$, where $P_wy_wd = f_cS_c$:

$$b_3 = \frac{P_wy_wd}{0.154f_cd^2} = \frac{22{,}000 \times 0.446 \times 18}{0.154 \times 1350 \times 18 \times 18} = 2.6 \text{ in.}$$

$$\text{final } b = b_1 + b_2 - b_3 = 17.1 + 5.2 - 2.6 = 19.7 \text{ in., say 20 in.}$$

$$\text{least flexural width } b_m = 17.1 - 2.6 = 14.5 \text{ in.}$$

$$A_s = 0.010 \times 14.5 \times 18 = 2.6 \text{ in}^2$$

Use $b = 20$ in., $d = 18$ in., $A_s = 6$ No. 6 bars.

If the computed width $b$ is unacceptable, it may also become necessary to limit $b$. For such cases, where both $b$ and $d$ are fixed, the solution becomes a trial-and-error solution to find a workable area of reinforcement, both for tensile steel and for compressive steel. The following example shows the procedure.

**Example 4-7**

*Working stress method*
*Design of a rectangular section for flexure in combination with a small axial compression*
*Both depth d and width b restricted*

*Given:*
$M_w = 96$ kip-ft
$P_w = 22$ kips
Grade 60 steel
$f'_c = 3000$ psi
Depth $d$ limited to 18 in. and $b$ to 15 in.

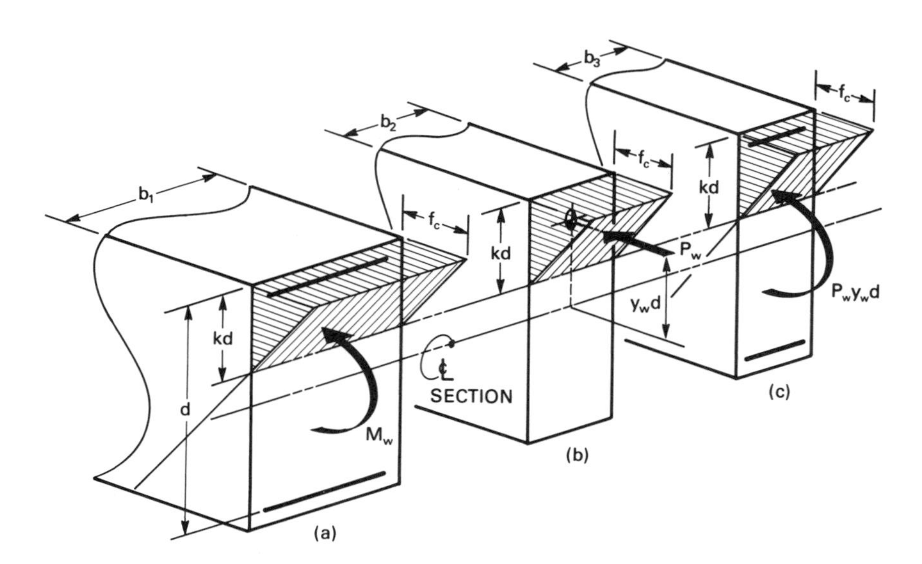


Since there are an infinite number of steel arrangements that would afford a satisfactory solution, the procedure is simply that of finding a steel ratio that meets the immediate requirements, a trial-and-error procedure. *Only the final trial is shown in the following calculations.*

From Table A-1, $f_c = 0.45f'_c = 1350$ psi, $f_s = 24{,}000$ psi.

Compute $f_s/f_c = 24{,}000/1350 = 17.8$.

From Table A-5, try $\rho = 0.014$, $\rho' = \frac{1}{2}\rho$, $S_c = 0.221b_1d^2$, $k_w = 0.343$, $y_w = 0.448$.

Solve for width $b_1$ required for flexure:

$$\frac{M_w}{f_c} = S_c \qquad \frac{96 \times 10^3 \times 12}{1350} = 853 \text{ in}^3 = 0.221b_1d^2 = 0.221b_1 \times 18 \times 18$$

Solve: $b_1 = 11.9$ in.

Add $b_2$ due to axial load, $P_w = \frac{1}{2}f_cb_2k_wd$:

$$b_2 = \frac{P_w}{\frac{1}{2}f_ck_wd} = \frac{22{,}000}{\frac{1}{2} \times 1350 \times 0.343 \times 18} = 5.3 \text{ in.}$$

Deduct $b_3$ due to eccentricity of axial load $P_wy_wd = f_cS_c$:

$$b_3 = \frac{P_wy_wd}{0.221f_cd^2} = \frac{22{,}000 \times 0.448 \times 18}{0.221 \times 1350 \times 18 \times 18} = 1.8 \text{ in.}$$

$$\text{required } b = b_1 + b_2 - b_3 = 11.9 + 5.3 - 1.8 = 15.4 \text{ in.} \quad \text{(O.K.—use)}$$

Note that the computed width $b$ is close enough to the allowable width of 15 in. to be an acceptable solution. For this case

$$A_s = 0.014 \times 15 \times 18 = 3.78 \text{ in}^2$$

$$A_s' = \tfrac{1}{2} \times 0.014 \times 15 \times 18 = 1.89 \text{ in}^2$$

Use $d$ = 18 in., $b$ = 15 in., $A_s$ = 5 No. 8 bars, and $A_s'$ = 2 No. 9 bars.

## INVESTIGATION OF ELASTIC FLEXURE*

On occasion, the reverse of the design problem occurs: It is the size and reinforcement of a member that is known and it is the loads or stresses that must be found. This type of solution is called "investigation" or "review" of a section. Where the axial load is zero, the investigation of a section is usually quite direct, as shown in the following examples.

**Example 4-8**

*Working stress method*

*Investigation of a rectangular section subject to a flexural loading to determine the allowable moment*

*Given:*

Section as shown

Grade 60 steel

$f_c' = 4000$ psi

Stresses $f_s$ and $f_c$ may not exceed the allowables

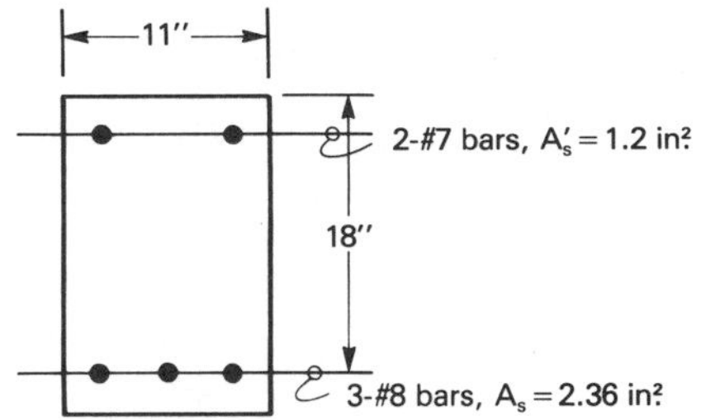


Note that neither $f_s$ nor $f_c$ are known. Also, since it is not known whether the stress in the steel or the stress in the concrete will limit the capacity of the section, it will be necessary to investigate both. With neither $f_s$ nor $f_c$ known, the usual access into the design tables using $f_s/f_c$ is not possible. Alternatively, however, the value of $\rho$ can be found, affording a second means to enter the tables.

Compute $\rho = A_s/bd$:

$$\rho = \frac{2.36}{11 \times 18} = 0.012 \qquad \text{and} \qquad \rho' = 0.50\rho \ \text{(approximately)}$$

Enter Table A-6: $S_c = 0.189bd^2$, $f_s/f_c = 17.7$, $k_w = 0.314$, $y_w = 0.458$.

With $f_c = 0.45f_c' = 1800$ psi, $f_s = 17.7 \times 1800 = 31{,}860$ psi (no good—maximum allowable $f_s$ is 24,000 psi).

With $f_s = 24{,}000$ psi, $f_c = 24{,}000/17.7 = 1356$ psi $< 1800$ psi.

Use $f_s = 24{,}000$ psi, $f_c = 1356$ psi.

$$\text{maximum moment} = f_c S_c = 1356 \times 0.189 \times 11 \times 18 \times 18.$$

$$\text{maximum } M_w = 913 \text{ kip-in.} = 76 \text{ kip-ft}$$

*This section is optional in a limited design course.

**Example 4-9**

*Working stress method*
*Investigation of a rectangular section subject to flexural loading to find the maximum stress in the steel and concrete*

*Given:*
Section as shown
$M_w$ = 64 kip-ft
Grade 60 steel
$f'_c$ = 4000 psi

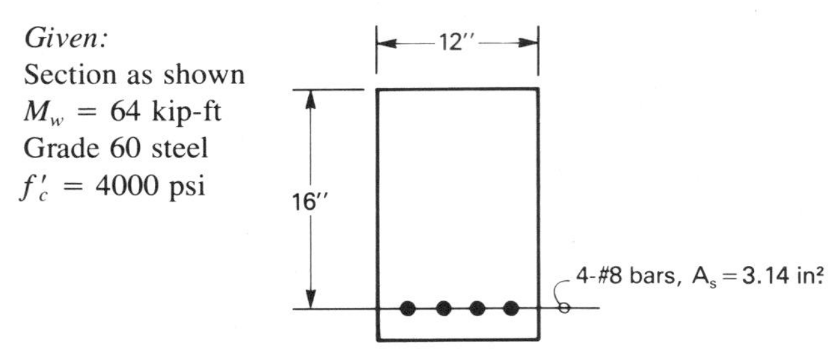


Compute $\rho = A_s/bd$:

$$\rho = \frac{3.14}{12 \times 16} = 0.0164$$

Enter Table A-6: $S_c = 0.173bd^2$, $f_s/f_c = 12.12$.

Compute $S_c = 0.173 \times 12 \times 16 \times 16 = 531 \text{ in}^3$.

Compute the maximum stress in the concrete:

$$f_c = \frac{M_w}{S_c} = \frac{64 \times 10^3 \times 12}{531} = 1446 \text{ psi}$$

Compute the maximum stress in the steel:

$$f_s = \frac{f_s}{f_c} f_c = 12.12 \times 1446 = 17{,}526 \text{ psi}$$

maximum stress in concrete = 1450 psi

maximum stress in steel = 17,500 psi

In Example 4-9 there is no concern whether the flexural stresses are high or low. The problem was simply to find the magnitudes of the stresses. It is noted, however, that the design of the beam is fairly well balanced, since both the steel and concrete will be stressed near their maximum allowable values simultaneously.

When the axial load is not zero, as it was in the last two examples, the procedure becomes more complex. As long as the final values of $f_s$ and $f_c$ are prescribed or known, however, there is a means of access into the design tables and the solution is still quite direct. The following example will illustrate the point.

**Example 4-10**

*Working stress method*
*Investigation of a rectangular section subject both to flexure and a small axial compression to determine $M_w$ and $P_w$ when stresses are maximum*

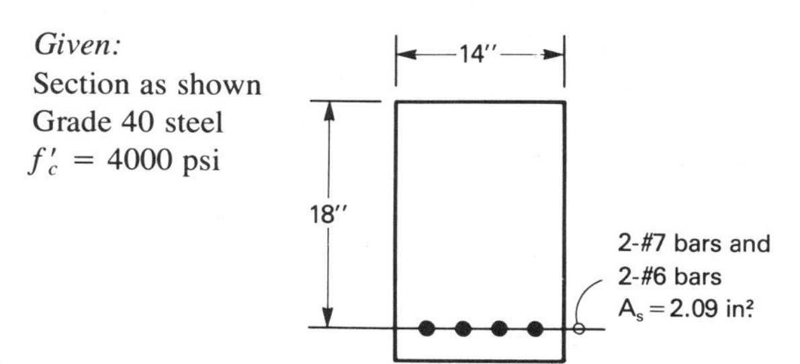


Compute the maximum allowable values for $f_s$ and $f_c$:

$$f_c = 0.45f'_c = 0.45 \times 4000 = 1800 \text{ psi}$$

$$f_s = 20{,}000 \text{ psi}$$

Compute the stress ratio $f_s/f_c = 20{,}000/1800 = 11.1$.

From Table A-6: $S_c = 0.181bd^2$, $\rho = 0.019$, $k_w = 0.420$, $y_w = 0.422$.

Compute the least flexural width $b_m$, where $\rho = A_s/b_m d$:

$$b_m = \frac{A_s}{\rho d} = \frac{2.09}{0.019 \times 18} = 6.1 \text{ in.}$$

With $b_m$ known, compute $b_2 = b - b_m$:

$$b_2 = 14 \text{ in.} - 6.1 \text{ in.} = 7.9 \text{ in.}$$

Compute the axial force $P_w = \frac{1}{2}f_c b_2 k_w d$:

$$P_w = \tfrac{1}{2} \times 1800 \times 7.9 \times 0.422 \times 18 = 53.8 \text{ kips}$$

Compute the allowable moment on the least flexural width:

$$M_w - P_w y_w d = f_c S_c = 1800 \times 0.181 \times 6.1 \times 18 \times 18 = 644 \text{ kip-in.}$$

Compute the allowable moment $M_w = 644 + P_w y_w d$:

$$M_w = 644 + 53.8 \times 0.422 \times 18 = 1053 \text{ kip-in.} = 88 \text{ kip-ft}$$

Working loads: $P_w = 53.8$ kips; $M_w = 88$ kip-ft.

It should be remembered that the maximum value recommended for $P_w$ should be less than about $0.06f'_c bh$ or $0.06 \times 4000 \times 14 \times (18 + 3) = 71$ kips. This value is seen to be greater than the applied load of 54 kips, so the solution is acceptable. If $P_w$ were greater than $0.06f'_c bh$, the member would have to be designed as a column.

A more complex problem occurs when both the moment $M_w$ and the axial load $P_w$ are knowns and the stresses $f_s$ and $f_c$ are the unknowns. For this case, the ratio $f_s/f_c$ cannot be determined and the steel ratio $\rho = A_s/b_m d$ cannot be computed since the least flexural width $b_m$ is also unknown. There is thus no direct access into the tables of section constants.

A trial-and-error solution can be used to find the least flexural width $b_m = b_1 - b_3$. As with most trial-and-error procedures, an ordered approach greatly simplifies the work. The following approach at guessing the final value of $b_m$ has been found to be reasonably direct.

For a first iteration, assume that the least flexural width is half the total width $b$. With this assumption, however incorrect, one-half of the total width of the section then becomes the first approximation for the least flexural width, $b_m$. The steel ratio $\rho = A_s/b_m d$ can then be computed, affording access to the tables of section constants.

With a set of section constants thus obtained, the additional width $b_2$ required to carry $P_w$ can be computed. (Note that this computed value of $b_2$ will *not* be the other half of the width.) The total required width for this first iteration is then $b_{i1} = b_{m1} + b_{21}$ and will be somewhat different from the given width $b$; the value of $b_m$ must be corrected accordingly.

For the second iteration, assume that although the first value of $b_m$ is known to be incorrect, its proportion of the total computed width $b_{i1}$ is approximately correct. By ratio, therefore, $b_m$ can be corrected proportionately such that it corresponds to the actual width $b$. The second approximation of $b_m$ is then:

$$b_{m2} = \frac{b_{m1}}{b_{i1}} b \tag{4-16}$$

With this second approximation for $b_m$, the procedure is repeated: the new steel ratio $\rho$ is computed, the new width $b_2$ is found, and the new total width $b_{i2}$ is found.

For the third iteration, it is again assumed that the earlier ratio of $b_m$ to the total iterated width $b_{i2}$ is becoming more correct (which it is). By ratio, again correct $b_m$ proportionately:

$$b_{m3} = \frac{b_{m2}}{b_{i2}} b \tag{4-17}$$

The computations are again repeated for these corrected values of $b_{m3}$.

Usually, the second or third iteration will yield a value for $b_m$ that no longer changes and is thus an acceptable solution. Any successive iterations that are required, however, would follow the established pattern. A change in $b_m$ of less than 5 to 10% should be considered satisfactory.

For very heavily reinforced sections the first approximation for $b_m$ using half the total width may produce a value for the steel ratio that is outside the tables. For such cases the first approximation for $b_m$ can be $0.75b$ or even $0.95b$. The remainder of the procedure still applies, regardless of what the first approximation is.

There are more direct methods in existence for solving this case, involving separate analysis, charts, tables, and so on. For the few times in a designer's working life that he or she encounters this problem, the trial-and-error solution described above, however clumsy, will yield satisfactory results. Note that no new concepts have been introduced beyond those used for design. Some examples will illustrate the procedure. In these examples, the sections contain compressive reinforcement.

**Example 4-11**

*Working stress method*

*Investigation of a rectangular section subject both in flexure and a small axial compression to determine the stresses in steel and concrete.*

*Given:*
Section as shown
$M_w = 83$ kip-ft
$P_w = 26$ kips
Grade 60 steel
$f'_c = 3000$ psi

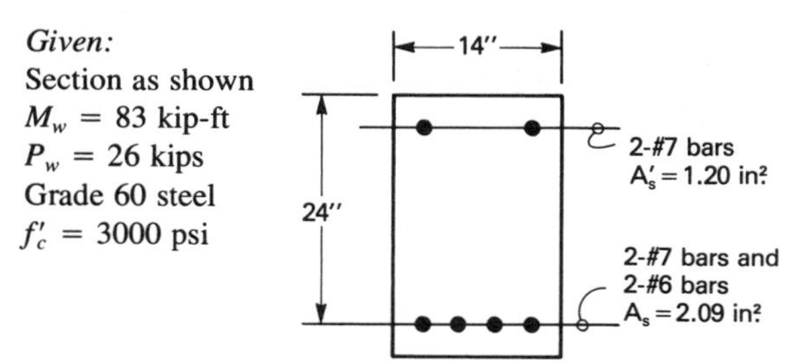


**First Iteration**

Assume that $b_m$ is one-half the given width, or $b_m = 7$ in. Compute the steel ratios to permit entry into Table A-5:

$$\rho = \frac{A_s}{b_m d} = \frac{2.09}{7 \times 24} = 0.012 \qquad \rho' = 0.5\rho \quad \text{(approximately)}$$

From Table A-5, $S_c = 0.204bd^2$, $k_w = 0.328$, $y_w = 0.453$.

Compute the stress in concrete corresponding to this $b_m$:

$$f_c = \frac{M_w - P_w y_w d}{S_c} = \frac{83 \times 10^3 \times 12 - 26 \times 10^3 \times 0.453 \times 24}{0.204 \times 7 \times 24 \times 24} = 867 \text{ psi}$$

Compute the additional width required for $P_w = \frac{1}{2} f_c b_2 k_w d$:

$$b_2 = \frac{P_w}{\frac{1}{2} \times f_c k_w d} = \frac{26{,}000}{\frac{1}{2} \times 867 \times 0.328 \times 24} = 7.6 \text{ in.}$$

The total width required for this iteration:

$$b_i = b_m + b_2 = 7 + 7.6 = 14.6 \text{ in.}$$

The assumed value for $b_m$ is thus seen to be too high. Its proportion of the total computed width, however, is seen to be

$$\frac{b_m}{b_i} = \frac{7}{14.6} = 0.479 \quad \text{or} \quad 48\% \text{ of the total}$$

Adjust $b_m$ to be 48% of the actual width of 14 in.

$$b_m = 0.48 \times 14 = 6.7 \text{ in.}$$

**Second Iteration**

(Same procedure as first iteration.) Assume that $b_m = 6.7$ in. Compute the steel ratios to permit entry into Table A-6:

$$\rho = \frac{A_s}{b_m d} = \frac{2.09}{6.7 \times 24} = 0.0130 \qquad \rho' = 0.5$$

From Table A-5, $S_c = 0.212bd^2$, $k_w = 0.336$, $f_s/f_c = 18.5$, $y_w = 0.451$ (interpolated).

Compute the stress in the concrete corresponding to this $b_m$:

$$f_c = \frac{M_w - P_w y_w d}{S} = \frac{83 \times 10^3 \times 12 - 26 \times 10^3 \times 0.451 \times 24}{0.212 \times 6.7 \times 24 \times 24} = 873 \text{ psi}$$

Compute the additional width required for $P_w = \frac{1}{2} f_c b_2 k_w d$:

$$b_2 = \frac{P_w}{\frac{1}{2} f_c k_w d} = \frac{26{,}000}{\frac{1}{2} \times 873 \times 0.336 \times 24} = 7.4 \text{ in.}$$

The total required width for a $b_m$ of 6.7 in.:

$$b = b_m + b_2 = 6.7 + 7.4 = 14.1 \text{ in.}$$

The assumed value of $b_m$ is seen to be well within the nominal 5% error. A third iteration is therefore unnecessary. The proportion of this $b_m$ to the computed width is seen to be

$$\frac{b_m}{b_i} = \frac{6.7}{14.1} = 0.475 \quad \text{or} \quad 47.5\% \text{ of the total}$$

Stress in concrete: $f_c = 873$ psi

Stress in steel: $f_s = 18.5 \times 873 = 16{,}150$ psi

The investigation thus reveals that both the concrete and steel are understressed under the given loads.

*Check:* The maximum recommended value for $P_w$ is

$$\text{maximum } P_w = 0.06 f'_c bh = 0.06 \times 4000 \times 14(24 + 3.5) = 92 \text{ kips}$$

The applied load of 26 kips is well within this maximum.

A second example will demonstrate the procedure when the given conditions specify one stress and one load. When one stress is known, it is a relatively simple matter to find the other.

**Example 4-12**

*Working stress method*

*Investigation of a rectangular section subject both to flexure and a small axial compression to find $P_w$ and $f_c$, given $M_w$ and $f_s$*

*Given:*
Section as shown
$M_w = 85$ kip-ft
Grade 60 steel
$f'_c = 4000$ psi
$f_s = 21{,}000$ psi

16″
22″
5-#6 bars
$A'_s = 2.21$ in.²
5-#6 bars
$A_s = 2.21$ in.²

Note that the stress in the steel, $f_s$, is less than the allowable stress of 24,000 psi.

**First Iteration**

Assume that $b_m$ is one-half the given width, or $b_m = 8$ in. Compute the steel ratios to permit entry into Table A-6:

$$\rho = \frac{A_s}{b_m d} = \frac{2.21}{8 \times 22} = 0.013 \qquad \rho' = \rho$$

From Table A-6, $S_c = 0.230\, b_m d^2$, $k_w = 0.290$, $f_s/f_c = 19.9$, $y_w = 0.466$ (interpolated).

Compute whichever of the two stresses is not prescribed:

$$f_c = \frac{f_s}{f_s/f_c} = \frac{21{,}000}{19.9} = 1055 \text{ psi}$$

Compute the moment acting on the width $b_m$:

$$M_w - P_w y_w d = f_c S_c = 1{,}055 \times 0.230 \times 8 \times 22 \times 22 = 940 \text{ kip-in.}$$

Compute the axial load $P_w$ (this is the axial load that must accompany the given moment if $b_m$ is to be 8 in.):

$$P_w = \frac{M_w - f_c S_c}{y_w d} = \frac{85 \times 10^3 \times 12 - 940{,}000}{0.466 \times 22} = 7.8 \text{ kips}$$

Compute the required width $b_2$ for this $P_w = \frac{1}{2} f_c b_2 k_w d$:

$$b_2 = \frac{P_w}{\frac{1}{2} f_c k_w d} = \frac{7800}{\frac{1}{2} \times 1055 \times 0.290 \times 22} = 2.3 \text{ in.}$$

$$\text{total required width} = b_m + b_2 = 8 + 2.3 = 10.3 \text{ in.}$$

The value of $b_m$ is seen to be low. Its proportion of the total computed width is

$$\frac{b_m}{b_i} = \frac{8}{10.3} = 0.777 \quad \text{or} \quad 77.7\% \text{ of the total}$$

Adjust $b_m$ to be 77.7% of the given width $b$:

$$b_m = 0.777 \times 16 = 12.4 \text{ in.}$$

**Second Iteration**

Assume that $b_m = 12.4$ in. Compute the steel ratios: $\rho = A_s/b_m d$.

$$\rho = \frac{2.21}{12.4 \times 22} = 0.008 \qquad \rho' = \rho$$

From Table A-6, $S_c = 0.171bd^2$, $k_w = 0.255$, $f_s/f_c = 23.69$, $y_w = 0.478$.

Compute the stress in the concrete:

$$f_c = \frac{f_s}{f_s/f_c} = \frac{21{,}000}{23.69} = 886 \text{ psi}$$

Compute the moment acting on the least flexural width:

$$M_w - P_w y_w d = f_c S_c = 886 \times 0.171 \times 12.4 \times 22 \times 22 = 909 \text{ kip-in.}$$

Compute the load $P_w$:

$$P_w = \frac{M_w - f_c S_c}{y_w d} = \frac{85 \times 10^3 \times 12 - 909{,}000}{0.478 \times 22} = 10.6 \text{ kips}$$

Compute the width required to carrying $P_w$: $P_w = \frac{1}{2} f_c b_2 k_w d$:

$$b_2 = \frac{P_w}{\frac{1}{2} f_c k_w d} = \frac{10{,}600}{\frac{1}{2} \times 886 \times 0.255 \times 22} = 4.3 \text{ in.}$$

$$\text{total width } b = b_m + b_2 = 12.4 + 4.3 = 16.7 \text{ in.} \quad \text{(O.K.—use)}$$

(This result is close enough to the given 16 in.)

Concrete stress: $f_c = 886$ psi

Axial load: $P_w = 10.6$ kips

**Example 4-13**

*Working stress method*

*Investigation of a rectangular section subject both to flexure and a small axial compression to find $M_w$ and $f_s$, given $P_w$ and $f_c$*

*Given:*
Section as shown
$P_w = 10$ kips
Grade 60 steel
$f'_c = 4000$ psi
$f_c = 900$ psi

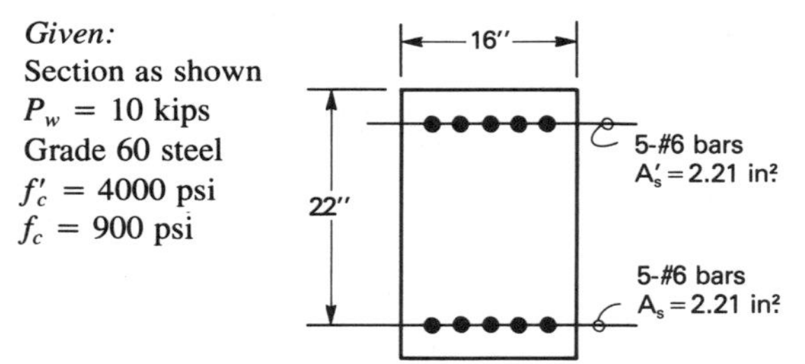


This example is essentially the same as Example 4-12, except that the given loads and stresses are reversed.

**First Iteration**

Compute steel ratios to permit entry into Table A-6, assuming that $b_m$ is half the total width, or 8 in.

$$\rho = \frac{A_s}{b_m d} = \frac{2.21}{8 \times 22} = 0.013 \qquad \rho' = \rho$$

From Table A-6, $S_c = 0.230bd^2$, $k_w = 0.290$, $f_s/f_c = 19.9$, $y_w = 0.466$.

Compute the additional width to carry $P_w$, $P_w = \frac{1}{2} f_c b_2 k_w d$:

$$b_2 = \frac{P_w}{\frac{1}{2} f_c k_w d} = \frac{10{,}000}{\frac{1}{2} \times 900 \times 0.290 \times 22} = 3.5 \text{ in.}$$

$$\text{total required width } b = b_m + b_2 = 8 + 3.5 = 11.5 \text{ in.}$$

The value of $b_m$ is seen to be low. Its proportion of the total computed width is

$$\frac{b_m}{b_i} = \frac{8}{11.5} = 0.696 \quad \text{or} \quad 69.6\% \text{ of the total width}$$

Adjust $b_m$ to be 69.6% of the given width of 16 in.:

$$b_m = 0.696 \times 16 = 11.1 \text{ in.}$$

This value of $b_m$ will be used in the second iteration. At this time it is well to observe that the moment $M_w$ has not appeared in the calculations. In assuming a value for $b_m$, the need for the value of $M_w$ has been obviated. It could easily be computed from the flexure formula, using the assumed value of $b_m$, but there is no need for it.

### Second Iteration

Assume that $b_m = 11.1$ in. Compute the steel ratios:

$$\rho = \frac{A_s}{b_m d} = \frac{2.21}{11.1 \times 22} = 0.009 \qquad \rho' = \rho$$

From Table A-6, $S_c = 0.183bd^2$, $k_w = 0.263$, $f_s/f_c = 22.73$, $y_w = 0.475$.

Compute the additional width $b_2$ to carry $P_w$, $P_w = \frac{1}{2}f_c b_2 k_w d$:

$$b_2 = \frac{P_w}{\frac{1}{2}f_c k_w d} = \frac{10{,}000}{\frac{1}{2} \times 900 \times 0.263 \times 22} = 3.8 \text{ in.}$$

$$\text{total required width } b = b_m + b_2 = 11.1 + 3.8 = 14.9 \text{ in.}$$

Although this value of $b$ is within 10% of the given value of 16 in., a third iteration will be made. For this third iteration, the ratio of $b_m$ to the total width is

$$\frac{b_m}{b_i} = \frac{11.1}{14.9} = 0.745 \quad \text{or} \quad 74.5\% \text{ of the total width}$$

Adjust $b_m$ to be 74.5% of the given width of 16 in.:

$$b_m = 0.745 \times 16 = 11.9 \text{ in.}$$

### Third Iteration

Use $b_m = 11.9$ in. Compute the steel ratios:

$$\rho = \frac{A_s}{b_m d} = \frac{2.21}{11.9 \times 22} = 0.008 \qquad \rho' = \rho$$

From Table A-6, $S_c = 0.171bd^2$, $k_w = 0.255$, $f_s/f_c = 23.69$, $y_w = 0.478$.

Add

$$b_2 = \frac{P_w}{\frac{1}{2}f_c k_w d} = \frac{10{,}000}{\frac{1}{2} \times 900 \times 0.255 \times 22} = 4.0 \text{ in.}$$

$$\text{total required width } b = b_m + b_2 = 11.9 + 4.0 = 15.9 \text{ in.}$$

This result is judged to be close enough to 16 in. The moment $M_w$ and the stress $f_s$ can now be computed:

$$M_w = f_c S_c + P_w y_w d = 886 + 105 = 991 \text{ kip-in} = 83 \text{ kip-ft}$$

$$f_s = f_c \frac{f_s}{f_c} = 900 \times 23.2 = 20{,}900 \text{ psi}$$

These results are seen to be very close to the starting values given in Example 4-12, verifying that two or three iterations are usually enough to yield a satisfactory result.

## ULTIMATE FLEXURE

The preceding sections dealt with design and investigation of members using the elastic method of analysis. The following sections deal with the same problems using the strength method. Although the final concern of the strength method is the ultimate load (rather than stress), the analysis must start with the usual stress distribution and develop that into the corresponding ultimate load.

The analysis for the ultimate moment on a concrete section is stringently prescribed by Code (10.2). To begin, it is required by Code (10.2.3) that the maximum strain in the concrete shall be 0.003 at ultimate moment, which includes all effects of shrinkage and creep. To accompany this prescribed strain in the concrete, the strain in the steel must take whatever value is required to produce equilibrium.

Code (10.2.2) also requires that the variation in strains across the section shall be assumed to follow a straight-line variation. The variations in stress should then follow the stress–strain curve of Fig. 1-4, but it does not matter since Code (10.2.7) also specifies an empirically derived "stress block" to be used in computing the compressive force. In all derivations it is assumed that both the concrete and the tensile steel are well into yield, but that compressive steel may or may not be in yield.

The straight-line variation in strain is shown graphically in Fig. 4-3b. The symbol $\epsilon_{sy}$ denotes the yield strain for the steel; the maximum strain in concrete is 0.003 in./in. Line $OA$ indicates the strains just as the tensile steel starts to enter yield, $OB$ just as the

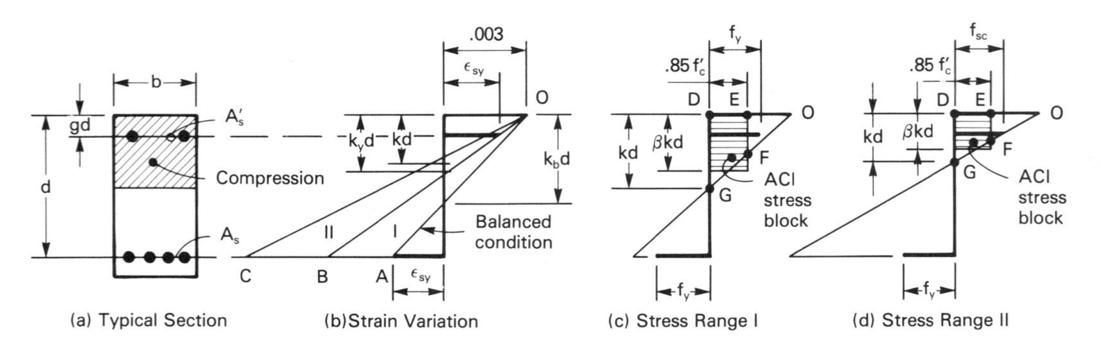


**Figure 4-3** Ultimate Conditions.

compressive steel starts to emerge from yield, and $OC$ at some arbitrarily chosen maximum rotation of the section.

Before proceeding, it is important to recognize the stress–strain relationships in Fig. 4-3. As the rotation progresses up to line $OA$, stress in the tensile steel increases as the strain increases. When the rotation progresses beyond $OA$ toward $OB$, strain in the tensile steel continues to increase but the stress in the steel cannot increase further; the steel simply yields and stays at $f_y$ regardless of how much additional strain occurs. Similarly, on the compression side, the concrete is also considered to be in yield whenever the stress exceeds the idealized yield stress of $0.85f'_c$, producing the trapezoidal stress block $DEFG$ shown in Fig. 4-3c and d.

The variations in stress that accompany the strain diagrams of Fig. 4-3b are shown in Fig. 4-3c and d. In Fig. 4-3c, the compressive steel is still in yield; in Fig. 4-3d it has emerged from yield and is in its elastic range. In both cases, the tensile steel remains in yield, denoted $f_y$.

The value of $k$ that occurs just as the compressive steel emerges from yield, designated $k_y$, can be computed by straight-line ratios from the strain diagram of Fig. 4-3b:

$$\frac{0.003}{k_y d} = \frac{\epsilon_{sy}}{k_y d - gd} \qquad \text{where } \epsilon_{sy} = \frac{f_y}{E_s} \tag{4-18}$$

The solution for $k_y$ is, with $E_s = 29{,}000$ psi,

$$k_y = \frac{0.003E_s g}{0.003E_s - f_y} = \frac{87{,}000g}{87{,}000 - f_y} \qquad f_y \text{ in psi} \tag{4-19}$$

For all values of $k$ greater than $k_y$ the stress diagram of Fig. 4-3c applies; for values of $k$ less than $k_y$, the stress diagram of Fig. 4-3d applies. Taken together, these two stress diagrams define the state of stress in concrete beams throughout the range of interest. The ultimate section modulus (plastic section modulus) can readily be developed from these diagrams.

The value of $\beta$ shown in the stress diagrams for the ACI stress block is prescribed by Code (10.2.7.3). Although the Code permits approaches other than this, the empirical Code value of $\beta$ is known to provide a close correlation to test data. Using the prescribed value of $\beta$, the force in the compression stress block is $0.85f'_c b\beta kd$.

In the form of an equation, $\beta$ is given by Code (10.2.7.3):

$$\beta = 0.85 - 0.05(f'_c - 4000) \qquad f'_c \text{ in psi} \tag{4-20}$$

In addition, $\beta$ may not be greater than 0.85 nor less than 0.65.

From Fig. 4-4, the sum of horizontal forces yields an expression for $k$, where $k > k_y$ and $f_s = f_y$:

$$\sum H = 0 = 0.85f'_c b\beta kd + f_y\rho' bd - f_y\rho bd \tag{4-21}$$

The solution for $k$ is, where $k > k_y$,

$$k = \frac{f_y(\rho - \rho')}{0.85f'_c\beta} \tag{4-22}$$

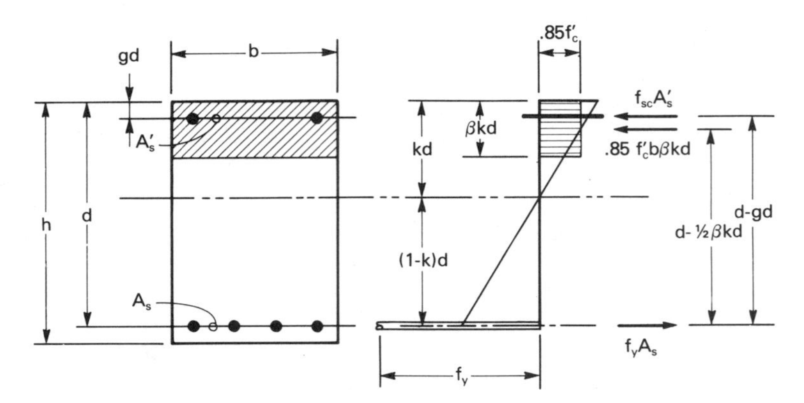


**Figure 4-4** Stresses at Ultimate Load.

Note that since $\beta$ is prescribed by Code, it is not necessary to deduct the compressive steel area from the stress block.

The sum of moments about the tensile steel yields

$$M_n = 0.85f'_c b\beta kd(d - \tfrac{1}{2}\beta kd) + f_y A'_s(d - gd) \tag{4-23}$$

Solve Eq. (4-23) for $M_n/0.85f'_c$, substitute $f_y$ from Eq. (4-22), and solve for the plastic section modulus $Z_c$. For $k > k_y$,

$$\frac{M_n}{0.85f'_c} = Z_c = k\beta\left(1 - \frac{k\beta}{2} + \rho'\,\frac{1-g}{\rho - \rho'}\right) bd^2 \tag{4-24}$$

where $k$ is the value given by Eq. (4-22).

Equations (4-22) and (4-24) are the values of $k$ and $Z_c$ when the compressive steel is in yield. When the compressive steel emerges from yield and is in the elastic range, its elastic strain $\epsilon_s$ can be found from Fig. 4-3d by ratio:

$$\frac{\epsilon_s}{.003} = \frac{kd - gd}{kd}, \text{ hence } f_{sc} = \epsilon_s E_s = 87{,}000\,\frac{k-g}{k} \tag{4-25a,b}$$

For values of $k < k_y$, the sum of horizontal forces shown in Fig. 4-4 yields an expression for $k$:

$$\sum H = 0 = 0.85f'_c b\beta kd + f_{sc}\rho' bd - f_y\rho bd \tag{4-26}$$

Equation (4-25b) is now substituted and the result solved for $k$, where $k < k_y$.

$$k = \sqrt{\left[\frac{f_y\rho - 87{,}000\rho'}{2 \times 0.85f'_c\beta}\right]^2 + \frac{87{,}000g\rho'}{0.85f'_c\beta}} + \frac{f_y\rho - 87{,}000\rho'}{2 \times 0.85f'_c\beta} \tag{4-27}$$

Moments are now summed about the tensile steel (Fig. 4-4),

$$M_n = 0.85f'_c b\beta kd(d - \tfrac{1}{2}\beta kd) + f_{sc}\rho' bd(d - gd) \tag{4-28}$$

Equation (4-25b) is substituted and the result solved for $M_n/0.85f'_c$. For $k < k_y$, the plastic section modulus $Z_c$ is then

$$\frac{M_n}{0.85f'_c} = Z_c = \left[k\beta\left(1 - \frac{k\beta}{2}\right) + \frac{87{,}000\rho'}{0.85f'_c}(1 - g)\frac{k - g}{k}\right]bd^2 \qquad (4\text{-}29)$$

where $k$ is the value given by Eq. (4-27).

Equations (4-22), (4-24), (4-27), and (4-29) provide the plastic section modulus for a concrete beam throughout its range of interest. As with the elastic analysis, it is not necessary to solve these rather lengthy equations manually. They can be solved by computer and the results tabulated for reference.

In addition to the plastic section modulus, the compressive area of concrete will also be used frequently. Its value is given by Code (10.2.7.1) as an equivalent area (see Fig. 4-5):

$$A_n = b\beta kd = (k\beta)bd \qquad (4\text{-}30)$$

Also, the distance from the centerline of the section to the center of the compressive force will be needed:

$$y_n d = \frac{h}{2} - \frac{k\beta d}{2} = \frac{1}{2}(1 + g - k\beta)d \qquad (4\text{-}31)$$

As observed earlier, this factor $y_n$ may sometimes be greater than 0.5 since it applies to $d$ rather than $h$.

Table A-5 through A-9 include the values of $Z_c$, $k\beta$, and $y_n$ given above. Similar to the elastic analysis, the tables are entered with the values of $\rho$ and $\rho'$, the ultimate strength of concrete and the grade of steel. The columns headed "at ultimate strength" are the tabulated values of Eqs. (4-24), (4-29), and (4-31).

The value of $k$ at ultimate load is not often of interest, but its value can be obtained from the listed value of $k\beta$ simply by dividing by $\beta$; the value of $\beta$ is given by Eq. (4-20) and is also given at the top of each beam table.

For ultimate load design, the Code places limits on both the minimum amount of steel that may be used and the maximum amount as well. Code (10.5.1) gives the minimum steel ratio by a simple formula:

$$\text{minimum } \rho = \frac{200}{f_y} \qquad (4.32)$$

The maximum steel ratio is somewhat more complicated. The maximum ratio is given by Code (10.3.3) as 75% of that required to produce the "balanced" condition (line *OA* of Fig. 4-3b). The balanced condition is also specified by Code (10.3.2). Again, by solving by ratios for this case and summing horizontal forces,

$$\text{maximum } \rho = 0.75\beta\,\frac{0.85f'_c}{f_y}\left(\frac{87{,}000}{87{,}000 + f_y}\right) \qquad f_y \text{ in psi} \qquad (4\text{-}33)$$

When a concrete beam is designed using the strength method, there is no requirement to check the beam for its day-to-day service stresses. It may be advisable to check these stresses, however, especially if brittle finishes or coatings are used. A rough check is provided in Tables A-5 through A-7, in the column headed "svc $f_c$." Values are listed first in $N/mm^2$, second in $lb/in^2$.

This rough check is readily derived. It was noted in Chapter 3 that an overall average load factor for concrete structures is about 1.52. With a value of $\phi$ of 0.9,

$$M_n = \frac{1.52}{0.9} M_w = 1.7M_w \quad \text{(average)} \tag{4-34}$$

It is noted that $M_n = 0.85f'_c Z_c$, and $M_w = f_c S_c$. These values are substituted into Eq. (4-34) and the result solved for $f_c$:

$$f_c = 0.5f'_c \frac{Z_c}{S_c} \tag{4-35}$$

The value of $f_c$ given by Eq. (4-35) is tabulated as "svc $f_c$" in the design tables.

Although rough, the tabulated value of service stress affords a guide to selecting a concrete section when using the strength method. When selecting a section, a value of should be chosen that keeps the service $f_c$ at about half of $f'_c$. It may be necessary to keep the service $f_c$ even lower if unusually brittle finishes are to be used.

The following examples illustrate the use of the foregoing equations. For the sake of simplicity, the examples are limited to flexure in members subjected only to gravity loads. Load factors are then 1.4 for dead load and 1.7 for live load. Strength reduction factor $\phi$ is taken at 0.9 for flexural members, as discussed in Chapter 3.

**Example 4-14**

*Strength method*
*Design of a rectangular section in flexure*
*No limitations on dimensions*

*Given:*
$M_{DL} = 36$ kip-ft
$M_{LL} = 42$ kip-ft
Load factors: 1.4 for DL and 1.7 for LL
$\phi = 0.9$
$f'_c = 3000$ psi
Grade 60 steel

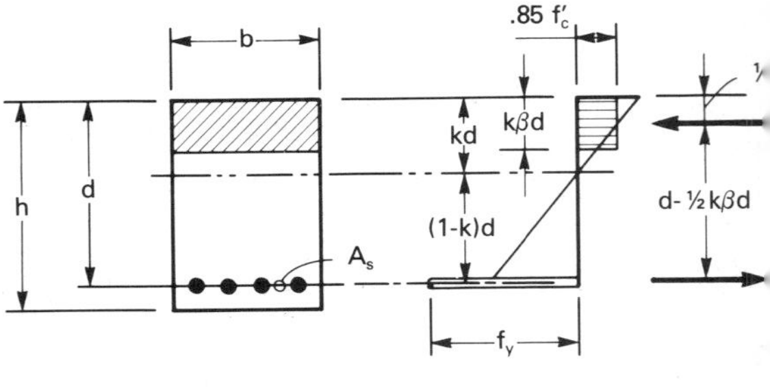

b
h
d
$A_s$
.85 $f'_c$
kd
$k\beta d$
(1-k)d
$f_y$
d- ½k$\beta$d

Compute the nominal ultimate moment $M_n$:

$$M_n = \frac{M_u}{\phi} = \frac{1.4M_{DL} + 1.7M_{LL}}{\phi} = \frac{1.4 \times 36 + 1.7 \times 42}{0.9}$$

$$= 135 \text{ kip-ft}$$

Enter Table A-5; avoid the use of compressive steel and try to keep "svc $f_c$" about half the ultimate strength of concrete. Select

$$\rho = 0.006 \qquad \text{and} \qquad Z_c = 0.131bd^2$$

Apply the flexure formula, where $b = 0.6d$.

$$\frac{M_n}{0.85f'_c} = Z_c \qquad \frac{135 \times 10^3 \times 12}{0.85 \times 3000} = 0.131bd^2 = 0.131 \times 0.6d \times d^2$$

Solve for $d$: $d = 20$ in., $b = 12$ in.

Compute the required area of steel:

$$A_s = \rho bd = 0.006 \times 12 \times 20 = 1.44 \text{ in}^2$$

From Table A-3, select $A_s = 5$ No. 5 bars, $A_s = 1.53$ in$^2$.

Use $d = 20$ in., $b = 12$ in., $A_s = 5$ No. 5 bars.

It should be emphasized that in selecting section constants from the tables of the Appendix, all sections listed in the tables are usable and are within all requirements of ACI. It is simply good practice to avoid the use of compressive steel (where possible) and to keep the steel ratio at about half the maximum value allowed. Nonetheless, a section having a steel ratio of 1.6% and its accompanying service stress of 2551 psi is an acceptable section under the Code.

Note that the Example 4-14 is comparable to Example 4-1. It should be recognized that the procedure used is directly parallel to that used ealier in the working stress method. The concept of using the flexure formula to find a section modulus is retained, even though the section is at ultimate load. The same general procedure still applies when the depth of the section is limited, as illustrated in the next example.

**Example 4-15**

*Strength method*
*Design of a rectangular section in flexure*
*Depth of section limited*

*Given*
$M_{DL} = 37$ kip-ft
$M_{LL} = 25$ kip-ft
Load factors: 1.4 for DL and 1.7 for LL
$\phi = 0.9$
Grade 50 steel
$f'_c = 4000$ psi
$d$ limited to 14 in.

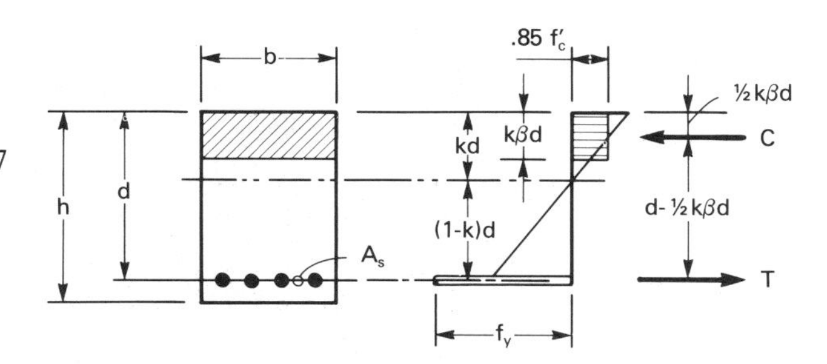


Compute the nominal ultimate moment $M_n$:

$$M_n = \frac{M_u}{\phi} = \frac{1.4M_{DL} + 1.7M_{LL}}{\phi} = \frac{1.4 \times 37 + 1.7 \times 25}{0.9} = 105 \text{ kip-ft}$$

Enter Table A-6; avoid compressive reinforcement and keep the service stress in concrete about half the ultimate strength. Select.

$$\rho = 0.011 \quad \text{and} \quad Z_c = 0.149bd^2$$

Apply the flexure formula, using $d = 14$ in.

$$\frac{M_n}{0.85f'_c} = Z_c \qquad \frac{105 \times 10^3 \times 12}{0.85 \times 4000} = 0.149bd^2 = 0.149 \times b \times 14 \times 14$$

Solve for $b$: $b = 12.7$ in.

Compute the required area of reinforcement:

$$A_s = \rho bd = 0.011 \times 12.7 \times 14 = 1.96 \text{ in}^2$$

From Table A-3, select $A_s = 2$ No. 9 bars.

Use $b = 13$ in., $d = 14$ in., $A_s = 2$ No. 9 bars.

When both the width $b$ and depth $d$ are limited, it may become preferable to use compressive steel, as in the next example.

**Example 4-16**

*Strength method*
*Design of a rectangular section in flexure*
*Both depth d and width b limited*

*Given:*
$M_{DL} = 62$ kip-ft
$M_{LL} = 35$ kip-ft
Load factors 1.4 for DL and 1.7 for LL
$\phi = 0.9$
Grade 40 steel
$f'_c = 3000$ psi
Effective depth $d$ limited to 17 in. and $b$ to 12 in.

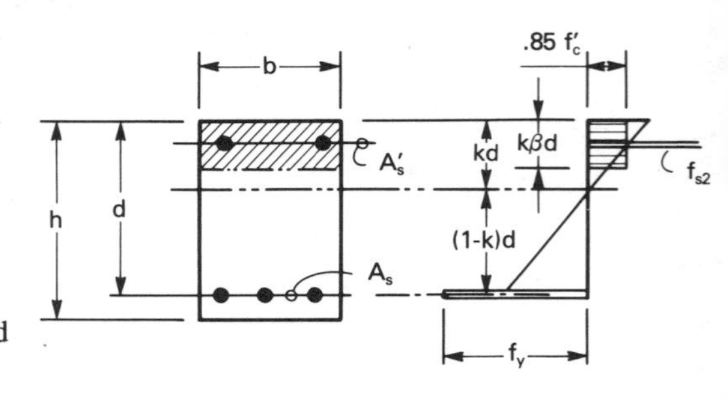


Since both $b$ and $d$ are prescribed, the only remaining problem is to select the reinforcement. Compute the nominal ultimate moment $M_n$:

$$M_n = \frac{M_u}{\phi} = \frac{1.4M_{DL} + 1.7M_{LL}}{\phi} = \frac{1.4 \times 62 + 1.7 \times 35}{0.9} = 163 \text{ kip-ft}$$

With both $b$ and $d$ known, the coefficient of the plastic section modulus can be computed, where $Z_c = \text{coeff.} \times bd^2$:

$$\frac{M_n}{0.85f'_c} = Z_c \qquad \frac{163 \times 10^3 \times 12}{0.85 \times 3000} = \text{coeff.} \times 12 \times 17 \times 17$$

Hence

$$\text{coeff.} = 0.221 \text{ (minimum allowable value)}$$

From Table A-5, if service stress is to be kept to less than half of ultimate, it is seen that compressive steel will be required. Select

$$\rho' = 0.5\rho \qquad \text{and} \qquad \rho = 0.016,\ Z_c = 0.227bd^2$$

Compute the required area of reinforcement:

$$A_s = \rho bd = 0.016 \times 12 \times 17 = 3.26 \text{ in}^2 \qquad \rho' = 0.5\rho = 1.63 \text{ in}^2$$

From Table A-3, select $A_s$ = 4 No. 7 and 2 No. 6 bars, $A_s'$ = 4 No. 6 bars.

Use $b$ = 12 in., $d$ = 17 in., $A_s$ = 4 No. 7 and 2 No. 6 bars, $A_s'$ = 4 No. 6 bars.

In Example 4-15, a section without compressive reinforcement could have been selected by allowing the service stress in concrete to go above 1500 psi. For example, from Table A-5, a section having a steel ratio of 0.016 would do, but the service stress would be around 1832 psi or about 60% of ultimate strength; lower service stresses are preferred.

## ULTIMATE FLEXURE PLUS AXIAL FORCE

When a small axial load $P_n$ is added to the flexural load $M_n$, the design procedure requires additional steps. The additional steps are directly parallel to those introduced earlier in the working stress method when a small axial force occurs. As in the working stress method, it is again assumed that the axial force acts at the centerline of the section.

It is also assumed that the magnitude of the factored axial compressive load $P_u$ is less than $0.10f_c'bt$, akin to the limit imposed on $P_w$ in the working stress method. For small values of $P_u$, however, the Code (9.3.2.2) requires a variable value to be assigned to $\phi$, with $\phi$ being 0.9 when $P_u = 0$ and 0.7 when $P_u$ is at its maximum value of $0.10f_c'bt$. Stated mathematically,

$$\phi = 0.9 - 0.2\frac{P_u}{0.10f_c'bt} \tag{4-36}$$

The correction of $\phi$ given by Equation (4-36) applies only when the axial load is in compression. When the axial load is in tension, the value of $\phi$ remains constant at 0.90; it applies for all values of the tensile axial load.

The general procedure developed for combined flexural and axial loading in the working stress method can also be applied to similar problems in the strength method. The following three steps apply to the strength method:

1. Using the procedures just developed for the strength method, find the width $b_1$ and the depth $d$ for a section subject only to the nominal moment $M_n$.
2. Expand the rectangular compression block laterally (without reinforcement) by a width $b_2$ such that the design load $P_n$ is exactly opposed, $P_n = 0.85f_c'k\beta b_2 d$.
3. Deduct the width $b_3$ to compensate for the moment induced by the eccentricity of the rectangular compression block.

The final required width is the algebraic sum of these three widths. When the nominal force $P_n$ is in tension, its sign is of course reversed. For this case, steps 2 and 3 become "deduct" and "expand," respectively.

The foregoing steps are illustrated in Fig. 4-5. The ultimate stress diagram is shown in Fig. 4-5a, the extended stress block is shown in Fig. 4-5b, and the stress diagram due to eccentric moment $P_n y_n d$ is shown in Fig. 4-5c. The angle of rotation of the section is held constant, such that $\beta$ and $k$ are constant throughout.

The width $b_1$ shown in Fig. 4-5a is found through ordinary methods of flexural design using the strength method. In this computation it is necessary to choose a steel ratio in order to obtain the section constants. It is not necessary to select the actual steel area $A_s$.

The force in the extended stress block of Fig. 4-5b is made equal to $P_n$:

$$P_n = 0.85 f'_c b_2 \beta k d \tag{4-37}$$

The width $b_2$ can be found from this equation, where $\beta k$ is given among the section constants just obtained for $b_1$.

The width $b_3$ shown in Fig. 4-5c is obtained using the same set of section constants; it is the flexural width required due to the eccentric moment $P_n y_n d$. When $P_n$ is compressive, the eccentric moment $P_n y_n d$ acts in the same direction as the nominal moment $M_n$, producing compression on the same side of the beam and allowing a reduction in width. When $P_n$ is in tension, an increase in width is required.

It is again emphasized that the actual amount of steel $A_s$ chosen for the final section must be computed for the least flexural width, $b_m = b_1 - b_3$:

$$A_s = \rho b_m d \tag{4-38}$$

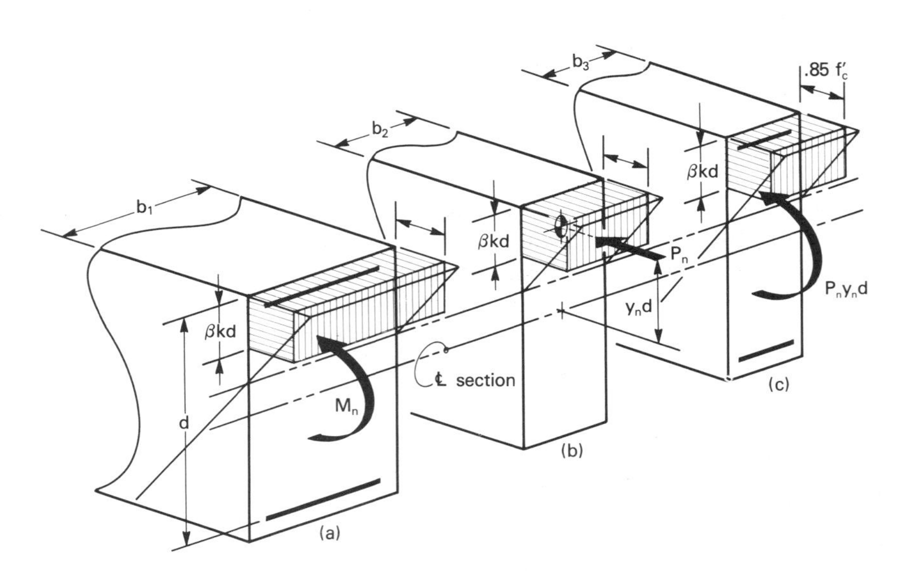


**Figure 4-5** Flexure Plus Axial Load.

The least flexural width $b_m$ will be used later in the investigation of known sections subject to both moment and axial load. For both design and investigation, however, from the time the axial load $P_n$ entered the calculations the area of steel $A_s$ could no longer be calculated using the full width $b$.

An example will illustrate the procedure in the strength method of design. In the first example, the load $P_n$ is compressive.

**Example 4-17**

*Strength method*
*Design of a rectangular section subject to both flexure and a small axial compression*
*No limitation on dimensions*

*Given:*
$M_{\text{DL}} = 44$ kip-ft
$P_{\text{DL}} = 16$ kips
$M_{\text{LL}} = 22$ kip-ft
$P_{\text{LL}} = 12$ kips
Load factors 1.4 for DL and 1.7 for LL
Grade 50 steel
$f'_c = 4000$ psi

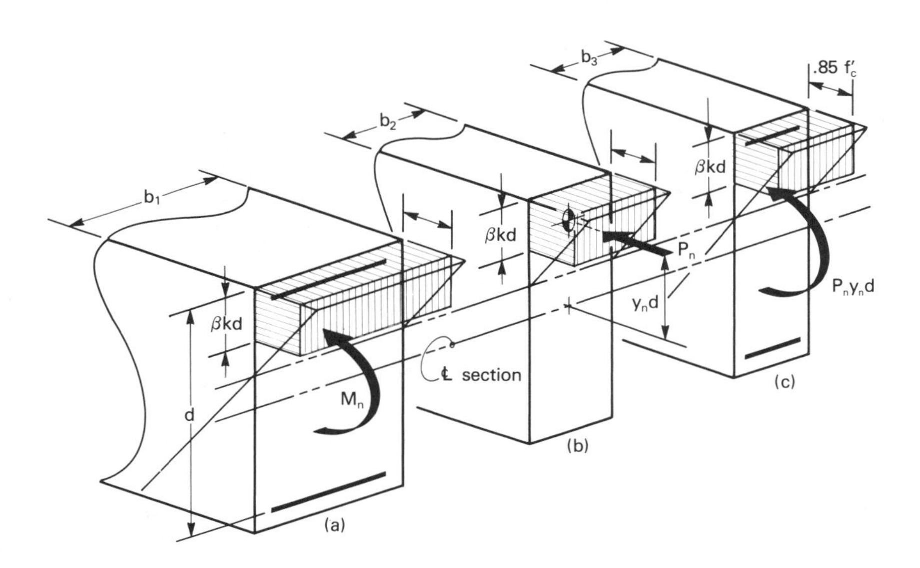


Estimate $\phi$. The exact value of $\phi$ cannot be determined until $b$ and $t$ are known [Eq. (4-34)]. Since $\phi$ will be somewhere between 0.7 and 0.9, a value of 0.8 is chosen. Compute the nominal ultimate loads $M_n$ and $P_n$:

$$M_n = \frac{M_u}{\phi} = \frac{1.4M_{\text{DL}} + 1.7M_{\text{LL}}}{\phi}$$

$$= \frac{1.4 \times 44 + 1.7 \times 22}{0.8} = 124 \text{ kip-ft}$$

$$P_n = \frac{P_u}{\phi} = \frac{1.4P_{\text{DL}} + 1.7P_{\text{LL}}}{\phi}$$

$$= \frac{1.4 \times 16 + 1.7 \times 12}{0.8} = \frac{42.8}{0.8} = 53.5 \text{ kips}$$

Select a plastic section modulus for flexure from Table A-6, keeping the service stress around 2000 psi. Try

$$\rho = 0.012 \qquad Z_c = 0.161bd^2 \qquad k\beta = 0.177 \qquad y_n = 0.474$$

Solve for initial sizes $b_1$ and $d$, assuming that $b_1 = 0.6d$:

$$\frac{M_n}{0.85f'_c} = Z_c \qquad \frac{124 \times 10^3 \times 12}{0.85 \times 4000} = 0.161b_1d^2 = 0.161 \times 0.6d \times d^2$$

Solve for $d$, $d = 16.5$ in., $b_1 = 9.9$ in.

Add the width $b_2$ required to sustain the axial load $P_n$, using the previously selected value of 16.5 in. for $d$:

$$P_n = 0.85f'_c b_2 \beta kd \qquad 53{,}500 = 0.85 \times 4000 \times b_2 \times 0.177 \times 16.5$$

Solve for $b_2$: $b_2 = 5.4$ in.

Deduct the width $b_3$ due to the eccentricity of the axial load $P_n$, which produces a moment $P_n y_n d$:

$$\frac{P_n y_n d}{0.85f'_c} = Z_c \qquad \frac{53{,}500 \times 0.474 \times 16.5}{0.85 \times 4000} = 0.161 \times b_3 \times 16.5 \times 16.5$$

Solve for $b_3$: $b_3 = 2.8$ in.

$$\text{final width } b = b_1 + b_2 - b_3 = 9.9 \times 5.4 - 2.8 = 12.5 \text{ in.}$$

Assume that overall height $h = d + 2\frac{1}{2}$ in. $= 19$ in.

Check the assumed value of $\phi$ by Eq. (4-35):

$$\phi = 0.9 - 0.2\frac{P_u}{0.1f'_c bh} = 0.9 - 0.2\frac{42{,}800}{0.1 \times 4000 \times 12.5 \times 19} = 0.81 \qquad \text{(O.K.)}$$

This value of $\phi$ is judged to be close enough to the assumed value of 0.8 to be acceptable.

Compute the required steel area:

$$A_s = \rho b_m d = 0.012 \times (9.9 - 2.8) \times 16.5 = 1.40 \text{ in}^2$$

From Table A-3, select $A_s = 2$ No. 6 and 2 No. 5 bars.

Use $b = 12.5$ in., $d = 16.5$ in., $A_s = 2$ No. 6 and 2 No. 5 bars.

If the computed value of $\phi$ in Example 4-17 had not been close enough to the estimated value to be acceptable, it would have been necessary to revise $d$ to the new value and to repeat all computations for this new value of $\phi$. However, since the factor

$\phi$ affects only loads and all other values are directly proportional to loads, the correction could be made simply by multiplying by the ratio $\phi_{old}/\phi_{new}$. For this example, carrying out the correction,

$$\text{ratio} = \frac{\phi_{old}}{\phi_{new}} = \frac{0.80}{0.814} = 0.9828$$

Now multiply the loads and the widths $b_1$, $b_2$, and $b_3$ by this ratio:

$$M_n = 122 \text{ kip-ft}$$

$$P_u = 42.1 \text{ kips} \qquad P_n = 52.6 \text{ kips}$$

$$b_1 = 9.7 \text{ in.} \qquad b_2 = 5.3 \text{ in.} \qquad b_3 = 2.75 \text{ in.} \qquad b = 12.25 \text{ in.}$$

Now recheck $\phi$, using $h = 20$ in.

$$\phi = 0.9 - 0.2\frac{P_u}{0.1f'_c bh} = 0.9 - 0.2\frac{42{,}100}{0.1 \times 4000 \times 12.25 \times 19} = 0.8096$$

This result is certainly close enough to 0.8140 to be accepted. The remaining calculations would then use these corrected values for $b$. If the axial load had been tensile rather than compressive, the signs of $b_2$ and $b_3$ change, as shown in the next example. Also, as noted earlier, the value of $\phi$ does not vary, but remains at 0.90 for all values of the tensile load.

**Example 4-18**

*Strength method*
*Design of a rectangular section subject to both flexure and a small axial tension*
*No limitations on dimensions*

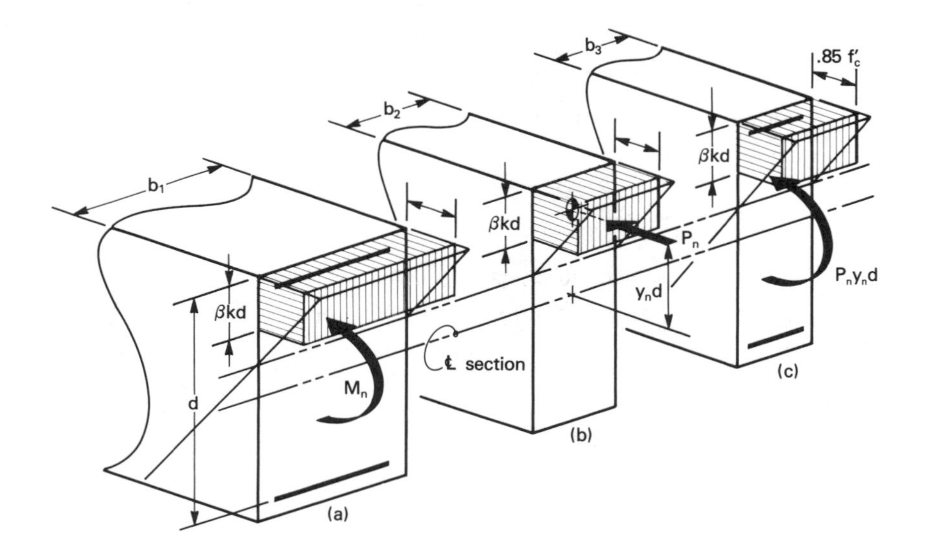

*Given:*
$M_{DL} = 32$ kip-ft
$T_{DL} = 9$ kips
$M_{LL} = 35$ kip-ft
$T_{LL} = 13$ kips
Load factors 1.4 for DL and 1.7 for LL
Grade 60 steel
$f'_c = 3000$ psi

Use $\phi = 0.90$ for members in tension, with or without flexure

Compute the nominal ultimate loads $M_n$ and $T_n$:

$$M_n = \frac{1.4M_{DL} + 1.7M_{LL}}{\phi}$$

$$= \frac{1.4 \times 32 + 1.7 \times 35}{0.9} = 116 \text{ kip-ft}$$

$$T_n = \frac{1.4T_{DL} + 1.7T_{LL}}{\phi}$$

$$= \frac{1.4 \times 9 + 1.7 \times 13}{0.9} = \frac{31}{0.9} = 39 \text{ kips}$$

Select a steel ratio from Table A-5 keeping the service stresses at about one-half the ultimate stress. Try

$$\rho = 0.0060 \qquad Z_c = 0.131bd^2 \qquad k\beta = 0.141 \qquad y_n = 0.492$$

Solve for initial sizes of $b_1$ and $d$, assuming that $b_1 = 0.8d$.

$$\frac{M_n}{0.85f'_c} = Z_c \qquad \frac{116{,}000 \times 12}{0.85 \times 3000} = 0.131b_1d^2 = 0.131 \times 0.8d \times d^2$$

Solve for $d$: $d = 17.3$ in., $b_1 = 13.9$ in.

Deduct the width $b_2$ due to tensile load $T_n$, using the previously selected size for $d$, $d = 17.3$ in.

$$T_n = 0.85f'_cb_2k\beta d \qquad 39{,}000 = 0.85 \times 3000 \times b_2 \times 0.141 \times 17.3$$

Solve for $b_2$: $b_2 = 6.3$ in.

Add the width $b_3$ due to eccentricity of the tensile load $T_n$ which produces a moment $T_ny_nd$:

$$\frac{T_ny_nd}{0.85f'_c} = Z_c \qquad \frac{39{,}000 \times 0.492 \times 17.3}{0.85 \times 3000} = 0.131 \times b_3 \times 17.3 \times 17.3$$

Solve for $b_3$: $b_3 = 3.3$ in.

$$\text{final width } b = b_1 - b_2 + b_3 = 13.9 - 6.3 + 3.3 = 11 \text{ in:}$$

$$\text{overall height } h = d + 2\tfrac{1}{2} \text{ in.} = 19.8 \text{ in., say 20 in.}$$

Compute the required steel area:

$$A_s = \rho b_m d = 0.006 \times (13.9 + 3.3) \times 17.3 = 1.79 \text{ in}^2$$

From Table A-3, select 3 No. 7 bars.

Use $b = 11$ in., $d = 17.5$ in., $A_s = 3$ No. 7 bars.

Neither Example 4-17 nor 4-18 had any restrictions on the size of the member. Where the depth $d$ is restricted, a common case around air-conditioning ductwork, the beam becomes wide and flat, as illustrated in the following example.

**Example 4-19**

*Strength method*
*Design of a rectangular section subject to both flexure and a small axial compression*
*Depth at section limited*

*Given:*
$M_{DL} = 80$ kip-ft
$P_{DL} = 21$ kips
$M_{LL} = 16$ kip-ft
$P_{LL} = 2$ kips
Load factors: 1.4 for DL and 1.7 for LL
Grade 60 steel
$f'_c = 3000$ psi
Depth $d$ limited to 18 in.

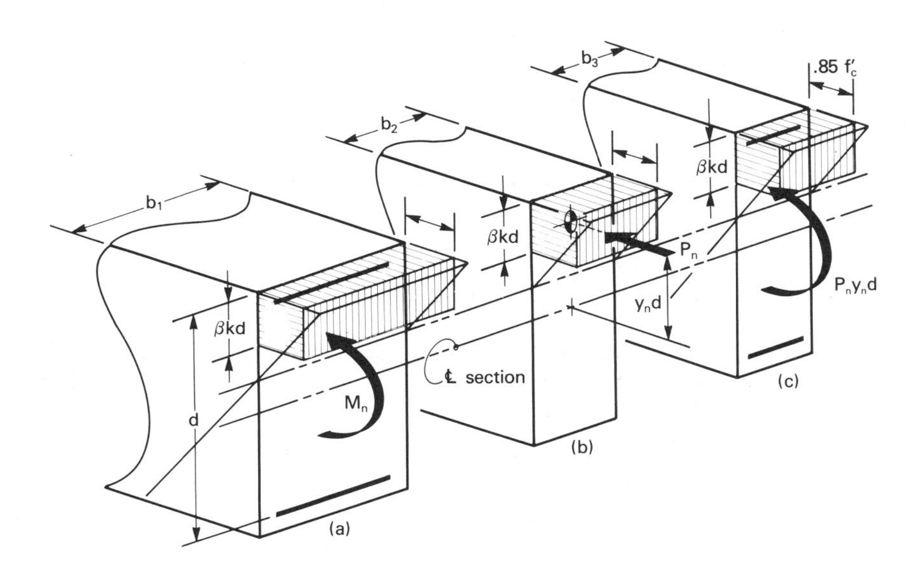

Estimate $\phi = 0.8$. Compute nominal ultimate loads $M_n$ and $P_n$:

$$M_n = \frac{1.4M_{DL} + 1.7M_{LL}}{\phi}$$

$$= \frac{1.4 \times 80 + 1.7 \times 16}{0.8} = 174 \text{ kip-ft}$$

$$P_n = \frac{1.4P_{DL} + 1.7P_{LL}}{\phi}$$

$$= \frac{1.4 \times 21 + 1.7 \times 2}{0.8} = \frac{33}{0.8} = 41 \text{ kips}$$

Select a steel ratio from Table A-5, trying to keep the service stress less than 1500 psi. Try

$$\rho = 0.0050 \qquad Z_c = 0.111bd^2 \qquad k\beta = 0.118 \qquad y_n = 0.504$$

Solve for initial size of $b_1$, holding $d$ to 18 in.

$$\frac{M_n}{0.85f'_c} = Z_c \qquad \frac{174{,}000 \times 12}{0.85 \times 3000} = 0.111b_1d^2 = 0.111b_1 \times 18 \times 18$$

Solve for $b_1$, $b_1 = 22.8$ in.

Add the width $b_2$ due to the axial load $P_n$:

$$P_n = 0.85f'_cb_2k\beta d \qquad 41{,}000 = 0.85 \times 3000 \times b_2 \times 0.118 \times 18$$

Solve for $b_2$: $b_2 = 7.6$ in.

Deduct the width $b_3$ due to eccentricity of axial load $P_n$, which produces a moment of $P_ny_nd$:

$$\frac{P_ny_nd}{0.85f'_c} = Z_c \qquad \frac{41{,}000 \times 0.504 \times 18}{0.85 \times 3000} = 0.111 \times b_3 \times 18 \times 18$$

Solve for $b_3$: $b_3 = 4.1$ in.

$$\text{final width } b = b_1 + b_2 - b_3 = 22.8 + 7.6 - 4.1 = 26.3 \text{ in.}$$

Assume that overall height $h = d + 2\frac{1}{2} = 18 + 2.5 = 20.5$ in.

*Check:*

$$\phi = 0.9 - 0.2\frac{P_u}{0.1f'_cbh} = 0.9 - 0.2\frac{33{,}000}{0.1 \times 3000 \times 26.3 \times 20.5} = 0.86$$

This value of $\phi$ is not close enough to be assumed value of 0.8 to accept. Revising $\phi$ to 0.86, the entire set of calculations to this point is revised (multiplying by the ratio 0.80/0.86), yielding

$$M_n = 162 \text{ kip-ft} \qquad P_u = 31 \text{ kips} \qquad P_n = 38 \text{ kips}$$

$$b_1 = 21.2 \text{ in.} \qquad b_2 = 7.1 \text{ in.} \qquad b_3 = 3.8 \text{ in.}$$

$$b = 24.5 \text{ in.} \quad d^2 = 18 \text{ in.} \qquad h = 20.5 \text{ in.}$$

*Recheck:*

$$\phi = 0.9 - 0.2\frac{P_u}{0.1f'_c bh} = 0.9 - 0.2\frac{33{,}000}{0.1 \times 3000 \times 24.5 \times 20.5} = 0.856$$

This value of $\phi$ is judged to be close enough to the estimated value of 0.86 to be acceptable.

Compute the steel area:

$$A_s = \rho b_m d = 0.005 \times (21.2 - 3.8) \times 18 = 1.57 \text{ in}^2$$

From Table A-3, select $A_s$ = 2 No. 8 bars.

Use $b$ = 25 in., $d$ = 18 in., $A_s$ = 2 No. 8 bars.

Rather than making the beam wider in Example 4-19, the width $b$ could have been kept narrow by using compressive steel. Compressive steel is usually avoided where possible, but its use is common in restricted areas. The next example considers a case where both $b$ and $d$ are restricted so much that the use of compressive steel becomes necessary.

**Example 4-20**

*Strength method*
*Design of a rectangular section subject to both flexure and a small axial compression*
*Both depth d and width b limited*

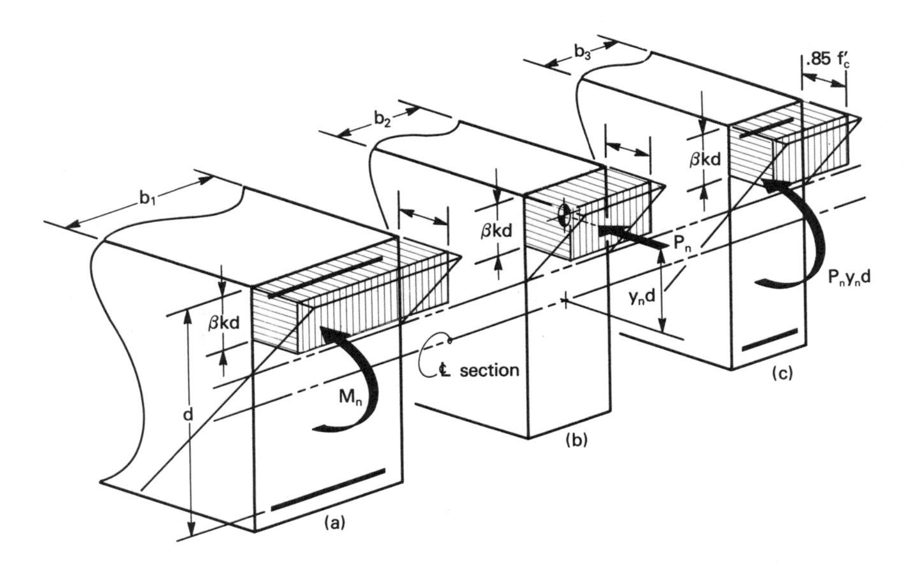


*Given:*
$M_{DL}$ = 67 kip-ft
$P_{DL}$ = 25 kips
$M_{LL}$ = 29 kip-ft

$P_{\mathrm{LL}} = 8$ kips
Load factors: 1.4 for DL and 1.7 for LL
Grade 60 steel
$f'_c = 4000$ psi
Effective depth $d$ limited to 18 in. and $b$ to 16 in.

As in Example 4-7, the solution to this problem is simply a trial-and-error solution to find a steel ratio that will provide the necessary magnitude for the section modulus while keeping the service stress within a desired range of values. To make this selection, a steel ratio is chosen by guess from the design table. The procedure from Example 4-19 is then followed to find the width $b$ required for this value of $\rho$. This computed value of $b$ is then compared to the original section to see if it is within about 10% of the given width. If not, another steel ratio either slightly higher or lower is chosen and the process repeated. In the following calculations, only the last trial for $\rho$ is shown; the incorrect trials were simply discarded.

Estimate $\phi = 0.825$ (final value). Compute nominal loads $M_n$ and $P_n$.

$$M_n = \frac{1.4M_{\mathrm{DL}} + 1.7M_{\mathrm{LL}}}{\phi}$$

$$= \frac{1.4 \times 67 + 1.7 \times 29}{0.825} = 173 \text{ kip-ft}$$

$$P_n = \frac{1.4P_{\mathrm{DL}} + 1.7P_{\mathrm{LL}}}{\phi}$$

$$= \frac{1.4 \times 25 + 1.7 \times 8}{0.825} = \frac{49}{0.825} = 59 \text{ kips}$$

Select a steel ratio $\rho$ from Table A-6, trying to keep the service stresses about half of the ultimate strength of the concrete. The first trial, using no compressive steel and a steel ratio of 0.008, yielded a value of $b$ far above the 16 in. allowed. For the second trial a steel ratio of 0.010 was tried with compressive steel half the tensile steel; the width $b$ was still more than the 16 in. allowed. For this third trial, a compressive steel ratio equal to the tensile steel ratio is chosen. Try

$$\rho' = \rho \qquad \rho = 0.010 \qquad Z_c = 0.162bd^2 \qquad k\beta = 0.130 \qquad y_n = 0.498$$

Compute width $b_1$ required for flexure, using $d = 18$ in.

$$\frac{M_n}{0.85f'_c} = Z_c \qquad \frac{173{,}000 \times 12}{0.85 \times 4000} = 0.162b_1d = 0.162 \times b_1 \times 18 \times 18$$

Solve for $b_1$: $b_1 = 11.6$ in.

Compute the additional width $b_2$ required for axial load $P_n$:

$$P_n = 0.85f'_cb_2k\beta d \qquad 59{,}000 = 0.85 \times 4000 \times b_2 \times 0.130 \times 18$$

Solve for $b_2$: $b_2 = 7.4$ in.

Compute the reduction in width $b_3$ due to the eccentricity of axial load $P_n$, which produces the moment $P_n y_n d$:

$$\frac{P_n y_n d}{0.85 f'_c} = Z_c \qquad \frac{59{,}000 \times 0.498 \times 18}{0.85 \times 4000} = 0.162 \times b_3 \times 18 \times 18$$

Solve for $b_3$: $b_3 = 3.0$ in.

$$\text{final width } b = b_1 + b_2 - b_3 = 11.6 + 7.4 - 3.0 = 16.0 \quad \text{(O.K.)}$$

$$\text{overall height } h = d + 2\tfrac{1}{2} = 18 + 2.5 = 20.5 \text{ in.}$$

*Check:*

$$\phi = 0.9 - 0.2\frac{P_u}{0.1 f'_c bh} = 0.9 - 0.2\frac{49{,}000}{0.1 \times 4000 \times 16 \times 20.5} = 0.825 \quad \text{(O.K.)}$$

Compute the steel areas:

$$A_s = \rho b_m d = 0.010(11.6 - 3)18 = .010(11.6 - 3)18 = 1.54 \text{ in}^2$$

From Table A-3, select $A_s$ = 2-No. 8 bars and $A'_s = A_s$.

Use $b = 16$ in., $d = 18$ in., $A_s$ = 2-No. 8 bars, $A'_s$ = 2-No. 8 bars.

Solutions such as that in Example 4-20 are admittedly tedious. Fortunately, this type of solution is not one that must be made frequently in the working life of an engineer. This type of solution has the advantage, however, that no new theory or development was necessary; the equations and the approach are directly those used in the other examples.

There are special approaches and procedures for such problems. Such specialized procedures are likely to be forgotten between uses when they are used infrequently. For that reason, only one approach is presented in this text, which, however clumsy, can be applied to a wide variety of problems.

## INVESTIGATION OF ULTIMATE FLEXURE*

All the preceding examples in the strength method were concerned only with the design of a concrete section, that is, the selection of the dimensions or reinforcement that are required if the beam is to carry the prescribed loads. The other general type of problem that arises occurs when the size and reinforcement of a section are known; the problem becomes that of finding how much load the section will carry. This "investigation" of a given section is treated next.

The investigation of a known section in the strength method uses essentially the same procedures and concepts as those developed earlier for the working stress method.

*This section optional in a limited design course.

The strength method, of course, uses ultimate loads and stresses with the plastic section modulus, whereas the working stress method used the working loads and stresses and elastic section modulus. As an advantage in the strength method, the stresses are always at yield, and are thus known, affording two fewer unknowns to be found than in the working stress method.

An example will illustrate the procedure for investigating a known section using the strength method. All terms and symbols are those used in the design procedures. The defining sketch for the terms and symbols is Fig. 4-4.

**Example 4-21**

*Strength method*
*Investigation of a rectangular section subject only to flexure to determine the nominal ultimate moment $M_n$*

*Given:*
Section as shown
Grade 60 steel
$f'_c = 4000$ psi

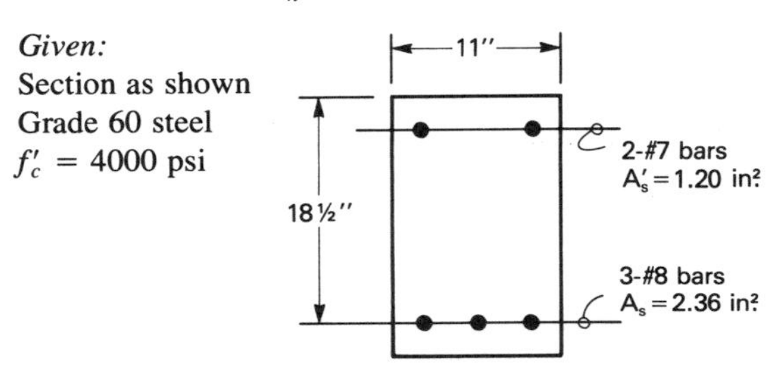


Compute the steel ratio $\rho$, where $\rho' = 0.5\rho$:

$$\rho = \frac{A_s}{bd} \qquad \rho = \frac{2.36}{11 \times 18.5} = 0.0116, \text{ say } 0.012$$

Enter Table A-6: $Z_c = 0.193bd^2$.

Compute the nominal ultimate moment from the flexure formula:

$$\frac{M_n}{Z_c} = 0.85f'_c \qquad M_n = 0.85f'_cZ_c = 0.85 \times 4000 \times 0.193 \times 11 \times 18.5 \times 18.5$$

Solve for $M_n$: $M_n = 2470$ kip-in. $= 206$ kip-ft.

The investigation of a given section subject to a known axial load $P_n$ as well as a known moment $M_n$ is usually undertaken only to see if the given section is adequate to sustain the two prescribed loads (rather than to determine exact capacities). At ultimate load, however, the loads $M_n$ and $P_n$ that can coexist on a given section are rigidly interrelated and the given loads are rarely in these required proportions. This required interrelationship between $M_n$ and $P_n$ at ultimate load is not complex; given one load and the least flexural width $b_m$, the related load can readily be computed.

The investigation is therefore a matter of applying one of the two loads to the section, almost always $P_n$, and computing the magnitude of the interrelated load, the moment $M_n$. This value is then compared to the original known moment $M_n$ to see if the section can support it. The investigation thus completely ignores the given moment until the final comparison.

In performing the actual computations, the approach is that used previously in the working stress method, where the value of the least flexural width $b_m$ was found by successive approximations. That approach can be applied in the strength method as well. Given the axial load and the section, therefore, the solution becomes that of successively approximating the least flexural width $b_m$ corresponding to the axial force $P_n$ and subsequently computing the interrelated moment $M_n$.

The trial-and-error solution for the least flexural width $b_m$ is described in detail with the working stress method earlier in this chapter. The following summary describes the procedure as adapted to the strength method.

1. *First iteration.* Assume that the least flexural width $b_m$ is half the total width of the section (see Fig. 4-5). Compute the steel ratio $\rho = A_s/b_m d$ for this value of $b_m$ and find the section constants from the Appendix tables. Compute the additional width $b_2$ required to sustain the axial load $P_n$ (see Fig. 4-5b), where $P_n = 0.85f'b_2k\beta d$. The total required width for this case is then $b_m + b_2$. If this is not equal to the given width $b$, go to a second iteration.
2. *Second and subsequent iterations.* Knowing $b_m$ to be incorrect, but assuming the proportion $r = b_m/(b_m + b_2)$ to be a close approximation, find a new $b_m$ in the same proportion to the actual width of the section: $b_m = r \times b$. Repeat the calculation above with this improved approximation of $b_m$ until the computed width $b_m + b_2$ is within a few percent of the actual width $b$.

The end result of this procedure in the strength method will be the determination of the axial load $P_n$ and all the section constants at ultimate load. The related moment is then found from the flexure formula:

$$0.85f'_c = \frac{M_n - P_n y_n d}{Z_c} \tag{4-39}$$

where $Z_c$ is determined for the least flexural width $b_m$. The following example illustrates the procedure.

**Example 4-22**

*Strength method*
*Investigation of a rectangular section subject to both flexure and a small compression to determine whether the section is adequate*

*Given:*
Section as shown
$M_n$ = 145 kip-ft
$P_n$ = 91 kips
Grade 60 steel
$f'_c$ = 5000 psi

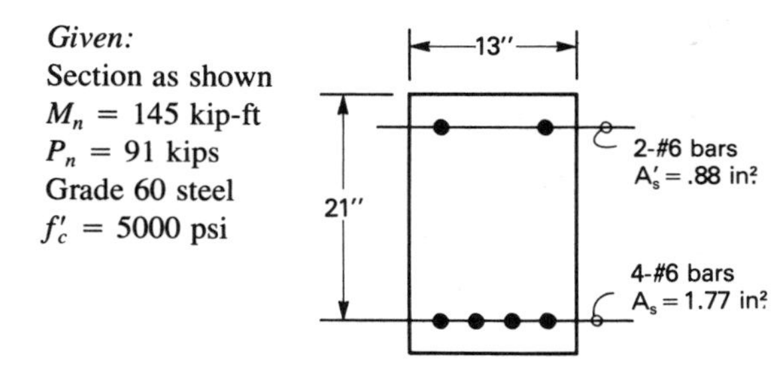

**First Iteration**

Estimate $b_m = 0.5b = 6.5$ in.

Compute

$$\rho = \frac{A_s}{b_m d} = \frac{1.77}{6.5 \times 21} = 0.013$$

Enter Table A-7 with $\rho = 0.013$, $\rho' = 0.5\rho$:

$$Z_c = 0.169bd^2 \qquad k\beta = 0.145 \qquad y_n = 0.491$$

Compute the additional width $b_2$ due to $P_n = 91$ kips:

$$P_n = 0.85f'_c b_2 k\beta d \qquad 91{,}000 = 0.85 \times 5000 \times b_2 \times 0.145 \times 21$$

Solve for $b_2$: $b_2 = 7.0$ in.

$$\text{final required width } b = b_m + b_2 = 6.5 + 7.0 = 13.5 \text{ in.}$$

Compute the corrected value for $b_m$:

$$b_m = \frac{6.5}{13.5} = 0.482 \quad \text{or} \quad 48.2\% \text{ of total width}$$

Try $b_m = 48.2\%$ of the actual given width of 13 in.

$$b_m = 0.482 \times 13 = 6.3 \text{ in.}$$

**Second Iteration:**

Estimate $b_m = 6.3$ in., $\rho = 1.77/6.3 \times 21 = 0.0134$, $\rho' = 0.5\rho$.

From Table A-7: $Z_c = 0.174bd^2$, $k\beta = 0.147$, $y_n = 0.489$.

Additional width $b_2$: $P_n = 0.85f'_c b_2 k\beta d = 0.85 \times 5000 \times b_2 \times 0.147 \times 21$.

Solve for $b_2$: $b_2 = 6.9$ in.

$$\text{final required width } b = b_m + b_2 = 6.3 + 6.9 = 13.2 \text{ in.} \quad \text{(O.K.)}$$

This value of $b$ is considered close enough to the given 13 in.

Compute the interrelated moment $M_n$ when $P_n = 91$ kips:

$$M_n - P_n y_n d = 0.85f'_c Z_c \qquad \text{[Eq. (4-39)]}$$

$$M_n = 91{,}000 \times 0.498 \times 21 + 0.85 \times 5000 \times 0.174 \times 6.3 \times 21 \times 21$$

$$= 2989 \text{ kip-in.} = 249 \text{ kip-ft}$$

Since this computed value of 249 kip-ft is much larger than the intended load of 145 kip-ft, the section is seen to be adequate.

It is well to point out that Example 4-22 omitted the use of the strength reduction factor $\phi$ to obtain $P_n$. It was omitted in this case for the sake of clarity. Depending on the particular problem, it could be necessary to determine $\phi$ in the same way that it was determined in Example 4-19.

For very heavily reinforced sections, the first approximation for $b_m$ at half the total width may produce a steel ratio $\rho$ that falls outside the tables. For such cases, the first approximation for $b_m$ can be made $0.75b$ or even equal to $b$. The remainder of the procedure is unchanged.

## COMPARISON OF STRENGTH METHOD TO WORKING STRESS METHOD

It is of interest to compare a rectangular section designed using the working stress method to one designed using the strength method, where both sections are deliberately made the same size. The design moments are chosen in an average range, $M_{DL} = 60$ kip-ft and $M_{LL} = 40$ kip-ft. The ultimate strength of concrete is chosen at 3000 psi and the yield stress of steel at 60,000 psi, a common combination in today's practice.

1. The section is designed first by the working stress method:

   $M_w = M_{DL} + M_{LL} = 100$ kip-ft $\quad f_c = 0.45f'_c = 1350$ psi $\quad f_s = 24{,}000$ psi

   From Table A-5, $S_c = 0.154bd^2$, $\rho = 0.0100$.
   Solve for section dimensions, using $b = 0.6d$:

   $$\frac{M_w}{f_c} = S_c \qquad \frac{100{,}000 \times 12}{1350} = 0.154 \times 0.6d\ x\ d^2 \qquad d = 21.3 \text{ in.}$$

   *Final results:*

   $$b = 12.8 \text{ in.} \qquad d = 21.3 \text{ in.} \qquad A_s = \rho bd = 2.7 \text{ in}^2$$

   Service stresses: $f_c = 1350$ psi

   $f_s = 24{,}000$ psi

2. The section is designed a second time by the strength method:

   $$M_n = \frac{1.4M_{DL} + 1.7M_{LL}}{0.9} = 169 \text{ kip-ft}$$

   Solve for coefficient of section modulus, $Z_c = \text{coeff} \times bd^2$

   $$\frac{M_n}{0.85f'_c} = Z_c \qquad \frac{169 \times 12}{0.85 \times 3} = \text{coeff} \times 12.8 \times 21.3 \times 21.3$$

   Solve for coeff.: coeff. $= 0.137$.
   From Table A-5, $\rho = 0.0063$, $S_c = 0.131bd^2$, $f_s/f_c = 22.9$.
   The actual service stresses are

   $$f_c = \frac{M_w}{S_c} = \frac{100{,}000 \times 12}{0.131 \times 12.8 \times 21.3 \times 21.3}$$

   $$f_c = 1577 \text{ psi} \qquad f_s = f_c(f_s/f_c) = 36{,}100 \text{ psi}$$

*Final results:*

$$b = 12.8 \text{ in.} \qquad d = 21.3 \text{ in.} \qquad A_s = \rho bd = 1.72 \text{ in}^2$$

$$\text{Service stresses: } f_c = 1577 \text{ psi}$$

$$f_s = 36{,}100 \text{ psi}$$

When the two designs are compared, the results demonstrate a general truth, that the service stresses in beams designed using the strength method can be expected to be higher than in beams designed using the working stress method. The difference is less marked when low-strength (grade 40) steel is used or when a higher-strength concrete is used ($f'_c > 3000$ psi). The difference is most marked when low-strength concrete is used with high-strength steels.

A second general observation can also be drawn: For the same size of member, the amount of steel required under the strength method is considerably less than that required under the working stress method. It is this difference that is probably the single most attractive feature in the use of the strength method. Although the overall difference cannot be expected to be as great as in this single example, the savings in steel tonnage can be significant, even on small buildings.

**TABLE 4-1 APPROXIMATE SERVICE STRESSES IN STEEL**

| Compressive steel ratio, $\rho^1/\rho$ | Steel ratio $A_s/bd$ | Service Stress, $f_s$ (psi) Grade 40 steel | Grade 50 steel | Grade 60 steel |
|---|---|---|---|---|
| 0.0 | 0.0025 | | | 36,600 |
| | 0.0050 | 24,800 | 30,700 | 36,400 |
| | 0.0100 | 24,500 | 30,000 | 35,200 |
| | 0.0150 | 24,100 | 29,100 | 33,700 |
| | 0.0200 | 23,400 | | |
| | 0.0250 | 22,000 | | |
| 0.5 | 0.0025 | | | 37,400 |
| | 0.0050 | 24,700 | 30,700 | 36,400 |
| | 0.0100 | 24,400 | 30,200 | 35,800 |
| | 0.0150 | 24,200 | 29,900 | 35,500 |
| | 0.0200 | 24,100 | | |
| | 0.0250 | 24,000 | | |
| 1.0 | 0.0025 | | | 37,600 |
| | 0.0050 | 25,100 | 30,800 | 36,500 |
| | 0.0100 | 24,400 | 30,100 | 35,900 |
| | 0.0150 | 24,100 | 29,900 | 35,700 |
| | 0.0200 | 24,000 | | |
| | 0.0250 | 23,700 | | |
| Allowable working stress: | | 20,000 | 20,000 | 24,000 |
| $0.6f'_y$ (for comparison): | | 24,000 | 30,000 | 36,000 |

To illustrate a third observation, it is necessary to tabulate the approximate service stress in steel over a wide range of steel ratios. Such a tabulation has been taken from Table A-5, shown in Table 4-1. The approximate service stress in steel has been obtained simply by multiplying the "svc $f_c$" column times the "$f_s/f_c$" column in the design tables.

It is apparent from Table 4-1 that the service stress in steel is remarkably consistent through a wide range of steel ratios. It is also apparent that Code permits the service stress in steel to go considerably higher in the strength method than in the working stress method. The significantly higher stress in steel suggests that proportionately more care must be exercised in controlling deflections and crack sizes when using the strength method.

Although not treated in this text, the measures prescribed by Code (Chap. 9) to achieve effective control of deflections and crack sizes are a major feature of the strength method of design. A thorough knowledge of Code requirements on these topics is recommended.

## FLOOR SLABS AND ROOF SLABS

Floor and roof slabs are a special case of rectangular sections. They are analyzed and designed as beams 1 ft. wide. A 1-ft strip in the interior of the slab is taken as being typical for the entire width, as shown in Fig. 4-6, and is designed as a rectangular beam.

Where negative moments occur over supports (producing tension on top), the reinforcement is placed at the top of the slab. At midspan, where moments are positive (producing tension on bottom), the reinforcement is placed at the bottom of the slab. A

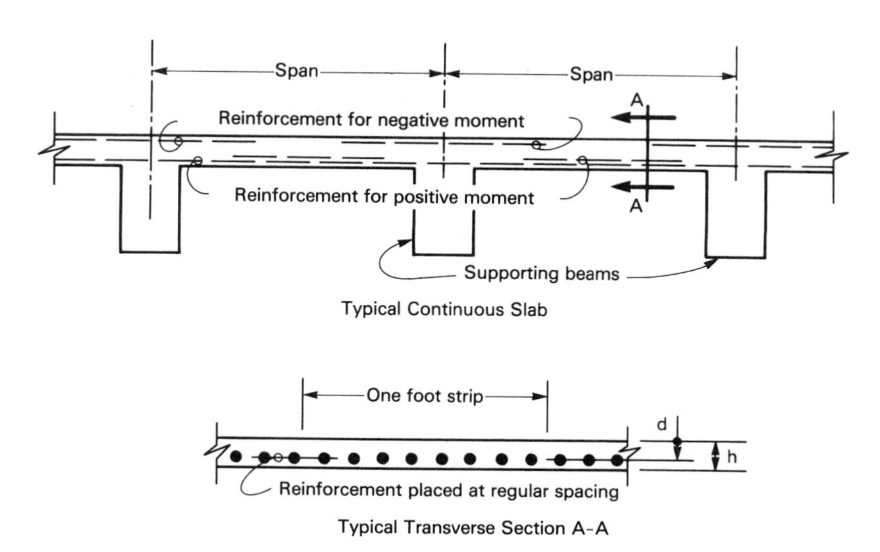


**Figure 4-6** Typical Structural Slab.

typical longitudinal section is included in Fig. 4-6; an actual design of continuous slabs, beams, and columns is treated in Chapter 9.

Reinforcing bars are placed at a fixed spacing across the width of the slab. For the most commonly used spacings and bar sizes, steel areas per foot of width are tabulated in Table A-2. Bars may be no closer together than one bar diameter (clear distance), no farther apart than three times the slab thickness $h$, with an absolute minimum clear distance of 1 in. and an absolute maximum spacing of 18 in.

In American practice, slab thicknesses are varied in $\frac{1}{2}$-in. increments. Forms, screeds, and accessories are manufactured in these dimensions and the tradition is so strong that change seems unlikely. In SI units, increments of 10 mm is common; reinforcement cannot reasonably be placed to closer tolerances than 10 mm.

Because the thickness is small, the reinforcement in slabs is similarly limited to smaller sizes. Bar sizes larger than No. 6 are not generally used in slabs; No. 4 is probably the most commonly used size, with No. 6 a close second. Bar sizes less than No. 4 are not used as structural reinforcement but may be used as temperature steel.

Slabs are never reinforced for compression. If a slab is made so thin that compressive reinforcement becomes necessary, the problems with deflections become almost insurmountable. It is far easier, cheaper, and more practical to add $\frac{1}{2}$ in. to the depth.

Where the shape of a slab is square, or nearly so, considerable savings in both concrete and steel can be effected by designing the slab for flexure in two directions. In such "two-way" slabs, part of the load is assumed to be carried in one direction, the remaining part in the other direction. The ACI Code includes a special section on the design of two-way slabs.

The following examples illustrate the design of one-way slabs. Since the procedure somewhat duplicates that used for beams, these examples are expanded to illustrate a means to estimate the dead load before the actual thickness of the slab is known. The rules of thumb given in Chapter 3 are helpful in estimating the thickness and computing the corresponding dead load.

**Example 4-23**

*Strength method*
*Design of a concrete floor slab*
*Simple supports, exposed to weather*

*Given:*
Slab as shown
Live load 100 psf
Grade 60 steel
$f'_c$ = 4000 psi
Normal-weight concrete
Exterior exposure

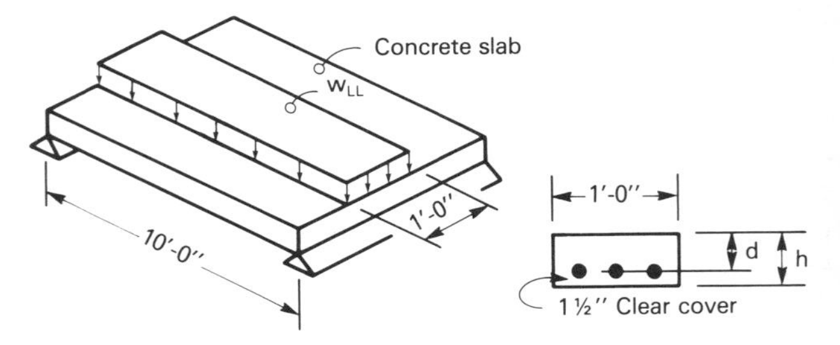


Compute the live-load moment at working levels. For a 1-ft strip, the uniform live load becomes a distributed load of 100 plf, and

$$M_{LL} = \frac{w_{LL}L^2}{8} = \frac{100 \times 10 \times 10}{8} = 1250 \text{ lb-ft/ft}$$

For control of deflections, overall slab thickness is limited (see Table 3-1). Overall thickness is limited to $L/20$ or 6 in.

Estimate thickness: $h = 6$ in. $\pm$, $d = 4$ in. $\pm$.

Compute the dead load of slab per foot of width:

$$\text{dead load} = (150 \text{ pcf})\left(\frac{6}{12}\right)$$

$$= 75 \text{ psf}$$

Compute the estimated dead-load moment

$$M_{DL} = \frac{w_{DL}L^2}{8} = \frac{75 \times 10 \times 10}{8} = 938 \text{ lb-ft/ft}$$

Compute the nominal ultimate moment:

$$M_n = \frac{1.4M_{DL} + 1.7M_{LL}}{\phi} = \frac{1.4 \times 938 + 1.7 \times 1250}{0.9}$$

$$= 3820 \text{ lb-ft.}$$

From Table A-6, try $\rho = 0.008$, $Z_c = 0.131bd^2$

Solve for $d$, using $b = 12$ in.

$$\frac{M_n}{0.85f'_c} = Z_c; \frac{3820 \times 12}{0.85 \times 4000} = 0.131 \times 12 \times d^2$$

$$d = 2.93 \text{ in.}, A_s = 0.008 \times 12 \times 2.93 = 0.281 \text{ in.}^2/\text{ft.}$$

Compare this value of $d = 2.93$ in. that is required for strength criteria against the value of $d$ given earlier for deflection criteria, where $d$ may be as much as 4 in. The deeper section will, of course, be provided. Recompute $\rho$ based on this deeper section, where $d = 4$ in. Both $b$ and $d$ are therefore known and the solution is then for the coefficient of $Z_c$:

$$\frac{M_n}{0.85f'_c} = Z_c \quad \frac{3820 \times 12}{0.85 \times 4000} = \text{coeff.} \times 12 \times 4 \times 4$$

From Table A-6, use $\rho = 0.0041$; hence

$$A_s = 0.0041 \times 12 \times 4 = 0.197 \text{ in}^2/\text{ft}$$

Note that the "svc $f_c$" for this value of $\rho$ is quite low, about 1300 psi. From the sketch, use $h = 6$ in., $d = 4$ in. In this case, deflection (rather than stress) is the controlling criterion.

From Table A-2, use No. 4 bars at 12 in. spacing.

Use $h = 6$ in., $d = 4$ in., $A_s =$ No. 4 bars at 12 in. o.c.

The investigation of a known section to find an allowable load is simply the reverse of the design procedure, as it was for rectangular sections. The following example will illustrate.

**Example 4-24**

*Strength method*
*Investigation of a section to determine the allowable live load on the given slab.*
*Simple supports, no exposure to weather*

*Given:*
Slab as shown
Grade 50 steel
$f'_c = 3000$ psi
Simple span, 12 ft

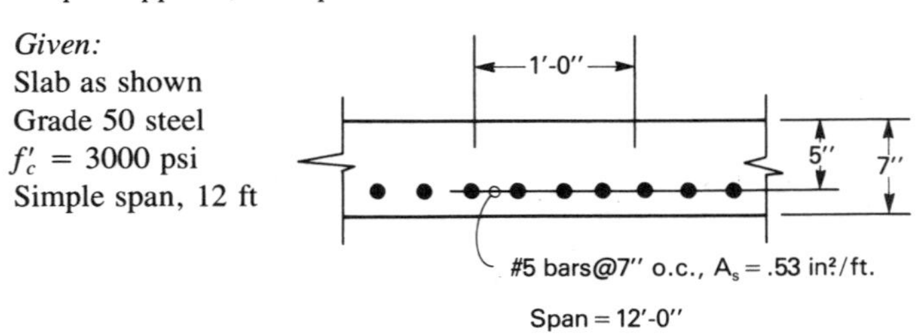


Compute $\rho$ and find the section modulus from tables.

$$\rho = \frac{A_s}{bd} = \frac{0.53}{12 \times 5} = 0.0088; \text{ use } 0.009$$

From Table A-5, $Z_c = 0.161bd^2$.

Compute the nominal ultimate moment:

$$M_n = 0.85f'_cZ_c = 0.85 \times 3000 \times 0.161 \times 12 \times 5 \times 5$$

$$= 123 \text{ kip-in.} = 10.3 \text{ kip-ft}$$

Compute the dead load and the dead-load moment:

$$w_{\text{DL}} = 150 \text{ pcf} \times \frac{7}{12} = 87.5 \text{ psf}$$

$$M_{\text{DL}} = \frac{w_{\text{DL}}L^2}{8} = \frac{87.5 \times 12 \times 12}{8} = 1575 \text{ lb-ft}$$

Compute the live-load moment:

$$M_n = \frac{1.4M_{\text{DL}} + 1.7M_{\text{LL}}}{\phi} \qquad 10.3 = \frac{1.4 \times 1.575 + 1.7M_{\text{LL}}}{0.9}$$

Solve for $M_{\text{LL}}$: $M_{\text{LL}} = 4.16$ kip-ft.

Solve for the uniform load:

$$M_{\text{LL}} = \frac{w_{\text{LL}}L^2}{8} \qquad 4.16 = \frac{w_{\text{LL}} \times 12 \times 12}{8}$$

Solve for $w_{\text{LL}}$: $w_{\text{LL}} = 231$ lb/ft$^2$.

Other investigations are similar to those already discussed with the rectangular sections.

## REVIEW QUESTIONS

1. Under what conditions might compressive reinforcement be required in a concrete beam?
2. At about what percentage of the ultimate stress should the service stress be held in a concrete beam?
3. Other than the examples given in the text, give an example of a flexural member subjected to a small axial load.
4. What is the maximum value of strain that is assumed to occur in concrete at ultimate load?
5. How is the design of a section different from the review (or investigation) of the section?
6. How is the value of the factor $\beta$ determined?
7. Compare the computed service stress in steel in Table 4-1 to the value of $0.6f_y$ for that steel. What is the maximum difference that occurs between these two values?
8. What is a section modulus?
9. Why are slabs rarely, if ever, reinforced in compression?
10. In Table 4-1, compare the working stress in steel in the working stress method to the service stress expected in steel in the ultimate strength method.

## PROBLEMS

Select a suitable beam section for the following conditions, using first the working stress method and then the ultimate strength method. Assume exterior exposure for all members

| Problem | Moment (k-ft) | | Axial load (kips) | | Steel grade | $f'_c$ psi | Limitations | |
|---|---|---|---|---|---|---|---|---|
| | DL | LL | DL | LL | | | $b$ (in.) | $d$ (in.) |
| **1** | 58 | 42 | 0 | 0 | 60 | 3000 | None | None |
| **2** | 40 | 49 | 0 | 0 | 60 | 4000 | None | 14 |
| **3** | 39 | 33 | 0 | 0 | 40 | 3000 | None | 16 |
| **4** | 68 | 54 | 0 | 0 | 40 | 4000 | 15 | 15 |
| **5** | 52 | 62 | 9 | 9 | 60 | 4000 | None | None |
| **6** | 34 | 41 | 14 | 5 | 60 | 5000 | None | None |
| **7** | 47 | 41 | 14 | 9 | 50 | 3000 | None | 19 |
| **8** | 38 | 57 | 23 | 7 | 50 | 3000 | None | 18 |
| **9** | 49 | 52 | 0 | 0 | 60 | 4000 | None | None |
| **10** | 41 | 47 | 0 | 0 | 40 | 4000 | None | 16 |
| **11** | 34 | 41 | 0 | 0 | 60 | 3000 | None | 15 |
| **12** | 55 | 38 | 0 | 0 | 40 | 4000 | 10 | 17 |
| **13** | 69 | 24 | 19 | 23 | 50 | 5000 | None | None |
| **14** | 55 | 32 | 21 | 16 | 50 | 4000 | None | None |
| **15** | 49 | 41 | 24 | 14 | 40 | 5000 | None | 15 |
| **16** | 58 | 49 | 22 | 15 | 60 | 3000 | None | 20 |

Investigate the following sections to find the allowable moment $M_w$ under working stress analysis and the nominal ultimate moment $M_n$ under ultimate strength conditions.

| Problem | $b$ (in.) | $d$ (in.) | $A_s$ (in$^2$) | $A'_s$ (in$^2$) | $f'_c$ (psi) | Steel grade |
|---|---|---|---|---|---|---|
| **17** | 16 | 28 | 3.93 | 0 | 3000 | 60 |
| **18** | 17 | 26 | 2.55 | 1.20 | 4000 | 40 |
| **19** | 18 | 24 | 3.45 | 3.45 | 4000 | 40 |
| **20** | 19 | 22 | 3.14 | 1.57 | 5000 | 50 |
| **21** | 20 | 20 | 4.34 | 0 | 4000 | 50 |
| **22** | 24 | 19 | 2.41 | 0 | 3000 | 60 |
| **23** | 23 | 18 | 7.07 | 7.07 | 3000 | 60 |
| **24** | 32 | 17 | 9.60 | 9.60 | 4000 | 60 |

**25.** Investigate the section of Problem 18 using the working stress analysis when an axial load $P_w = 46$ kips is applied. Find the allowable moment $M_w$ when concrete stress is at $0.45f'_c$.

**26.** Investigate the section of Problem 21 using the working stress analysis when an axial load $P_w = 41$ kips is applied. Find the stress in the tensile steel when the moment is at its maximum allowable value.

**27.** Investigate the section of Problem 22 using the ultimate strength method when an axial load $P_n = 72$ kips is applied. Determine the maximum moment $M_n$ that may accompany the axial load.

**28.** Investigate the section of Problem 23 using the ultimate strength method when an axial load $P_n = 27$ kips is applied. Determine the maximum moment $M_n$ that may accompany the axial load.

# 5

# Shear in Beams

## GENERAL

In Chapter 4 the subject of flexure in concrete beams was developed, together with methods for designing concrete beams to sustain the flexural loadings. It was found that concrete beams could be designed for flexure using the same concepts that were developed in elementary stength of materials for steel or timber. The lack of tensile strength in the concrete was overcome by adding reinforcement wherever tension was expected to occur. The procedures for designing concrete beams thus became somewhat more detailed than for steel or timber, but the concepts and procedures were familiar ones.

The lack of tensile strength in concrete also has a profound effect on the ability of concrete beams to resist shear. Even the relatively low levels of shear that are encountered in routine buildings can introduce serious tension fields in concrete beams. The design of concrete beams to sustain these shears has no counterpart in other common structural materials; no other material is routinely reinforced for the tensions produced by beam shear.

The procedures for designing concrete beams for shear are simple and well developed and their effectiveness has been well proven over the years. Since there are hundreds of combinations of loads that could vary the shear patterns, the design for shear has evolved into a semiempirical ''blanket'' method which assures that the member will be capable of carrying the extremes in shear. Although the extremes may not occur in all beams, Code requires that all beams be capable of sustaining them.

A detailed analysis of shear in beams is beyond the scope of this book. Even if it were included, however, it would provide only background information; the actual design is prescribed by Code without requiring a rational analysis. The following sections contain only brief discussions and explanations of shear patterns in beams, intended to identify the sources of the shear problem and the solutions currently being used.

## SHEAR AS A MEASURE OF DIAGONAL TENSION

When a concrete beam is subjected to high shearing forces, tension stresses develop in the beam. It should be noted that using "shear as a measure of diagonal tension" is an accurate summary of the approach used by ACI to predict the magnitude of these tension stresses.

A simply supported rectangular beam subject to applied loadings is shown in Fig. 5-1a. A section is removed and shown in Fig. 5-1b, showing the shearing force that occurs across the section. Shear on a section is taken as positive when the left side moves up with respect to the right.

The theory concerning the distribution of shear stress across a beam section is treated in elementary strength of materials texts. The typical distribution of shear stress in a rectangular beam is shown in Fig. 5-1c. Shear stress is seen to be maximum at the neutral axis and zero at the extreme fibers, just reverse to the distribution of flexure stress, shown for comparison in Fig. 5-1d.

Typical stressed particles are shown in Fig. 5-1b. Particle I is in the zero where the concrete is in flexural compression, particle II is at the neutral axis, and particle III is in the zone where the concrete is in flexural tension. In addition to these flexural stresses, the particles are also subject to shear stresses.

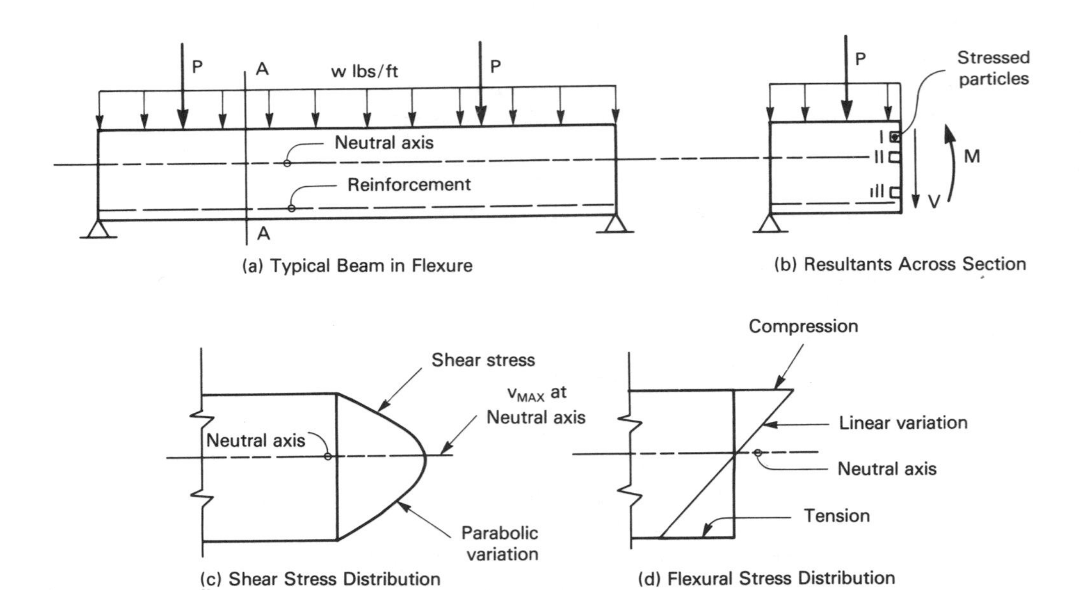


**Figure 5-1** Beam Under Load.

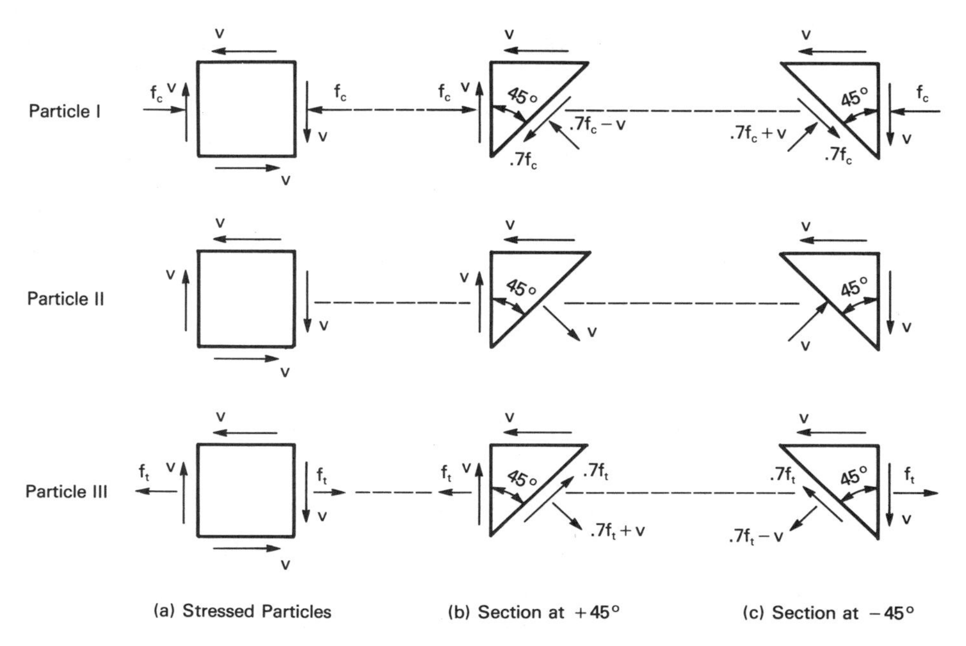


**Figure 5-2** Stressed Particles.

The three stressed particles of Fig. 5-1b are shown larger in Fig. 5-2a, with their flexural stresses $f_c$ and $f_t$ and shear stress $v$ shown on the faces of the particles. Also included as Fig. 5-2b and c are two subparticles cut from the original particle, taken at +45° and −45°. At the neutral axis (particle II) it is seen that the shear stress actually becomes a tensile stress at +45° and a compressive stress at −45°; there is no flexural stress at the neutral axis to influence these stresses.

When the tensile stress due to shear is combined with the compressive stress due to flexure (particle I, Fig. 5-2b), it is seen to reduce the size of the compressive stress. Depending on the magnitudes of these two stresses, the shear stress $v$ could actually cause some tension well into the normally compressive zone, although at some angle away from the flexural compressive stress $f_c$. For the common types of loadings, the magnitude of the shear stress will be significantly greater than the compressive stress only toward the ends of the beam, where shears are high and moments are low.

In particle III, however, the tensile stress due to shear at +45° adds to the tensile stress due to flexure (particle III, Fig. 5-2b). Their sum sharply influences the tension cracking in the concrete. It is this region of the cross section where the shear stresses may be investigated and reinforcement must be considered. Note that the worst case is always at +45°; the tensile stress is always less when the angle is −45°.

At the ends of a simply supported beam where there is no flexural stress and the shear stresses are highest, the shears acting alone will produce tension stresses acting at 45° from horizontal, as shown in Fig. 5-3a. Examining the stress pattern from left to

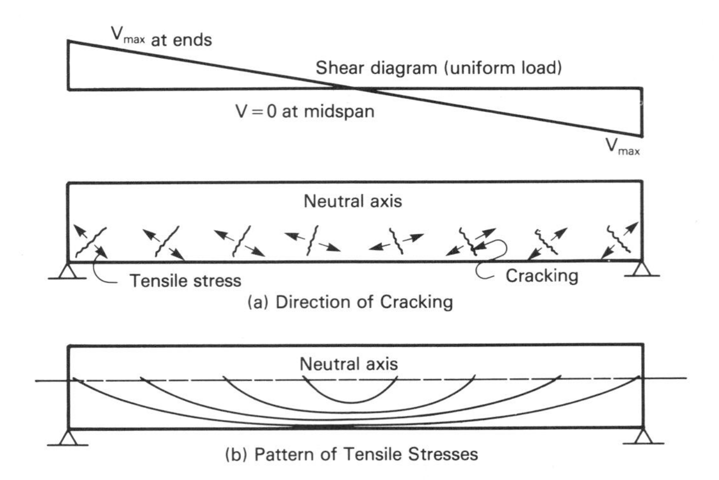


**Figure 5-3** Directions of Tensile Stresses.

right across the span, it is seen that the angle must slowly change across the span toward the center as the shear decreases to zero and the bending moment increases to maximum. At midspan, where there is no shear on the section, the moment acting alone produces a tension stress acting horizontally.

The tension cracks that can be expected to accompany the shear and moment stresses are also shown in Fig. 5-3a. As stated earlier, these cracks are usually quite small, to the point of being invisible to the naked eye. Regardless of how small, however, these cracks must exist if the steel reinforcement is to reach any significant levels of stress.

The general direction of tension stresses in a symmetrically loaded beam is shown in Fig. 5-3b. Because concrete is so weak in tension, reinforcement must be provided wherever the diagonal tension reaches significant levels. It should be noted that diagonal tension can occur wherever there is shear stress; it is toward the ends of the beam at the neutral axis where the shear stress is highest and where the problem is most serious.

## TYPES OF EFFECTIVE SHEAR REINFORCEMENT

Referring again to Fig. 5-3, it is apparent that the directions of the cracks shown in Fig. 5-3a will always be perpendicular to the direction of the tensile stresses shown in Fig. 5-3b. A potential crack would therefore form first at the bottom of the beam, where tensile stresses due to flexure would be highest (Fig. 5-1d). At the lowest point, the shear stresses would be zero (Fig. 5-1c), so the crack would start as a vertical crack. The crack would then progress upward, turning gradually toward 45° as it nears the neutral axis, where shear stress is maximum and flexural stress is zero. At all points, the direction of the crack would be perpendicular to the direction of the tensile stresses, as shown in Fig. 5-3b.

The crack pattern just described is typical of a shear crack in beams, as indicated in Fig. 5-4a. In this type of failure, the longitudinal steel is largely ineffective in resisting the separation of the beam into an upper piece and a lower piece. To be effective, any additional reinforcement would have to cross the crack and thereby keep the crack from becoming large enough to be detrimental. As noted before, the crack *must* form before the reinforcement can be stressed to any appreciable amount. The crack need be only the size of a hairline, however, for the steel to become effective.

Ideally, the shape of the tensile reinforcement in a beam should follow the general pattern of the tensile stresses, shown in Fig. 5-3b. Spaced at some relatively close spacing, the reinforcement would then cross any potential crack that might form, such as those shown in Fig. 5-3a. Such a pattern of steel bars would be difficult to bend and nest into such a shape, but an approximate shape can readily be configured.

The schematic pattern of reinforcement shown in Fig. 5-5a follows such an approximate configuration. When compared to the directions of stress shown in Fig. 5-3b, it is apparent that this approximation is reasonably close to that desired. The longitudinal flexural reinforcement has simply been bent upward across the web after the moment has decreased to the point that it is no longer needed as flexural reinforcement. (The ends of the bars are bent back into a horizontal plane to provide anchorage.)

Toward the middle of the span, shear stresses become much lower and little, if any, shear reinforcement may be required. The crack pattern in this area is shown in Fig. 5-3a and is almost vertical; such tension comes as a result of flexural tension rather than shear. Toward the ends of the beam, the diagonal steel of Fig. 5-5a intercepts the cracks almost at right angles and becomes effective reinforcement against diagonal tension.

Less efficient, but still effective, the vertical bars of Fig. 5-5b also intercept the potential shear cracks. In combination with the separate bars used for longitudinal reinforcement, the configuration of Fig. 5-5b provides only a rude approximation of the directions shown in Fig. 5-3b. However approximate, the crack patterns have nonetheless been intercepted and the desired purpose has been achieved.

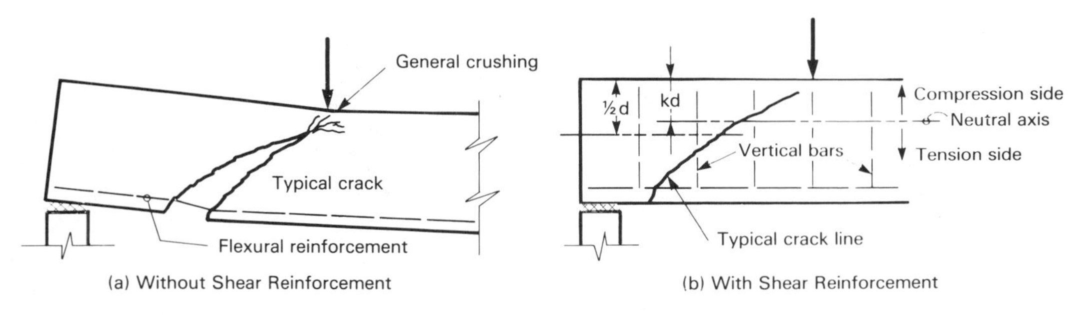


**Figure 5-4** Typical Shear Cracking.

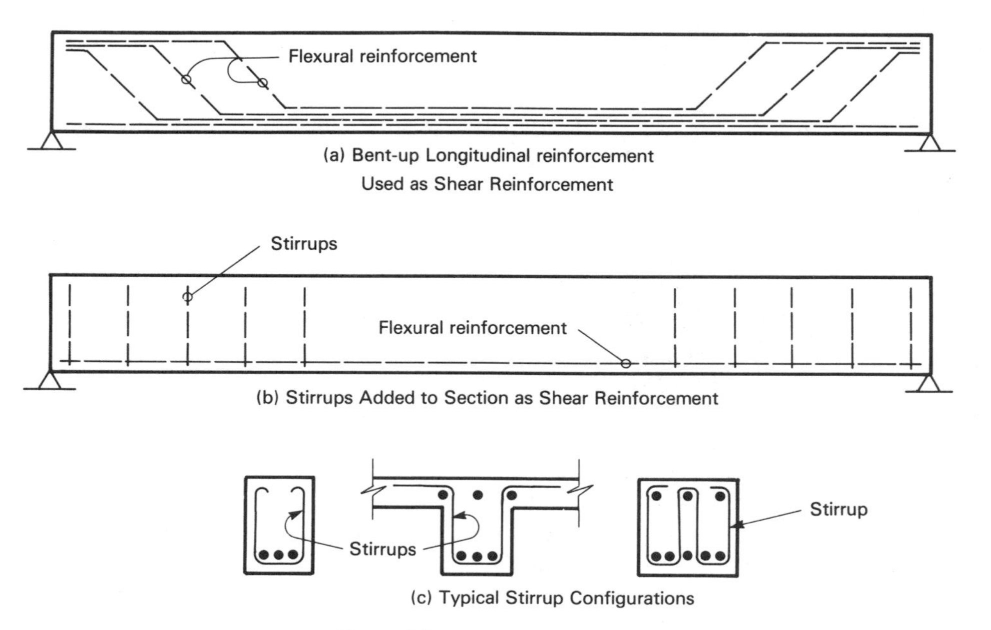


**Figure 5-5** Shear Reinforcement.

The pattern of diagonal shear reinforcement formed by bending the longitudinal reinforcement, as shown in Fig. 5-5a, has been in use for many years. Additional diagonal bars may be added to extend the pattern of diagonal bars both toward the center of the span and toward the ends. In recent years, however, labor costs have increased disproportionately in comparison to materials costs and this method has been replaced by methods requiring less labor.

With few exceptions, the pattern of separate vertical bars shown in Fig. 5-5b is used for shear reinforcement in today's practice. The cost of the small additional amount of steel is more than offset by lower labor costs and faster handling times. In this book, only this pattern of vertical shear reinforcement is presented.

Typical configurations of the separate vertical bars are shown in Fig. 5-5c. Called *stirrups*, these bars are spaced along the span according to the level of shear stress; where shear stresses are higher, the stirrups are placed closer together. Where shear stresses are nearly zero toward the center of the span, stirrups can theoretically be omitted.

Stirrups offer an additional benefit to the builder, in that they provide a sturdy means of tieing and holding the longitudinal steel in position while concrete is being cast. For that reason, the builder will frequently use lightweight stirrups even where none are required, or may extend the stirrups (at a wider spacing) across the middle part of the span even when the drawings show them omitted. The judicious use of stirrups in the design can be applied somewhat generously; they will usually be doubly beneficial.

## CODE REQUIREMENTS FOR STRENGTH METHOD

Where vertical stirrups are used for shear reinforcement, their design is stringently prescribed by Code (11.5). In the strength method, a simplified design is permitted which is based only on the ultimate shear force acting across the section; no consideration is given to any moments acting on the section. Code (11.3.2) also prescribes a more exact refinement of the method in which the moments are included; that refinement is not included in the following discussions.

Code (B.7) also permits the design of shear reinforcement using the alternate method, or working stress method, which will be discussed fully in a subsequent section. As usual in the alternate method, Code prescribes allowable stresses rather than allowable loads. Even so, both the strength method and the alternate method use essentially the same approach, but it is emphasized that one is not an exact multiple of the other.

Returning to the strength method, Code (11.1.1) recognizes that the concrete by itself will take a certain portion of the total shear force even at failure. The remainder of the shear force is, of course, taken by stirrups. That part taken by concrete alone, $V_c$, is computed by

$$V_c = 2\sqrt{f'_c}\, b_w d \qquad \text{(strength method)} \tag{5-1}$$

where $b_w$ is the width of the concrete *web* (or stem of a tee) and all other symbols are the same as defined earlier. The effective web width $b_w$ is taken as the diameter of circular sections.

The terms used in Eq. (5-1) are usually interpreted as an "average" ultimate shear stress of $2\sqrt{f'_c}$ acting across the somewhat arbitrary web area (or stem area of a tee) given by $b_w d$. Although this is a convenient interpretation and this "average" ultimate shear stress is a useful parameter, the Code makes no such claim. The equation is simply a highly simplified and proven means to compute the load taken by the concrete; it applies equally to rectangular beams and slabs as well as tee shapes, circular shapes, and irregular shapes.

The remainder of the nominal ultimate shear force $V_n$ is taken by the vertical steel stirrups. That force, $V_n - V_c$, is designated $V_s$ and is also prescribed by Code (11.5.6), but it is, of course, dependent on the spacing of the stirrups. The horizontal spacing of the stirrups, shown in Fig. 5-6, is specified as a portion of the effective depth, such as $\frac{1}{2}d$, $\frac{1}{4}d$, and so on. As the stirrups are spaced closer together, the force $V_s$ is permitted to go higher.

Code (11.5.4) prescribes a maximum spacing of $\frac{1}{2}d$ for stirrups and implies a minimum spacing of $\frac{1}{4}d$, but places no restrictions on the number of intermediate spacings that may be used. Considering the overall level of accuracy of the design method, a rigorous calculation using several intermediate stirrup spacings is difficult to justify. In this text, stirrup spacings will be limited to the two Code values, $\frac{1}{2}d$ and $\frac{1}{4}d$.

The force per space $V_s$ that may be taken by a single stirrup is given by Code (11.5.6.2) in terms of the spacing:

$$V_s = V_n - V_c = \frac{A_v f_y d}{s} \qquad \text{(strength method)} \tag{5-2}$$

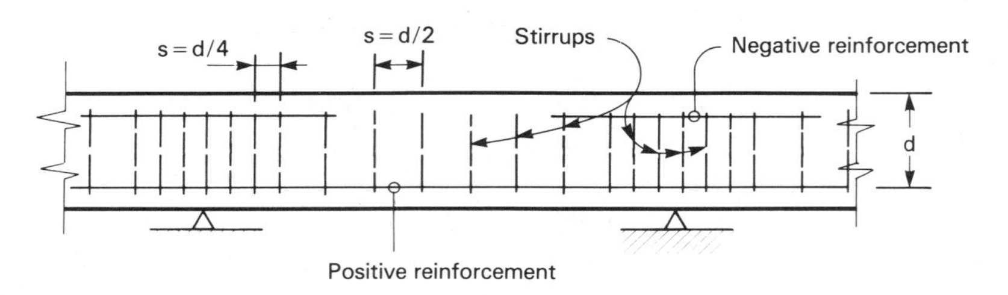


**Figure 5-6** Stirrup Spacing.

where $s$ is the stirrup spacing, shown in Fig. 5-6, and $A_v$ is the total cross-sectional area of steel in the stirrup; other symbols are those used previously.

The nominal ultimate shear force $V_n$ that the entire section can develop, to include benefits due to stirrups, can be stated as some hypothetical shear stress $v_n$ times the cross-sectional area:

$$V_n = v_n b_w d \qquad \text{(strength method)} \tag{5-3}$$

where $v_n$ is an overall "average" value of the ultimate shear stress on the section.

The maximum values of this average ultimate shear stress $v_n$ that can be developed at various levels of reinforcement are, when extracted from the Code:

$$\text{No stirrups:} \quad v_n = 2\sqrt{f'_c} \tag{5-4a}$$

$$\text{Stirrups at } \tfrac{1}{2}d\text{:} \quad v_n = 6\sqrt{f'_c} \qquad \text{(strength method)} \tag{5-4b}$$

$$\text{Stirrups at } \tfrac{1}{4}d\text{:} \quad v_n = 10\sqrt{f'_c} \tag{5-4c}$$

Code limitations do not permit the average ultimate shear stress to exceed $10\sqrt{f'_c}$.

When the values of Eqs. (5-4) are used to compute the shear forces, it is convenient to use the ratio of the stresses rather than the stresses themselves.

$$\text{For concrete alone:} \quad V_c = v_n\, b_w d = 2\sqrt{f'_c}\, b_w d \tag{5-5}$$

$$\text{Without stirrups:} \quad V_n = V_c \tag{5-5a}$$

$$\text{With stirrups at } \tfrac{1}{2}d\text{:} \quad V_n = 3V_c \tag{5-5b}$$

$$\text{With stirrups at } \tfrac{1}{4}d\text{:} \quad V_n = 5V_c \tag{5-5c}$$

These shear forces are shown schematically in Fig. 5-7.

Before proceeding with an example, it is well to list several other requirements and practices:

1. Sections located closer than a distance $d$ from face of support may be designed for a "critical" shear force $V_{cr}$ computed at a distance $d$ from face of support.

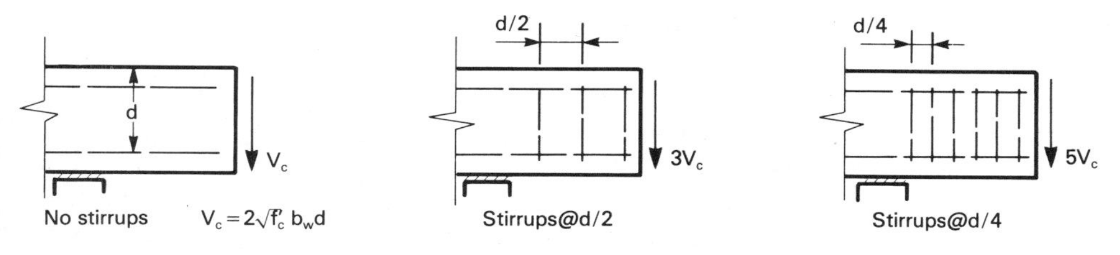


**Figure 5-7** Allowable Shear on a Section.

2. The first stirrup shall be placed within a distance of $\frac{1}{2}d$ from face of support where the required spacing is $\frac{1}{2}d$, and within a distance $\frac{1}{4}d$ from face of support where the required spacing is $\frac{1}{4}d$.
3. Minimum shear reinforcement shall be placed wherever the nominal ultimate shear force $V_n$ acting on the section exceeds one-half the shear strength of the concrete $V_c$.
4. Minimum total cross-sectional area of shear reinforcement in all legs is designated $A_v$ and is computed by

$$A_v = 50 \frac{b_w s}{f_y} \tag{5-6}$$

5. Strength reduction factor $\phi$ for shear is 0.85.

Requirement 3 is something of a curiosity. It states that shear reinforcement is not required at all unless the nominal ultimate shear force exceeds $V_c$, but when the nominal ultimate shear does exceed $V_c$, the shear reinforcement must begin back at $\frac{1}{2}V_c$ rather than at $V_c$.

In addition to the foregoing requirements, there is a practical matter to be considered concerning the placement of stirrups. Heavy stirrups at close spacing are most likely to be required at the ends of heavily loaded girders. At a module point, it can be expected that two such girders with their heavy negative reinforcement and their stirrups will intersect a column with its heavy vertical reinforcement and ties. The congestion of reinforcement in such locations is often formidable. Wherever reasonable, the size of the girders should be kept large enough that the minimum stirrup spacing of $\frac{1}{4}d$ is never required; a spacing not smaller than $\frac{1}{2}d$ should be maintained.

It should be recognized that the strength reduction factor $\phi$ is applied only to external loads, not to any of the computed capacities of the section. For that reason it is recommended that the factor $\phi$ always be included with the shear diagram when making shear computations; it need never be applied thereafter. By this means, there is no confusion later whether the factor $\phi$ should or should not be applied to any other loads that may be under consideration.

## EXAMPLES IN THE STRENGTH METHOD

The following examples illustrate the procedure; a relatively simple continuous rectangular beam is chosen as the first example. Shear and moment diagrams are given. A more general coverage is presented in Chapter 6 with tee beams, but it should be recognized that in designing shear reinforcement, all sections are considered to be rectangular; the shear capacity is based on a rectangular area $b_w d$ regardless whether the section is a tee or a rectangle or a circle.

**Example 5-1**

*Strength method*
*Design of shear reinforcement*
*Flexural design already completed*

Determine the shear reinforcement for the symmetrically loaded rectangular beam shown below. Reinforcement for flexure has already been selected as indicated. Shear and moment diagrams include dead load and are drawn for the nominal ultimate values of $V_n$ and $M_n$; they include the strength reduction factor $\phi$. Use grade 60 steel, $f'_c$ = 3000 psi.

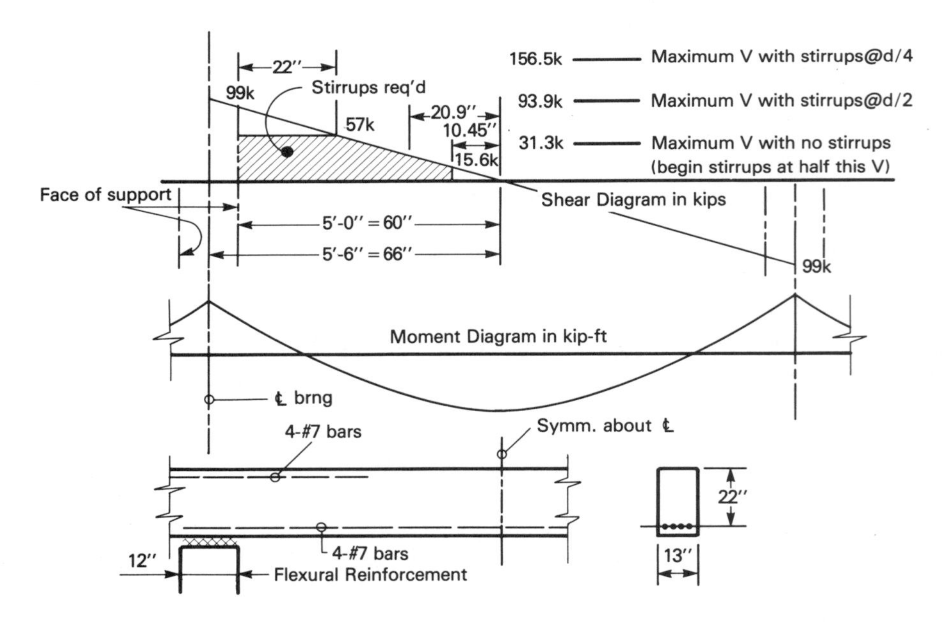


Compute (or scale) critical $V_{cr}$ at a distance $d$ from face of support, by ratio;

$$V_{cr} = \frac{60 - 22}{66} 99 = 57 \text{ kips (includes } \phi)$$

Compute the shears at three levels of reinforcement:

$$V_c = 2\sqrt{f'_c}\, b_w d = 2\sqrt{3000} \times 13 \times 22$$

No stirrups: $V_n = 31.3$ kips

Stirrups at $\frac{1}{2}d$: $V_n = 3 \times 31.3 = 93.9$ kips

Stirrups at $\frac{1}{4}d$: $V_n = 5 \times 31.3 = 156.5$ kips

Compute the horizontal distance to $V_{cr}$ (without stirrups) from the centerline of beam:

$$\text{distance} = \frac{31.3}{99} 66 = 20.9 \text{ in. from centerline of beam}$$

The foregoing shears and distances have been plotted on the shear diagram above. Note that a stirrup spacing of $\frac{1}{2}d$ (11 in.) will be required, beginning within a distance of $\frac{1}{2}d$ from face of support and extending to within 10.45 in. of the centerline of the beam. Since the stirrups extend so close to the center, it is chosen in this case to use stirrups across the entire span.

The chosen layout of stirrups is shown in the following sketch.

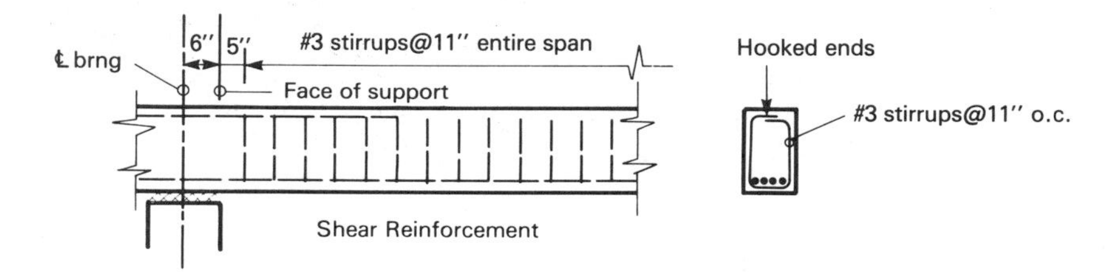


Compute the required size of the stirrups for the indicated layout, for grade 60 steel. The maximum shear to be carried by any section is 57 kips.

$$V_s = V_{cr} - V_c = \frac{A_v f_y d}{s} \qquad A_v = \frac{(V_{cr} - V_c)\, s}{f_y d}$$

$$A_v = \frac{(57 - 31.3) \times 10^3 \times 11}{60{,}000 \times 22} = 0.214 \text{ in}^2 \quad \text{(in two legs)}$$

Use stirrups No. 3 bars at 11 in. o.c. as shown in the sketch.

The stirrups chosen for Example 5-1 are shown with "hooks" at their upper end. The hooks are an anchorage requirement that will be discussed in Chapter 7. It is also an anchorage requirement that the stirrups must be anchored within the compression side of the beam. Theoretically, for those parts of the span where the bottom of the beam is in compression, the stirrups would have to be turned upside down, such that the hooks are on the bottom. Such a state of stress occurs at the supports, where the moment is negative, but turning the stirrups is rarely done.

Just as it is possible to vary the stirrup spacing across the span to suit the variations in the shearing force, so is it possible to vary the size of the bar used for stirrups where the load is small enough to justify it. Such refinements are rarely made in small buildings. The tonnage of steel that can be saved by such measures would rarely justify the time and effort spent in engineering, drafting, field layout, and in just keeping track of the additional mark numbers of stirrups.

Where a beam is not quite symmetrically loaded but is nearly so, the shear reinforcement is usually laid out symmetrically to avoid the additional labor hours in drafting and field layout. It also obviates any possibility of the stirrups being installed backwards in the beam. On occasion, however, where a beam is distinctly unsymmetrical, the shear reinforcement must be laid out to suit the actual shear diagram. The procedure is the same, just more complex, as illustrated in the next example.

**Example 5-2**

*Strength method*
*Design of shear reinforcement*
*Flexural design already completed*

Determine the shear reinforcement for the unsymmetrically loaded rectangular beam shown below. Reinforcement for flexure has already been selected as indicated. Shear and moment diagrams include dead load and are drawn for the nominal ultimate values of $V_n$ and $M_n$; the indicated loads include load factors as well as the strength reduction factor $\phi$. Use grade 60 steel and $f'_c = 3000$ psi.

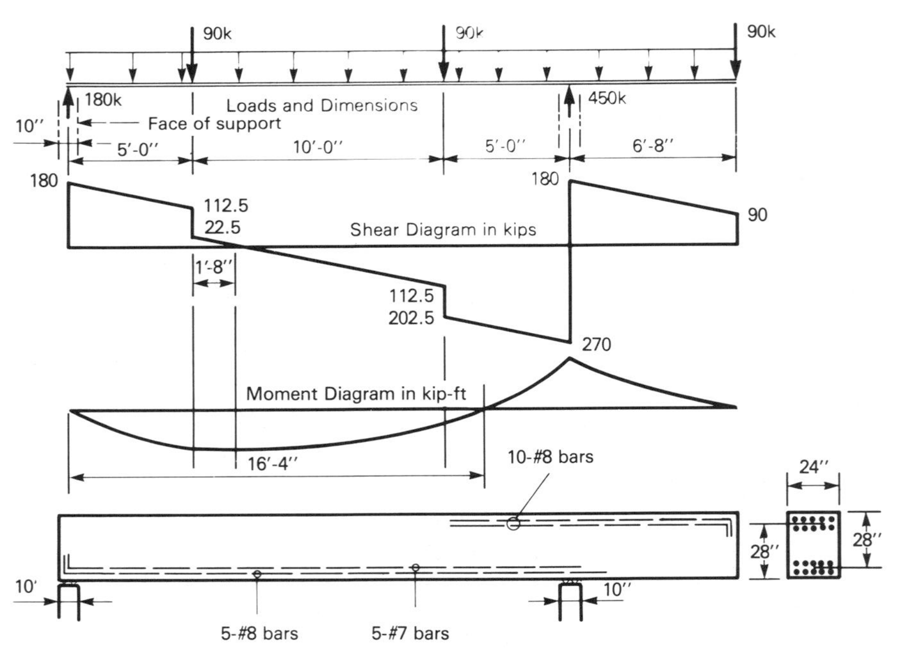

Flexural Reinforcement

Compute (or scale) $V_{cr}$ at a distance $d$ from face of support:

At right side of left support, 33 in. from centerline of support:

$$V_{cr} = 180 - (180 - 112.5) \times \frac{33}{60} = 143 \text{ kips}$$

At left side of right support, 33 in. from centerline of support:

$$V_{cr} = 270 - (270 - 202.5) \times \frac{33}{60} = 233 \text{ kips}$$

At right side of right support, 33 in. from centerline of support:

$$V_{cr} = 180 - (180 - 90) \times \frac{33}{80} = 143 \text{ kips}$$

Compute the shears for three levels of reinforcement:

No stirrups: $V_c = 2\sqrt{f'_c}\, b_w d = 2 \times \sqrt{3000} \times 24 \times 28$

$= 73.6 \text{ kips}$

Stirrups at $\frac{1}{2}d$: $V_n = 3 \times 73.6 = 221 \text{ kips}$

Stirrups at $\frac{1}{4}d$: $V_n = 5 \times 73.6 = 368 \text{ kips}$

Plot the foregoing values to scale on the shear diagram as shown below.

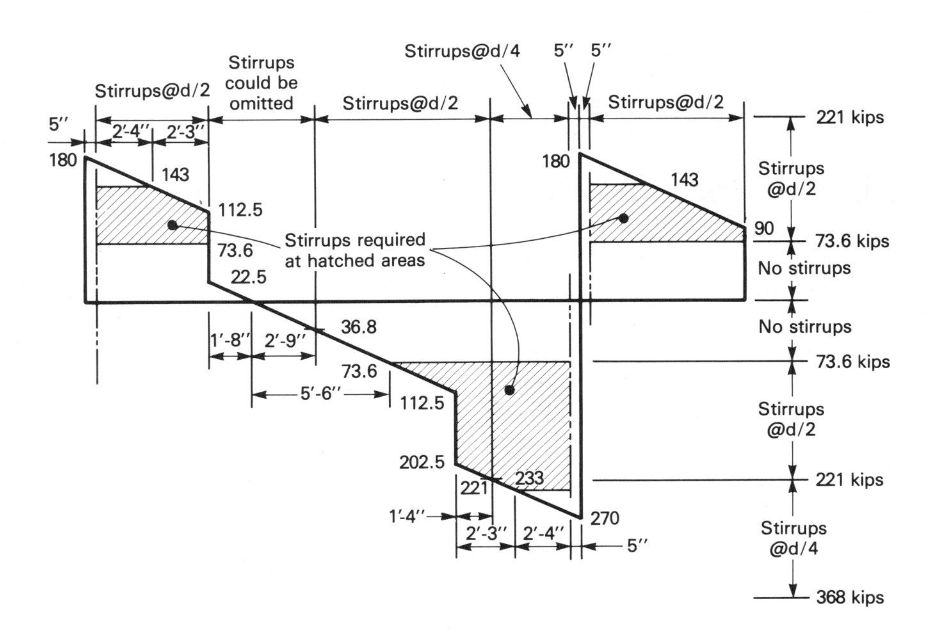

Compute the required size for the stirrups:

$$V_s = V_n - V_c = \frac{A_v f_y d}{s} \qquad A_v = \frac{(V_n - V_c)s}{f_y d}$$

$$\text{For spacing at } \tfrac{1}{4}d\text{:} \quad A_v = \frac{(233 - 73.6) \times 7}{60{,}000 \times 28}$$

$$= 0.664 \text{ in}^2$$

$$\text{For spacing at } \tfrac{1}{2}d\text{:} \quad A_v = \frac{(221 - 73.6) \times 14}{60{,}000 \times 28}$$

$$= 1.23 \text{ in}^2$$

Use four legs, No. 5 bars, $A_v = 1.23$ in$^2$.

Final shear reinforcement is shown in the following sketch.

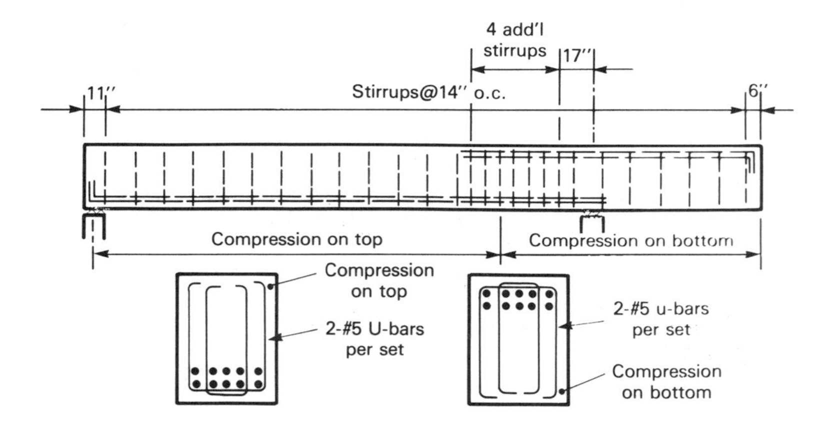


## CODE REQUIREMENTS FOR WORKING STRESS METHOD

The design of shear reinforcement using the working stress method can be handled almost identically to the approach used in the strength method. The major difference lies in the magnitudes of the "average" shear stresses used in the two methods. The two methods will not afford exactly the same final design, emphasizing once again that the two methods of design are not equivalent.

As before, Code (B.3) recognizes that if diagonal tension is high enough to cause cracking, the concrete will be taking only a portion of the total load and the stirrups will be taking the rest. The maximum allowable working stress on the concrete acting alone, $v_w$, is given by Code (B.3.1):

$$v_w = 1.1\sqrt{f'_c} \qquad \text{(working stress method)} \tag{5-7}$$

The units in Eq. (5-7) are psi; the equation is empirical and is not homogeneous.

Spacing of stirrups in the working stress method is the same as that specified earlier, with a maximum spacing of $\frac{1}{2}d$ at low shears to a minimum spacing of $\frac{1}{4}d$ at high shears. The same criteria will be applied—that only these two spacings, $\frac{1}{2}d$ and $\frac{1}{4}d$, will be used. No intermediate spacings will be considered.

The allowable force per space, $V_s$, that may be taken by one stirrup is found from the Code to be

$$V_s = V_w - V_c = \frac{A_v f_s d}{s} \qquad \text{(working stress method)} \qquad (5\text{-}8)$$

where $f_s$ = elastic or working stress in the steel
$V_w$ = total shear force on the section
$V_c$ = allowable shear force on the concrete

Following the approach used before, the maximum allowable shear across the section, to include any increases due to addition of stirrups, is

$$V_w = v_w b_w d \qquad \text{(working stress method)} \qquad (5\text{-}9)$$

where $v_w$ is an allowable "average" shear stress across the section. The maximum allowable values of $v_w$ are extracted from the Code for various levels of reinforcement:

$$\text{No stirrups:} \quad v_w = 1.1\sqrt{f'_c} \qquad (5\text{-}10a)$$

$$\text{Stirrups at } \tfrac{1}{2}d\text{:} \quad v_w = 3.1\sqrt{f'_c} \qquad \text{(working stress method)} \qquad (5\text{-}10b)$$

$$\text{Stirrups at } \tfrac{1}{4}d\text{:} \quad v_w = 5.5\sqrt{f'_c} \qquad (5\text{-}10c)$$

Code (B.3.1) limitations do not permit the allowable average shear stress to exceed $5.5\sqrt{f'_c}$.

The detailing requirements for the working stress method are almost exactly the same as listed earlier:

1. Sections located closer than a distance $d$ from face of support may be designed for the shear force computed at a distance $d$ from face of support.
2. The first stirrup shall be placed at a distance of $\frac{1}{2}d$ from face of support where the required spacing is $\frac{1}{2}d$ and within a distance $\frac{1}{4}d$ from face of support where the required spacing is $\frac{1}{4}d$.
3. Minimum shear reinforcement shall be placed wherever the average shear stress $v_w$ acting on the section is greater than one-half the allowable shear stress $v_c$ carried by the concrete alone, or where $v_w$ is greater than $0.55\sqrt{f'_c}$.
4. The minimum area of reinforcement is computed by

$$A_v = 50\frac{b_w s}{f_y} \qquad (5\text{-}11)$$

5. As stated earlier, it is highly desirable to keep the stirrup spacing not closer than $\frac{1}{2}d$, to reduce congestion of reinforcement.

The strength reduction factor $\phi$ that was used in the strength method does not apply in the working stress method.

The procedure for using the working stress method is identical to that for the strength method, except, of course, that the working stress method uses the allowable stresses and shear diagrams at working levels rather than at ultimate load. With these changes in the numbers, the procedure for the working stress method is identical to the strength method.

## REVIEW QUESTIONS

1. When a particle is subjected to a pure shear, how is it that a tension stress occurs?
2. Where in the cross section of a beam is flexure stress highest? Lowest? Where is shear stress highest? Lowest?
3. Where in the span of a simply supported beam is bending moment highest? Where are shearing forces highest?
4. In view of the answers to Questions 2 and 3, explain why the design for shear in a concrete beam may be treated generally independently from the design for flexure.
5. In view of the answers to Questions 2 through 4, explain why the direction of shear cracks will be close to 45° at the ends of a simple span, becoming more nearly vertical toward midspan.
6. How do vertical stirrups provide reinforcement for a diagonal tension stress?
7. On a beam having a circular cross section, what is the shear width $b_w$?
8. In view of the answer to Question 7, determine an effective width for $b_w$ for a hollow circular beam of reinforced concrete.
9. What is the effective width $b_w$ for an I shape, similar to that shown in Fig. 3-1? Justify your answer.
10. What is the strength reduction factor $\phi$ for shear?
11. How is the strength reduction factor $\phi$ applied when the strength method is used to design for shear?
12. How is the strength reduction factor $\phi$ applied when the working stress design method is used to design for shear?

## PROBLEMS

An interior span in a continuous rectangular beam has the shear and moment diagrams indicated below. The size of the member and its flexural reinforcement have already been selected as given. Select the shear reinforcement for the given section using the strength method. Use grade 60 steel, $f'_c$ = 3000 psi.

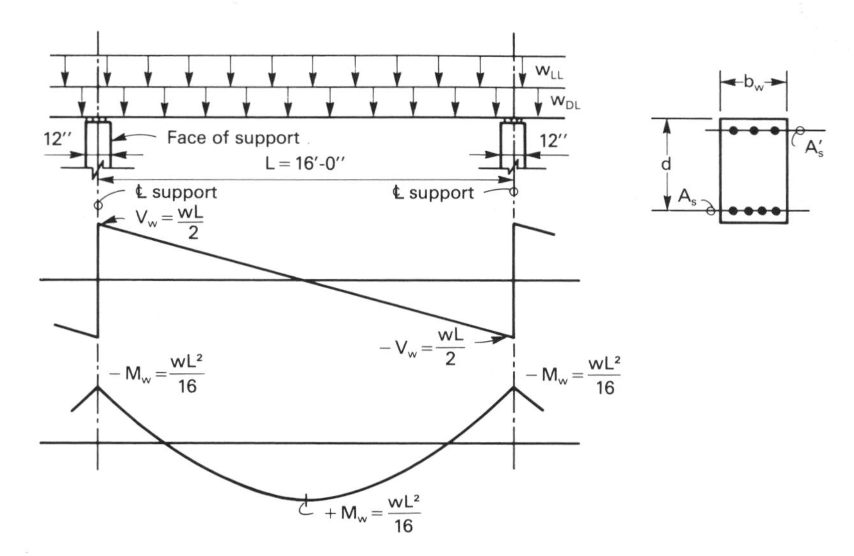


| Problem | Uniform load DL (kips/ft) | Uniform load LL (kips/ft) | $d$ (in.) | $b_w$ (in.) | $A_s$ (in$^2$) | $A'_s$ (in$^2$) |
|---|---|---|---|---|---|---|
| **1** | 3.5 | 15.75 | 36 | 18 | 3.14 | 1.50 |
| **2** | 5.25 | 15.75 | 34 | 20 | 4.00 | 2.09 |
| **3** | 7.00 | 14.00 | 32 | 22 | 4.68 | 2.53 |
| **4** | 8.75 | 14.00 | 30 | 24 | 5.57 | 2.88 |
| **5** | 10.50 | 12.25 | 28 | 26 | 6.33 | 3.12 |
| **6** | 12.25 | 12.25 | 28 | 28 | 7.33 | 3.76 |
| **7** | 14.00 | 10.50 | 26 | 28 | 7.57 | 3.88 |
| **8** | 15.75 | 10.50 | 24 | 30 | 8.78 | 4.10 |
| **9** | 17.50 | 8.75 | 24 | 30 | 9.37 | 5.07 |
| **10** | 19.50 | 8.75 | 24 | 32 | 10.3 | 5.57 |

# 6

# Tee Beams and Joists

## GENERAL

The topics treated in this chapter utilize the theoretical concepts introduced in Chapters 4 and 5. Since there is no development of any new theory, this chapter might be viewed as a ''how-to'' chapter on the design of tee-beam floor and roof systems. Joists are included also, but it will be seen that they are simply tee beams that meet special conditions, thereby gaining some design advantages under the Code.

Tee beams are one of the more important features that may be used in concrete construction, important enough that an entire chapter devoted to their design is justified. The advantage of having a concrete deck working monolithically with its concrete beams offers a significant increase in efficiency of materials. For comparison, designing a steel deck to be stressed monolithically with its steel supporting member would rarely be feasible outside aircraft or ship design.

A typical concrete tee beam is shown in Fig. 6-1. A certain width of the concrete floor slab becomes the compression flange whenever the beam is in positive moment (tension on bottom). This compressive width, designated $b'$ in Fig. 6-1, is the effective width for flexure only; the width of the web, $b_w$, is still the effective width for shear.

A system of repetitive tee beams might well be considered a thick slab that has had its reinforcement lumped together at regular intervals and in addition has had the concrete in the unreinforced areas removed to save weight. Since all concrete in the tension zone ceases to exist, insofar as the flexural analysis is concerned, it does not really matter whether these ''dead'' areas are solid or void. The tee beam is thus equivalent to a thick slab with its dead weight considerably reduced.

Although the removal of the recessed areas makes little difference in the capacity to resist flexure, it causes a serious reduction in the capacity to resist shear. Tee beams

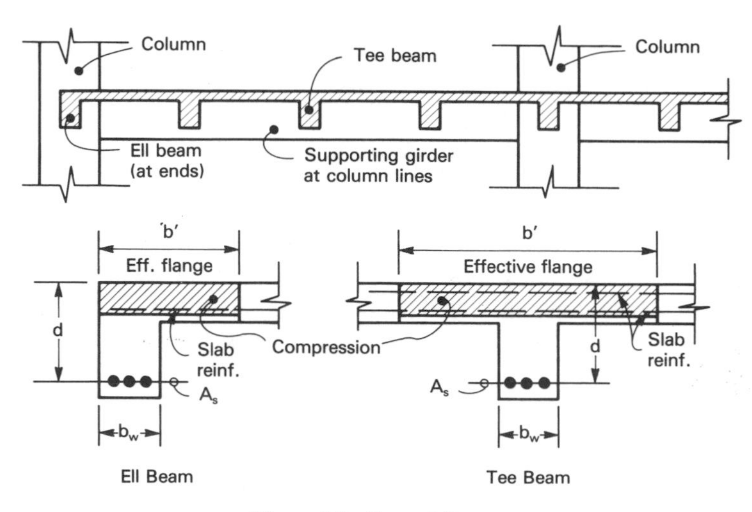


**Figure 6-1** Formed Recess.

are quite weak in shear compared to slabs and usually require shear reinforcement. The spacing of repetitive tee beams should therefore be set with due regard to the shear capacity of each tee beam.

Concrete tee sections. Exposed concrete beams are often used as a design feature, such as these roof beams in a contemporary design. (Courtesy Portland Cement Association).

## TEE BEAMS SUBJECT TO NEGATIVE MOMENT

Design for flexure and design for shear were developed independently of each other in Chapters 4 and 5. Both may now be utilized to design a typical tee-beam stem subject to a large shear and a small negative moment. Such cases occur commonly where spans are short and loads are heavy.

Under negative moment, the compressive area of a tee beam is at the bottom of the stem, as shown in Fig. 6-2. The top of the section is in tension and therefore disappears as far as the flexural analysis is concerned. The section responds to load exactly like an upside-down rectangular section.

For continuous rectangular beams, the negative moment and the high shears at the supports can be expected to be larger than the positive moment and low shears at midspan; conditions at the supports will therefore usually govern the design. For continuous tee beams, where the compressive area is so much smaller at the support than at midspan, the conditions at the supports will almost always govern the size of the tee-beam stem.

The size of the section is usually determined first for flexural considerations; in this solution the section modulus is found and solved for $b_w$ and $d$. Then, from the shear criteria, a separate and independent solution for $b_w$ and $d$ can be found, where $b_w$ and $d$ are deliberately kept large enough that stirrups are not needed, or at best, a spacing closer than $d/2$ is not required. These two solutions for $b_w$ and $d$ are then compared and the final choice is made.

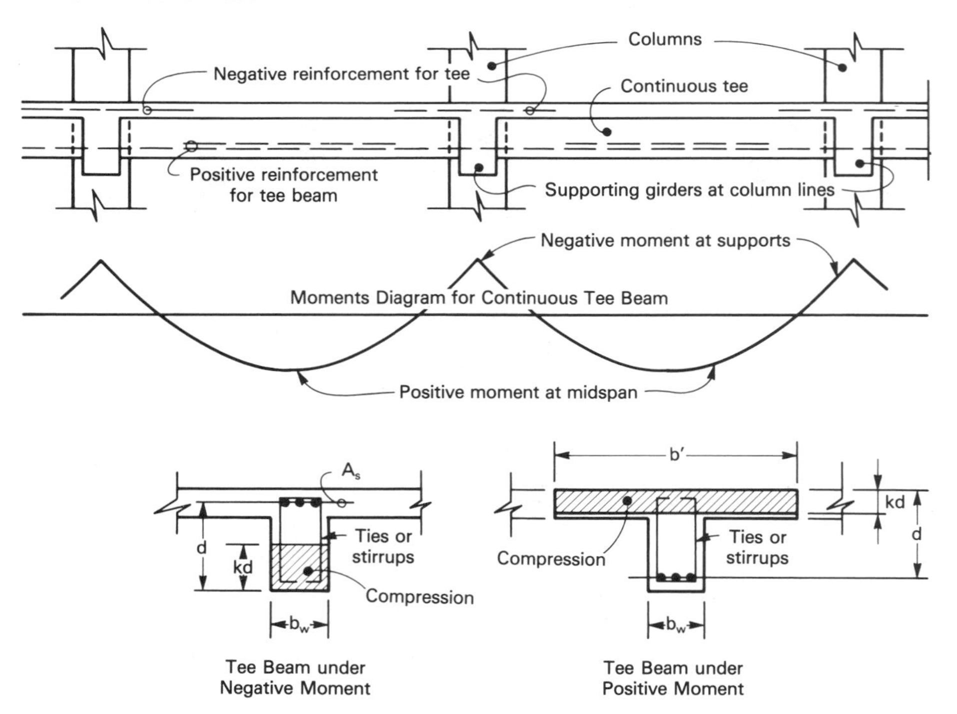


**Figure 6-2** Tee Beam Subject to Moments.

An example will illustrate the procedure. The example is presented using the strength method. The working stress method would be directly parallel.

**Example 6-1**

*Strength Method*
*Selection of size of a tee-beam stem*
*Stem subject to shear and to negative moment*

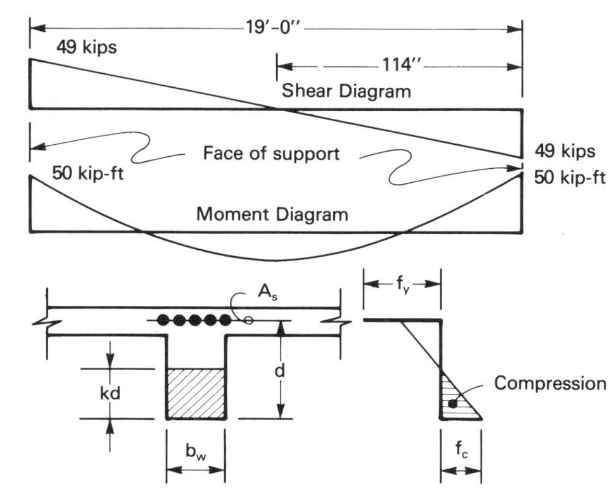


*Given:*
Beam as shown
Grade 60 steel
$f'_c = 3000$ psi
At face of support, $M_n = 50$ kip-ft and $V_n = 49$ kips

From Table A-5, for "svc $f_c$" of less than 1500 psi, select $Z_c$. Try

$$Z_c = 0.111bd^2 \qquad \rho = 0.005$$

Compute $d$, using $b_w = 0.5d$:

$$\frac{M_n}{0.85f'_c} = Z_c \qquad \frac{50 \times 10^3 \times 12}{0.85 \times 3000} = 0.111b_wd^2 = 0.111 \times 0.5d \times d^2$$

For flexure, required depth $d = 16.2$ in., $b = 8.1$ in.

An independent solution for $b$ and $d$ is now obtained from shear criteria.

Compute the maximum average ultimate shear stress with stirrups spaced at $\frac{1}{2}d$:

$$v_n = 6 \times \sqrt{f'_c} = 6\sqrt{3000} = 329 \text{ psi}$$

Compute the shear force at a distance $d$ from face at support, using $d = 16.2$ in. as a first approximation for $d$:

$$V_{cr} = \frac{114 - 16.2}{114} \times 49 = 42 \text{ kips}$$

Compute the required size to keep stirrup spacing $\geq \frac{1}{2}d$:

$$V_{cr} = v_nb_wd \qquad 42{,}000 = 329 \times 0.5d \times d$$

For shear, required depth $d = 16.0$ in.; $b_w = 8$ in.

This solution for $b$ and $d$ is compared to the earlier solution using flexure criteria. The larger values for $b$ and $d$ are chosen for the final design.

From flexural analysis, use $d = 16\frac{1}{2}$ in., $b_w = 9$ in., $\rho = 0.005$. Place stirrups at $\frac{1}{2}\,d$.

It is again emphasized that the first calculation for the required size of the stem is made for flexure. The second calculation is made for shear. It was noted earlier that for continuous beams, such as this example, the negative moment can be expected to be larger than the positive moment and will therefore govern the size of the stem. The shears are then reviewed to see that shear reinforcement is acceptable.

The foregoing solution would be identical if the beam had been specified as a rectangular section rather than a tee section. Such rectangular beams occur in structural systems where a cast-in-place rectangular concrete frame is used to support flooring made of precast prestressed "planks." In these systems, which are in relatively common use, the supporting member is obviously a rectangular section throughout its length.

When the moment on a tee beam becomes positive, the tee section suddenly gains a great deal more compressive material, as shown in Fig. 6-3. When the neutral axis falls within the top slab, or flange, the beam still responds to load as does any other rectangular section, but as one having a compressive width of $b'$. All the design procedures discussed previously are still valid; the section simply happens to be wide and shallow.

A complication arises when the neutral axis falls below the soffit of the slab, as shown in Fig. 6-3c. For this case the compressive area is no longer rectangular. The amount of error between this case and the truly rectangular case will be seen later to be quite small at service levels and in some cases can be ignored.

## CODE REQUIREMENTS FOR TEE SHAPES

Before examining a tee section subject to positive moment, it is first necessary to define the effective dimensions of a tee beam. A typical view of a floor system is shown in Fig. 6-1, with a section taken through the system showing the beams. At midspan the moment is positive and the floor slab is in compression; the section is recognized as a tee beam.

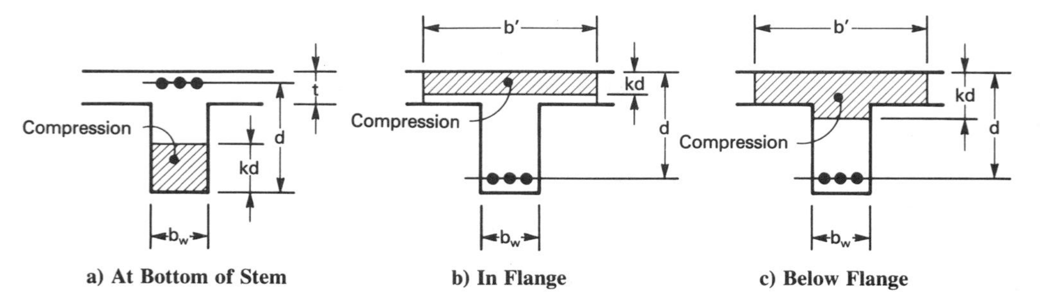


**Figure 6-3** Compression Zones in Tee Sections

Not all of the slab will be effective as a compression flange, however. Code (8.10.2) prescribes the width of the slab that may be considered to be effective in compression; these limits and requirements are based on extensive performance tests. Test results indicate that the effective flange will not extend halfway to the next stem unless the stems are spaced fairly close together.

Code (8.10) limitations concerning the effective compression area are given in the following requirements.

1. The total width of slab effective as a tee-beam flange (dimension $b'$ in Fig. 6-1) shall not exceed one-fourth the span length of the beam.
2. The effective overhanging flange width on each side of the web (or stem) shall not exceed eight times the slab thickness nor one-half the clear distance to the next web.
3. For beams having a slab on one side only (termed ell beams), the effective overhanging flange width shall not exceed one-twelfth the span length of the beam nor six times the slab thickness nor one-half the clear distance to the next web.
4. Isolated beams in which the tee shape is used to provide a flange for additional compressive area shall have a flange thickness not less than one-half the width of the web and an effective flange width not more than four times the width of the web.

Requirement 4 is applicable to the mixed precast "planks" and cast-in-place rectangular frames mentioned in the preceding section. Where an isolated framing member is required, a tee section offers a great deal more lateral stability than does a rectangular section.

The foregoing requirements are obviously somewhat arbitrary, since there can be no sharp break between the stress in the effective flange and the stress in the immediately adjacent parts of the slab. The rather rough approximation does not matter, however, since the total area of concrete is so high that concrete stresses at service levels are invariably quite low. The service stress in the concrete is rarely even checked in practice, but will be checked here as a matter of academic interest.

Occasionally, compressive reinforcement is used in a tee beam. Compressive reinforcement can be effective in reducing deflections. Tee beams, often being comparatively shallow, invite problems with deflections.

One additional point of the Code deserves repeating. The minimum steel ratio $\rho$ is given by Code (10.5.1):

$$\text{minimum } \rho \geq \frac{200}{f_y}$$

In checking for the minimum steel ratio in tee beams, the width of the stem is used (see Fig. 6-1):

$$\text{minimum } \rho = \frac{A_s}{b_w d} \geq \frac{200}{f_y}$$

This artificial steel ratio $\rho_{\min}$, based on $b_w$, must be applied to tee sections. The real steel ratio, when based on the compressive width $b'$ (see Fig. 6-1), may be considerably smaller than this artificial value. This point will be illustrated in Examples 6-2 and 6-3 when the steel ratio based on $b'$ must be used when selecting a tee section.

## TEE BEAMS SUBJECT TO POSITIVE MOMENT

Two common cases will be considered for tee-beam design:

1. *Continuous tee beam with negative moment at supports.* The tee beam is subject to negative moment at its ends and to positive moment at midspan.
2. *Simple span.* The tee beam is never subject to negative moment.

For case 1, the size of the stem is invariably set by shear and moment considerations at the support. At midspan, since the size of the stem is fixed, the only remaining calculation is for the area of reinforcement to take the positive moment. This calculation is demonstrated in Example 4-16, where both $b_w$ and $d$ are fixed, but will be repeated in the next example to show the effects of the wider compression area.

**Example 6-2**

*Strength method*
*Design of a tee beam subject to positive moment*
*Size of stem fixed by conditions at supports*

*Given:*
Tee section as shown
$M_{DL} = 57$ kip-ft
$M_{LL} = 61$ kip-ft
Grade 60 steel
$f'_c = 3000$ psi
Span = 23 ft 0 in.
Tees at 10 ft 0 in. o.c.

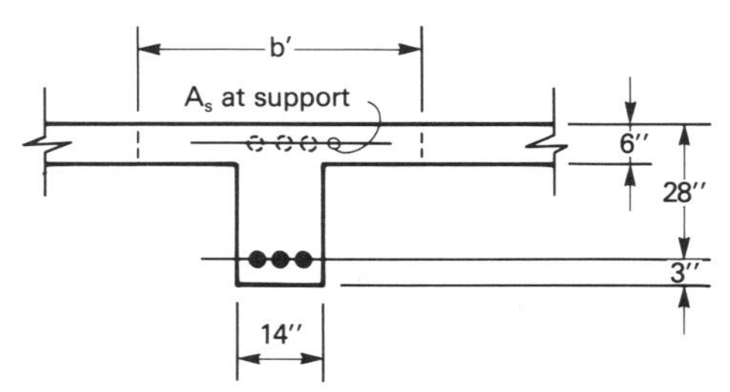


Determine the effective width of the flange (in inches):

$$b' < \frac{\text{span}}{4} \qquad b' < \frac{23 \times 12}{4} \text{ or } 69 \text{ in.} \quad \text{(use)}$$

$$b' < 8 \times \text{flange} + b_w \qquad b' < 8 \times 6 \times 2 + 14 \text{ or } 110 \text{ in.}$$

$$b' < \text{tee spacing} \qquad b' < 10 \times 12 \text{ or } 120 \text{ in.}$$

Compute the nominal ultimate moment:

$$M_n = \frac{1.4M_{\text{DL}} + 1.7M_{\text{LL}}}{\phi} = \frac{(1.4 \times 57 + 1.7 \times 61)}{0.9}$$

$$= 204 \text{ kip-ft}$$

Since both $b'$ and $d$ are known, compute the coefficient of $Z_c$:

$$\frac{M_n}{0.85 f'_c} = Z_c \quad \frac{204{,}000 \times 12}{0.85 \times 3000} = \text{coeff.} \times b'd^2 = \text{coeff.} \times 69 \times 28 \times 28$$

coeff. = 0.018.

From Table A-5, select $Z_c = 0.023b'd^2$ at $\rho = 0.0010$, or, interpolating, $Z_c = 0.018b'd^2$ at $\rho = 0.00075$.

Select the steel area:

$$A_s = \rho b'd = 0.00075 \times 69 \times 28 = 1.45 \text{ in}^2$$

From Table A-3, select 2 No. 8 bars.

Use $d = 28$ in., $b' = 69$ in., $b_w = 14$ in., 2 No. 8 bars.

Example 6-2 is quite direct. It emphasizes that the minimum steel requirement given by Code, $\rho_{\text{min}} = 200/f_y$, must always be computed using the stem width $b_w$ rather than the actual compressive width $b'$. The actual design, however, is still based on the actual compressive width $b'$.

Example 6-2 could also have been solved using the working stress method. In this application, however, the working stress method is tedious and clumsy and is not recommended. In a subsequent section a simple approximate method is presented that yields somewhat conservative but satisfactory results with much less effort.

For the sake of demonstration, however, the next example presents the solution to Example 6-2 using the working stress method. Since the stress in the concrete is not known, there is no direct access to the design tables, neither by computing $f_s/f_c$ nor by computing the coefficient of $S_c$. With no direct access to the design tables, the solution is one of simply trying various values of steel ratios until a satisfactory one is found.

**Example 6-3**

*Working stress method*
*Design of a tee beam subject to positive moment*
*Size of stem fixed by conditions at supports*

Using the working stress method, determine the required steel area of the tee beam of Example 6-2. Compute the working moment $M_w$:

$$M_w = M_{\text{DL}} + M_{\text{LL}} = 57 + 61 = 118 \text{ kip-ft.}$$

Starting with the lower steel ratios of Table A-5, try various steel ratios (and associated concrete stresses) until one is found such that the steel is stressed near its allowable value of 24,000 psi. $d = 28$ in., $b' = 69$ in.

Try $\rho = 0.005$, $S_c = 0.045b'd^2$, $f_s/f_c = 92.18$:

$$f_c = \frac{M_w}{S_c} = \frac{118{,}000 \times 12}{0.045 \times 69 \times 28 \times 28} = 582 \text{ psi}$$

Compute $f_s = f_c \times (f_s/f_c) = 582 \times 92.18 = 53{,}600$ psi (no good—steel stress too high).

Try $\rho = 0.0010$, $S_c = 0.061b'd^2$, $f_s/f_c = 63.89$:

$$f_c = \frac{M_w}{S_c} = \frac{118{,}000 \times 12}{0.061 \times 69 \times 28 \times 28} = 429 \text{ psi}$$

Compute $f_s = f_c(f_s/f_c) = 429 \times 63.89 = 27{,}400$ psi (no good—steel stress still too high).

Try $\rho = 0.0015$, $S_c = 0.073b'd^2$, $f_s/f_c = 51.37$:

$$f_c = \frac{M_w}{S_c} = \frac{118{,}000 \times 12}{0.073 \times 69 \times 28 \times 28} = 358 \text{ psi}$$

Compute $f_s = f_c(f_s/f_c) = 358 \times 51.37 = 18{,}400$ psi (O.K.—steel stress less than 24,000 psi).

Interpolate the results between $\rho = 0.0010$ and $\rho = 0.0015$. Try $\rho = 0.00125$, $S_c = 0.067b'd^2$, $f_s/f_c = 57.63$:

$$f_c = \frac{M_w}{S_c} = \frac{118{,}000 \times 12}{0.067 \times 69 \times 28 \times 28} = 391 \text{ psi}$$

Compute $f_s = f_c(f_s/f_c) = 391 \times 57.63 = 22{,}500$ psi (O.K.—steel stress near maximum allowable).

Compute the required steel area:

$$A_s = \rho b'd = 0.00125 \times 69 \times 28 = 2.42 \text{ in}^2$$

From Table A-3, select 4 No. 7 bars.

Use $d = 28$ in., $b' = 69$ in., $b_w = 14$ in., 4 No. 7 bars.

Note that the steel area in Example 6-3 is not equal to that of Example 6-2, indicating once again that the two methods are not equivalent.

The second case to be considered where tee beams are subject to positive moment is the case where the beam is simply supported and is therefore subject to positive moment throughout the span. For this case the size of the stem has not been set by moment conditions at the support as it was in the first case. However, the shear at the support must still be considered in selecting the size of the stem.

Other considerations must also enter the selection. Depending on the magnitude of loads and length of span, the shear conditions might not be serious and may not therefore require a very large section. Consequently, a small shallow section might suffice if the end shear were to be the only consideration.

Unusually shallow tee sections, however, can introduce problems in deflections and may require excessive amounts of flexural reinforcement; they should be used with caution. To satisfy most of the routine requirements, the overall depth of a tee section should be about four to five times the thickness of the slab it supports and it should be roughly equal to one-tenth to one-twelfth the span length. Obviously, such rules must vary considerably with span length, intensity of load, and spacing of tee stems.

Such considerations are not usually a formal part of the calculations. All such limitations are usually juggled mentally as the designer makes the final choice. The following example, however, includes these items as formal considerations for the sake of demonstration.

**Example 6-4**

*Strength method*
*Design of a tee beam subject to positive moment*
*Simply supported span, slab thickness fixed.*

*Given:*
Grade 60 steel
$f'_c = 4000$ psi
Concrete weight 145 pcf
$\phi = 0.90$ for flexure and 0.85 for shear

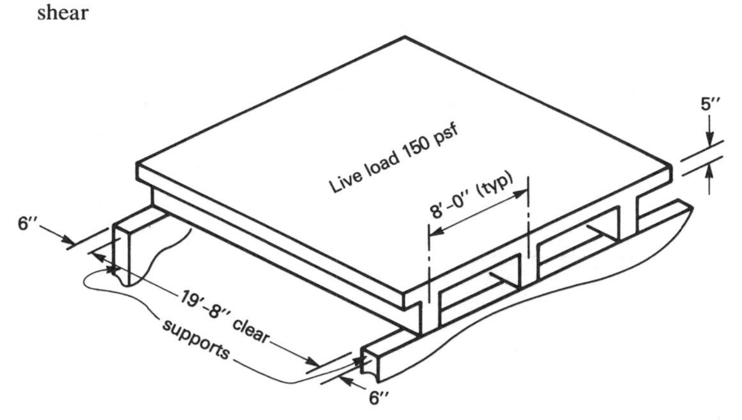


Estimate the stem size (for dead-load calculations):

$$\text{approximate } h = \frac{\text{span}}{10} \text{ to } \frac{\text{span}}{12} \qquad h = 20 \text{ to } 24 \text{ in.}$$

$$\text{approximate } h = 5 \times \text{slab thickness} \qquad h = 20 \text{ to } 25 \text{ in.}$$

$$\text{minimum} \quad h = \frac{\text{span}}{16} \text{ (Table 3-1)} \qquad h = 15 \text{ in.}$$

Try $h = 22$ in., $d = 19$ in., $b = 10$ in.

$$\text{weight per foot} = b(h - t) \times \frac{145}{144} = 171 \text{ plf}$$

Compute the dead load and live load for shear and moment diagrams:

$$\text{uniform live load} = 150 \text{ psf} \times 8 = 1200 \text{ plf}$$

$$\text{uniform dead load} = \left(\frac{5}{12}\right) \times 145 \times 8 = 483 \text{ plf}$$

$$\text{uniform stem load} = 171 \text{ plf}$$

$$w_{\text{DL}} = 654 \text{ plf}$$

Draw the shear and moment diagrams:

$$M_{\text{LL}} = \frac{w_{\text{LL}} \times L^2}{8} = \frac{1200 \times 19.67^2}{8} = 58 \text{ kip-ft}$$

$$M_{\text{DL}} = \frac{w_{\text{DL}} \times L^2}{8} = \frac{654 \times 19.67^2}{8} = 32 \text{ kip-ft} \qquad M_n = 159 \text{ kip-ft}$$

$$V_{\text{LL}} = \frac{w_{\text{LL}} \times L^2}{2} = \frac{1200 \times 19.67}{2} = 12 \text{ kips}$$

$$V_{\text{DL}} = \frac{w_{\text{DL}} \times L^2}{2} = \frac{654 \times 19.67}{2} = 6.5 \text{ kips} \qquad V_n = 35 \text{ kips}$$

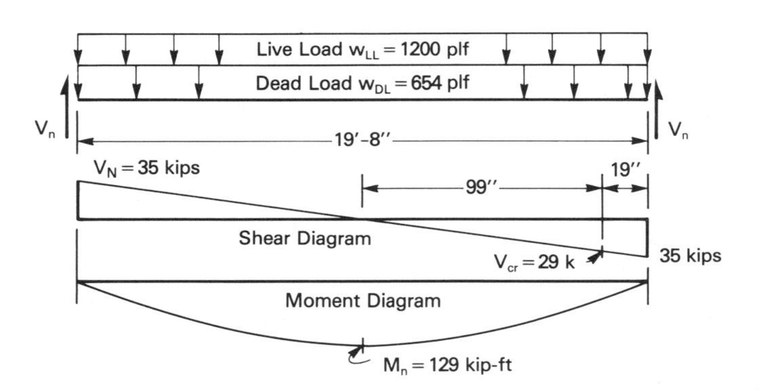


With no negative moment, the stem size is not controlled by flexure; set the stem size to suit shear requirements. The dimensions $b$ and $d$ are computed first for the case where no stirrups are needed and second for the case where stirrups are placed at $\frac{1}{2}d$.

$$\text{With no stirrups:} \quad v_n = 2\sqrt{f'_c} = 126 \text{ psi}$$

$$V_n = v_n b_w d \qquad 29{,}000 = 126 \times 0.5d \times d$$

Solve for $d$: $d = 21$ in., $b = 10\frac{1}{2}$ in.

$$\text{With stirrups at } d/2: \quad v_n = 6\sqrt{f'_c} = 380 \text{ psi}$$

$$V_n = v_n b_w d \qquad 29{,}000 = 380 \times 0.5d \times d$$

Solve for $d$: $d = 12.4$ in., $b = 6.2$ in.

For a final choice between the two cases, it is elected to use the section that does not require stirrups. Critical shear is corrected for $d = 21$ in. and the increase in dead load; the error is found to be negligible. Hence use $h = 24$ in., $d = 21$ in., $b = 12$ in. (The increase of dead load above the initial estimate is considered to be negligible.) Compute the effective width of the compression flange.

$$b' < \frac{\text{span}}{4} \quad \text{or} \quad \frac{242}{4} \quad \text{or} \quad 61 \text{ in.} \quad \text{(use)}$$

$$b' < 8 \times t \times 2 + \text{stem} \quad \text{or} \quad 8 \times 5 \times 2 + 12 \quad \text{or} \quad 92 \text{ in.}$$

$$b' < \text{tee spacing} \quad \text{or} \quad 96 \text{ in.}$$

Select reinforcement by finding the required section modulus:

$$Z_c = \frac{M_n}{0.85 f'_c} \qquad \text{coeff.} \times b'd^2 = \frac{M_n}{0.85 f'_c}$$

$$\text{coeff.} \times 61 \times 21 \times 21 = \frac{159{,}000 \times 12}{0.85 \times 4000}$$

Solve for coeff.: coeff = 0.021.

From Table A-6, $\rho = 0.0012$, $k\beta = 0.021$.

Select steel: $A_s = \rho b'd = 0.0012 \times 61 \times 21 = 1.94 \text{ in}^2$.

Use 2 No. 7 and 2 No. 4 bars, no stirrups.

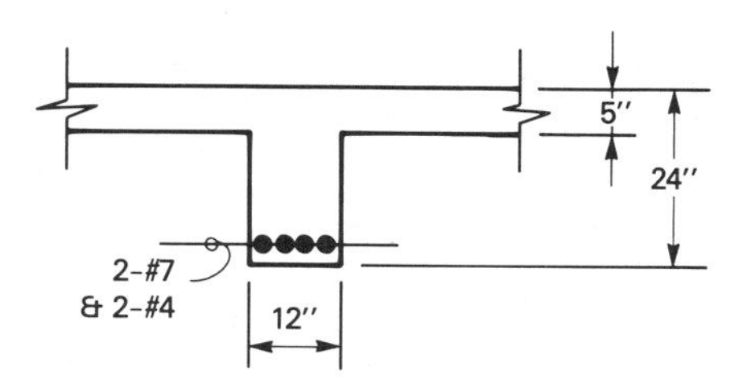


In Example 6-4 there were no limitations imposed on the overall height of the beam. Had such a limitation existed, a shallower section might have been required. A shallower section would, in turn, require shear reinforcement, probably stirrups at a spacing of $d/2$.

Example 6-4 might also have been solved using the working stress method. The procedure would be directly parallel to the procedure shown, but with working loads and working stresses being used instead of ultimate loads and ultimate stresses. It should be expected that a significantly higher steel ratio would be required under the working stress method.

## NEUTRAL AXIS BELOW THE FLANGE

It was recognized earlier that a complication arises in the analysis when the neutral axis falls in the stem of the tee beam, below the soffit of the slab. Such a case is shown in Fig. 6-4. The stress diagrams at service levels are also shown, one through the stem and one through the overhang at some distance from the stem.

It is evident from Fig. 6-4 that the compressive area is now irregular; it had been assumed to be rectangular in the flexure analysis of Chapter 4. The effect of the new void areas, however, is quite small where the penetrations of the neutral axis into the stem is small.

Now consider the loss in compressive force, $C_v$, shown in the stress diagram of Fig. 6-4. This loss in compressive force is the product of a relatively low stress acting on a relatively thin strip of area. The loss in moment is then this small force $C_v$ acting over a reduced moment arm $a_v$. In the elastic range, the total reduction in moment, $C_v a_v$, is obviously small when compared to the total elastic moment.

As load increases, the rotation of the section continues and the ultimate moment is reached; the stress diagrams shown in Fig. 6-5 then become the final state of stress. Note here that the loss in moment is the full ultimate stress acting over the void area, but again with a reduced moment arm. The loss in moment can be judged to be higher than at working levels and that it would be a small loss only as long as the penetration of the neutral axis into the stem is small.

A cursory examination of Tables A-5 through A-7 reveals that for a given stress ratio $\rho$, the depth of the compressive area of ultimate load, $k\beta d$, is significantly less than the depth at elastic levels, $k_w d$. The difference is more marked at lower steel ratios, where tee beams would normally fall. It may be concluded that as rotation of the section progresses from the elastic range into the ultimate state, the neutral axis shifts significantly upward and the area in compression becomes thinner.

For tee beams, this shift means that many beams having the neutral axis in the stem at elastic levels will have the neutral axis back into the slab at ultimate load. Con-

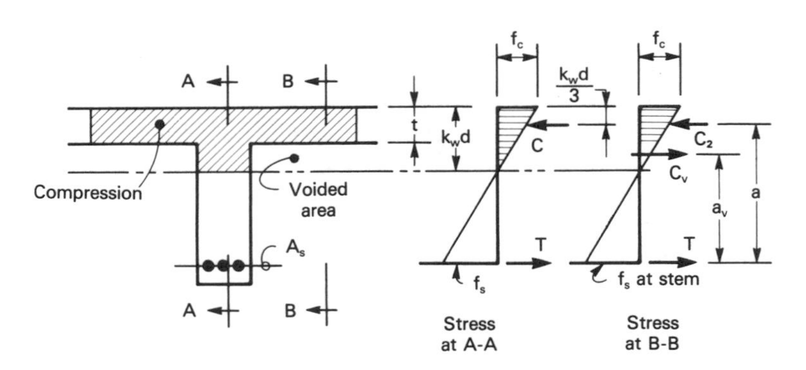


**Figure 6-4** Neutral Axis of Tee Shape.

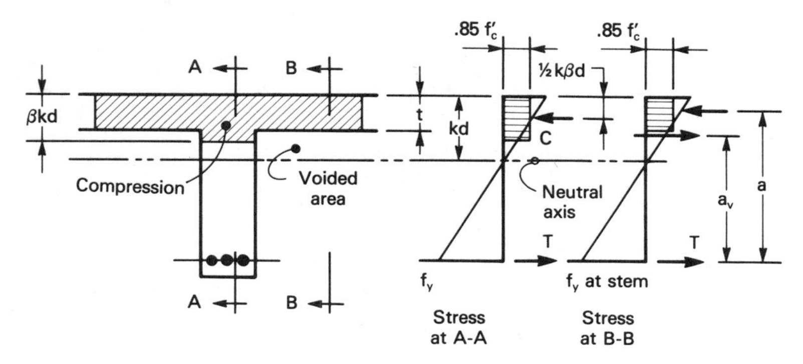


**Figure 6-5** Tee Beam at Ultimate State.

sequently, the strength method of design is more likely to keep its compression block out of the stem than is the working stress method.

The incidence of the neutral axis falling below the slab into the stem of the tee beam is somewhat unusual. It occurs where circumstances permit a very thin slab to be used, but also requires that the stems be deep, closely spaced, and heavily reinforced. Its treatment here would require additional design tables, which, since this case is not common, are not included in this limited work. Since this case is not included in this book, the following limits provide a means to recognize when the neutral axis is in the stem of the tee and some means to avoid such a design.

Again referring to the elastic stress diagrams of Fig. 6-4, it is seen that the neutral axis stays within the slab where the depth of the compression block is deliberately kept less than the slab thickness $t$. Similarly, for the ultimate stress diagrams of Fig. 6-5, the compression block stays rectangular where $k\beta d$ is kept less than the slab thickness $t$. Summarizing these observations:

At working levels, select $\rho$ low enough that $k_w < t/d$.

At ultimate loads, select $\rho$ low enough that $k\beta < t/d$.

If $k\beta$ or $k_w$ exceeds these values of $t/d$, the neutral axis will be in the stem of the tee. For the more common values of $t$ and $d$, the problem is rarely encountered.

## APPROXIMATE ANALYSIS OF TEE SHAPES

The conclusions of the preceding section suggest that an approximate analysis can be made, based on the stress in the tensile steel. Refer to Fig. 6-6 for the state of stress due to working loads. The sum of moments about the compressive force $C$ yields

$$M_w = f_s A_s \frac{d - k_w d}{3} \tag{6-1a}$$

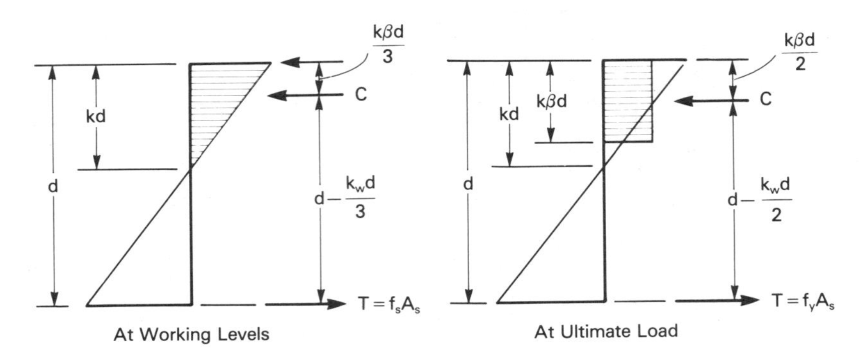


**Figure 6-6** State of Stress.

Similarly, for the ultimate load condition shown in Fig. 6-6,

$$M_n = f_y A_s \frac{d - k\beta d}{2} \tag{6-1b}$$

It should be recognized that an approximate value of $k\beta$ or $k_w$ in these equations would afford some useful analytical equations.

To establish approximate values of $k\beta$ and $k_w$, the values of $k\beta$ and $k_w$ given in Tables A-5 through A-7 are examined. It is recognized that the lower range of values for steel ratios are the values that would normally be applicable for tee beams. In this range it is observed that the value of $k_w$ is not higher than about 0.30 and the value of $k\beta$ is not higher than about 0.10. These values are consistent with the limits assumed for $t/d$ by the general rules of the preceding section, where $d$ is taken at about $4t$ to $5t$. Substitute these values into Eqs. (6-1) to find:

$$\text{At working levels:} \quad M_w = f_s A_s (0.90d) \tag{6-2a}$$

$$\text{At ultimate loads:} \quad M_n = f_y A_s (0.95d) \tag{6-2b}$$

These equations provide a means to make a fast manual check on steel stresses or steel areas. This check does not require a reference to the design tables.

For a typical application, refer to Example 6-2. The nominal ultimate moment $M_n$ is shown as 204 kip-ft and depth is fixed at 28 in. The area of grade 60 steel can be calculated directly from Eq. (6-2b):

$$M_n = f_y A_s (0.95d) \qquad 204{,}000 \times 12 = 60{,}000 \times A_s \times 0.95 \times 28$$

Hence $A_s = 1.53 \text{ in}^2$. This result compares to an area of 1.45 $\text{in}^2$ obtained in the example, derived from the design tables of the Appendix. Note that it was not necessary to know the effective width of compression, $b'$, before obtaining this result. Neither was it necessary to have access to a set of design tables.

For another application, refer to the rather tedious solution of Example 6-3. The working moment in this example is 118 kip-ft and the effective depth is 28 in. Use Eq. (6-2a) for working levels and an allowable working stress in steel of 24,000 psi to find

$$M_w = f_s A_s(0.90d) \qquad 118{,}000 \times 12 = 24{,}000 \times A_s \times 0.90 \times 28$$

Hence $A_s = 2.34\ \text{in}^2$. This result compares to the computed area of $2.42\ \text{in}^2$ in the exact solution.

A third application may be taken from Example 6-4. In that example, the nominal ultimate moment is 159 kip-ft, the steel is grade 60, and the selected value for $d$ is 21 in. Solve for area of steel using Eq. (6-2b):

$$M_n = f_y A_s(0.95d) \qquad 159{,}000 \times 12 = 60{,}000 \times A_s \times 0.95 \times 21$$

Hence $A_s = 1.59\ \text{in}^2$. This area compares to an area of $1.54\ \text{in}^2$ in the exact solution.

The approximate equations (6-2) are seen to be useful in making rough calculations and in making a quick independent check on existing calculations. They may be readily committed to memory by remembering that the moment arm for steel in the elastic range is about 90% of $d$, shifting to 95% of $d$ at ultimate load. The equations are also quite useful when there are no design tables immediately available.

When compression steel is used, its center of gravity will necessarily be close to the top face of the tee. The resultant force on this compressive reinforcement will therefore be quite close to $C$, the resultant force on the concrete (Fig. 6-6). As a consequence, the approximate equations (6-2) lose but little accuracy in beams reinforced in compression.

## ONE-WAY JOIST CONSTRUCTION

When tee beams are spaced at close intervals, the stems can be set just wide enough to develop the necessary strength in shear without requiring stirrups. At this close spacing, the effective flange width $b'$ will always be taken halfway to the two adjacent stems; the other requirements never govern. In addition, the floor slab can be made quite thin, although practical considerations usually will not permit a slab less than 2 in. thick to be used. Such a floor system, called a one-way joist system, is shown in Fig. 6-7a.

To build such a configuration, the filled spaces between stems can be formed by any available block or building tile that can be held in position while the concrete is being cast around them, such as the tiles shown in Fig. 6-7b. Or special steel forms for forming void space can be rented that can be handled easily, placed manually, and subjected to many reuses. These joists are cast in place at the same time and in the same plane as their supporting girders (without having to sit on top of their supports), thereby reducing the story height by a considerable amount (Fig. 6-7c).

A floor or roof system of this type is also called a *pan-joist* system or a *tile-filler* system of floor construction. It is one of the more popular floor systems in the industry and represents a major advantage in the use of concrete construction. Other materials currently being used to form the joists are such things as fiberglass pans, corrugated paper blocks, foamed concrete blocks, Styrofoam blocks, and kraft paper tubes. When fillers are used, the bottom surface is usually just sprayed to form the finished ceiling.

There are many innovations in common use. The ends of the joists can be widened or tapered as shown in Fig. 6-7d to accept unusually heavy shears (or negative mo-

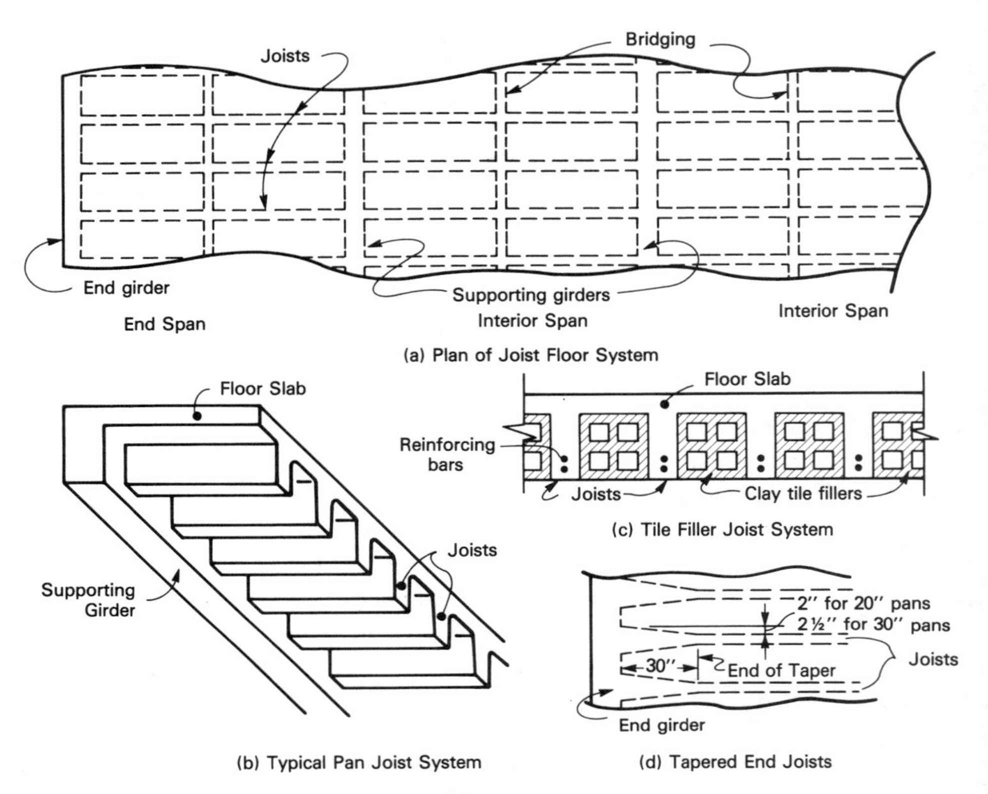


**Figure 6-7** Joist Systems.

ments). Or, the tile filler blocks can simply be omitted over short distances where shears and moments are too high for the thin stem. Shallower filler tiles or pans can be placed where a plumbing fitting requires extra slab thickness or where special structural attachments are to be placed.

Distribution members, or bridging, are required by Code at the interior of longer spans, as shown in Fig. 6-7a. Such members are easily formed simply by leaving a space between the tiles or pans when they are located and fixed in the forms. An individual cross-rib for attaching special ceiling-mounted equipment or balcony hangers can be formed the same way.

The design of joist systems is the same as that for other tee-beam systems, but with some additional Code (8.11) requirements. There are, however, some benefits to be had, such as a higher allowable shear on the section. The Code requirements that are of immediate interest are:

1. Ribs shall be not less than 4 in. in width.
2. Ribs shall have a depth not less than 3.5 times the least width.
3. Clear spacing between ribs shall not exceed 30 in.

4. When removable pans are used, slab thickness shall be not less than one-twelfth the clear distance between ribs but not less than 2 in.; slab reinforcement normal to the ribs shall be provided as required for flexure but shall be not less than the minimum amount required for shrinkage and temperature.
5. When permanent clay tile fillers are used that have a compressive strength at least as great as that of the concrete, the vertical shells of fillers in contact with ribs may be included in the flexural calculations; slab thickness over permanent fillers may not be less than one-twelfth the clear distance between ribs nor less than $1\frac{1}{2}$ in.
6. Shear stress in the working stress method may not exceed $1.2\sqrt{f'_c}$.
7. Joists not meeting the requirements shall be designed as ordinary slabs and tee beams.

In the United States, the "pans" used to form the void space have become reasonably well standardized (Concrete Reinforcing Steel Institute, 1984). The standardization has been made in Imperial units and does not seem likely to change in the near future. The dimensions shown in Fig. 6-8 are generally accepted standard dimensions in the industry.

In countries using metric units, there seems to be little standardization. American-manufactured pans are usually available in most countries; design calculations simply adjust these dimensions to the nearest few millimeters. Slab thicknesses are commonly 60, 80, or 100 mm and rib widths are commonly 120, 140, 160, 180, and 200 mm.

Depths of joist systems can be considerably less (about one-third less) than that of widely spaced tee-and-slab systems. Deflection problems with joist systems are common, however, and deflection limitations should always be checked.

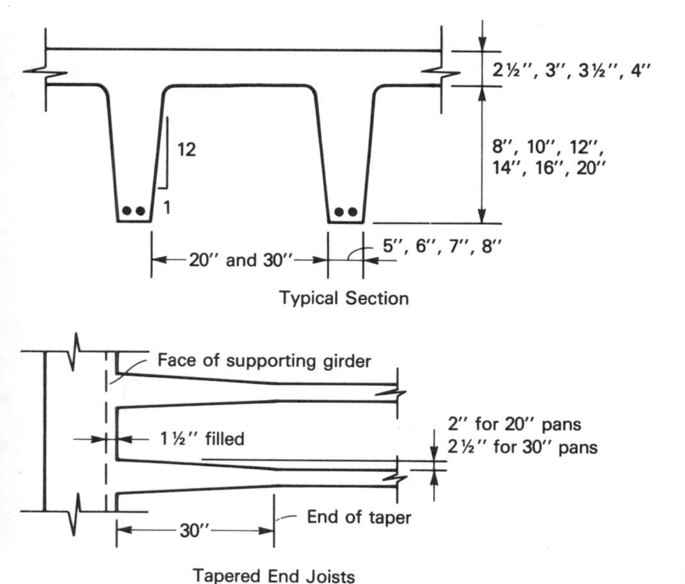


**Figure 6-8** Standard Joist Dimensions.

Code permits shear reinforcement to be used with joist systems as shown in Fig. 6-9, but the deep, narrow stems make placement of such reinforcement difficult. It is far simpler to increase the stem width slightly where shears are excessive than to place numerous stirrups in each of the many tall, thin stems. It was noted earlier that one of the attractive features of joist systems was that it becomes possible to eliminate shear reinforcement; to use shear reinforcement with a joist system effectively discards one of the attractive features of the system.

Where shears are light, it may be desirable to use the minimum widths of joist. In such cases, stacking the reinforcement vertically as shown in Fig. 6-9c is a reasonable alternative. The minimum widths of a section for various bar sizes placed in horizontal layers are given in Table A-4; the column headed "with no stirrups" is intended to apply particularly to joist systems.

Solid blocking, or bridging, perpendicular to the joists is always placed at support lines and where partitions are located in the span. Lines of bridging to stabilize the slender joists are also placed at the interior of the span where spans are long or loads are heavy. Interior bridging is not usually used with roof joists. For floor joists:

No interior lines of bridging for spans less than 20 ft

One interior line of bridging (midspan) for spans 20 feet to 30 ft

Two interior lines of bridging (third points) for spans longer than 30 ft

Where bridging is required, its dead load may be estimated at about 2 psf.

Due to the large number of combinations of slab thicknesses, stem widths, stem heights, and joist spacings, the dimensions of joist systems are almost always selected from handbooks or, in unusual configurations, designed using one of the many computer programs available. Deflection criteria are always included in these standard design aids.

For the sake of demonstration, the design of a joist system is presented here. Before beginning the example, it will be necessary to have a table of the average uniform dead weights of a joist system for various configurations. Such a tabulation is presented in Table 6-1 for common dimensions.

It should be noted in Table 6-1 that joist systems are omitted where the total height

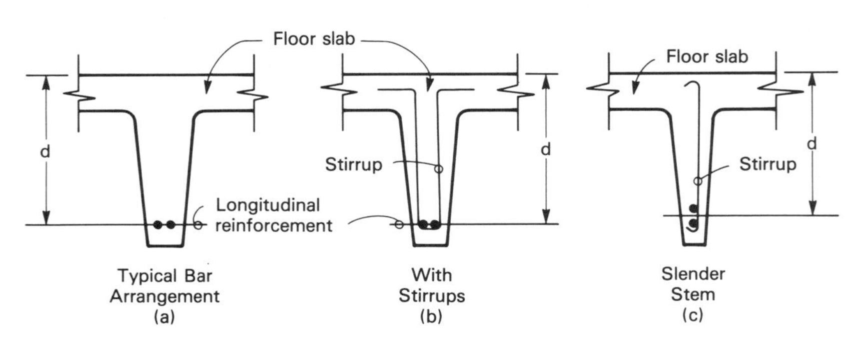


**Figure 6-9** Shear Reinforcement in Tee Shapes.

**TABLE 6-1 AVERAGE DEAD LOAD OF JOIST SYSTEMS IN POUNDS PER SQUARE FOOT**

| | | | Pan widths (in.) 20 | | | | | Pan widths (in.) 30 | | | | |
|---|---|---|---|---|---|---|---|---|---|---|---|---|
| | | | Stem widths (in.) 4 | 5 | 6 | 7 | 8 | 4 | 5 | 6 | 7 | 8 |
| Pan height (in.) | Slab thickness (in.) | Total height (in.) | Joist spacing (in.) 24 | 25 | 26 | 27 | 28 | 34 | 35 | 36 | 37 | 38 |
| 8 | 2½ | 10½ | 51 | 54 | 57 | 60 | 62 | | | | | |
| | 3 | 11 | 57 | 60 | 63 | 66 | 68 | 51 | 54 | 56 | 58 | 60 |
| | 3½ | 11½ | 63 | 66 | 69 | 72 | 75 | 57 | 60 | 62 | 64 | 67 |
| | 4 | 12 | 69 | 73 | 76 | 78 | 81 | 64 | 66 | 69 | 71 | 73 |
| 10 | 2½ | 12½ | 56 | 60 | 64 | 68 | 71 | | | | | |
| | 3 | 13 | 63 | 67 | 70 | 74 | 77 | 55 | 58 | 61 | 64 | 67 |
| | 3½ | 13½ | 69 | 73 | 77 | 80 | 83 | 62 | 65 | 67 | 70 | 73 |
| | 4 | 14 | 75 | 79 | 83 | 86 | 89 | 68 | 71 | 74 | 76 | 79 |
| 12 | 2½ | 14½ | 63 | 67 | 72 | 76 | 79 | | | | | |
| | 3 | 15 | 69 | 74 | 78 | 82 | 86 | 60 | 63 | 67 | 70 | 73 |
| | 3½ | 15½ | 75 | 80 | 84 | 88 | 92 | 66 | 69 | 73 | 76 | 79 |
| | 4 | 16 | 81 | 86 | 90 | 94 | 98 | 72 | 76 | 79 | 82 | 86 |
| 14 | 3 | 17 | 75 | 81 | 86 | 90 | 95 | 64 | 68 | 72 | 76 | 80 |
| | 3½ | 17½ | 81 | 87 | 92 | 97 | 101 | 70 | 75 | 79 | 82 | 86 |
| | 4 | 18 | 88 | 93 | 98 | 103 | 107 | 77 | 81 | 85 | 89 | 92 |
| 16 | 3½ | 19½ | 88 | 94 | 100 | 105 | 110 | 75 | 80 | 84 | 89 | 93 |
| | 4 | 20 | 94 | 101 | 106 | 112 | 117 | 81 | 86 | 91 | 95 | 99 |
| 20 | 4 | 24 | | | 124 | 130 | 136 | | | 103 | 109 | 114 |

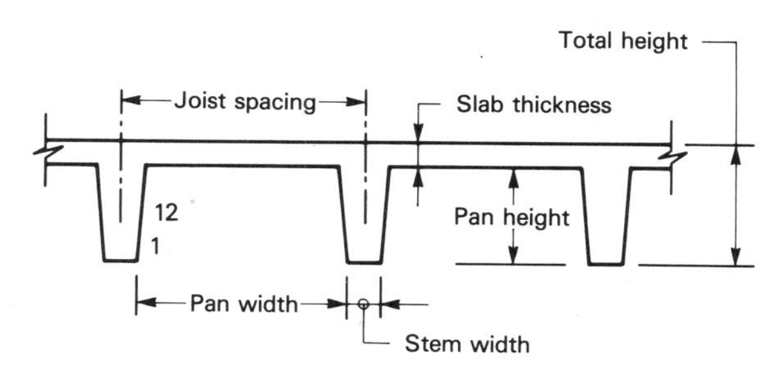


of stem is more than five times the slab thickness. Also, joist systems are omitted where the pan width (clear span of the slab) is more than 12 times the thickness of the slab and where pan height is greater than 3.5 times the least width of the stems. The systems listed in Table 6-1 are therefore permissible under all the limitations imposed earlier in this chapter.

The following example uses a typical system. The span is long, 32 ft, and the live load is relatively heavy at 160 psf. It should be noted that the entire system must be selected before the calculations can be started. The calculations are thereafter limited to selection of reinforcement and a review for shear.

**Example 6-5**

*Strength method*
*Design of a pan joist system*
*Simple span, no limitations on dimensions*
*Given:*
Simple span, 31 ft clear
Seated on 12 in. bearing wall
Live load 160 psf
Grade 60 steel
$f'_c = 4000$ psi
Floor tile 4 psf
Exterior exposure

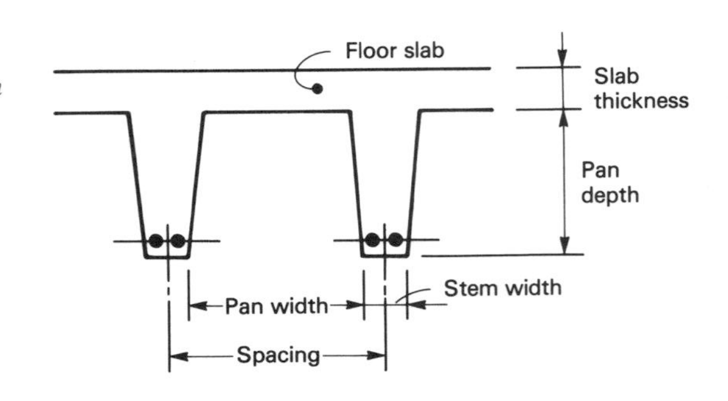


Assume that two lines of bridging will be required for a span more than 30 ft; assume that the dead load of bridging is 2 psf. For deflection control, minimum $h = L/16$ or 24 in. (Table 3-1). Try depth about 66% of the usual depth: approximate depth $0.66 \times L/12$ or 21 in.; use 24 in. Try pans 30 in. wide, 20 in. deep; slab 4 in. thick; stem 7 in. wide; joist spacing 37 in.; dead load = 109 psf.

Compute the shear and moment diagrams:

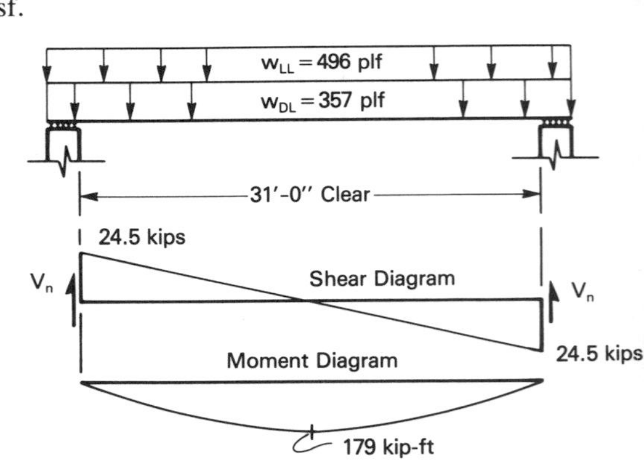


Dead load (psf)

| | |
|---|---|
| Joist system | 109 |
| Floor tile | 4 |
| Bridging | 2 |
| Total | 115 psf |

For joist spacing of 37 in.:

$$w_{DL} = 115 \times 3.1 = 357 \text{ plf}$$

$$w_{LL} = 160 \times 3.1 = 496 \text{ plf}$$

Compute the working shear and moment:

$$V_{DL} = \frac{w_{DL}L}{2} = \frac{357 \times 31}{2} = 5.5 \text{ kips}$$

$$V_{LL} = \frac{w_{LL}L}{2} = \frac{496 \times 31}{2} = 7.7 \text{ kips}$$

$$M_{DL} = \frac{w_{DL}L^2}{8} = \frac{357 \times 31 \times 31}{8} = 42.9 \text{ kip-ft}$$

$$M_{LL} = \frac{w_{LL}L^2}{8} = \frac{496 \times 31 \times 31}{8} = 59.6 \text{ kip-ft}$$

Compute the nominal ultimate shear and moment:

$$V_n = \frac{1.4V_{\text{DL}} + 1.7V_{\text{LL}}}{0.85} = 24.5 \text{ kips/ft}$$

$$M_n = \frac{1.4M_{\text{DL}} + 1.7M_{\text{LL}}}{0.90} = 179 \text{ kip-ft/ft}$$

Compute $V_{cr}$ at a distance $d$ from face of support. Assume that $d = h - 3 = 21$ in.; $V_{cr}$ at 21 in. from face of support.

$$V_{cr} = \frac{165}{186} \times 24.5 = 21.7 \text{ kips}$$

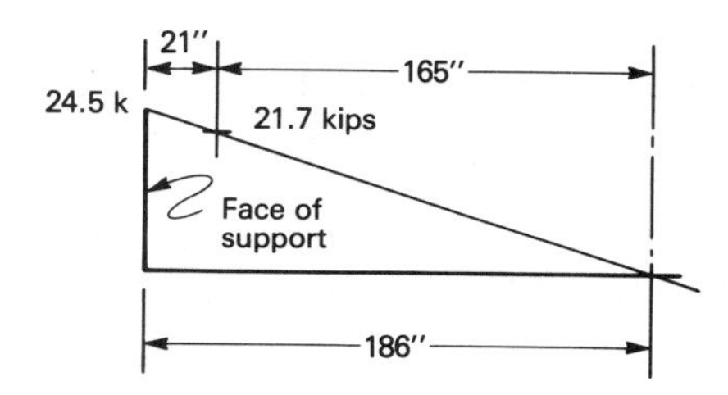


Compute the average stem width and flange width:

$$b_w = 7 + \frac{20}{12} = 8.7 \text{ in.}$$

$$b' = \text{joist spacing} = 37 \text{ in.}$$

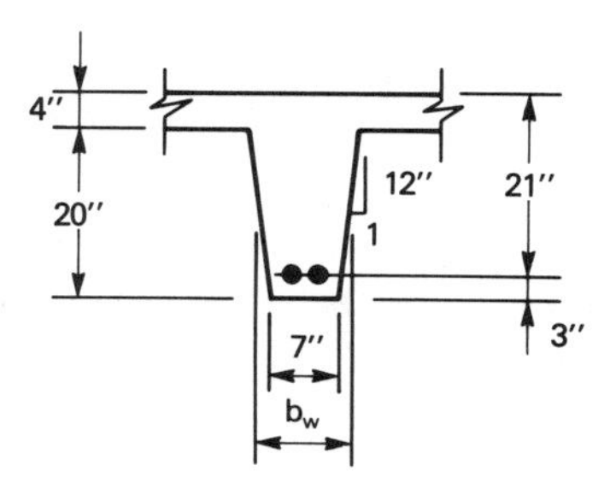


Compute the coefficient of section modulus:

$$Z_c = \frac{M_n}{0.85f'_c} \quad \frac{179{,}000 \times 12}{0.85 \times 4000} = \text{coeff.} \times b'd^2 = \text{coeff.} \times 37 \times 21 \times 21$$

$$\text{coeff.} = 0.039$$

From Table A-6, select $\rho = 0.0023$, $Z_c = 0.039b'd^2$.

Compute the steel area:

$$A_s = b'd = 0.0023 \times 37 \times 21 = 1.79 \text{ in}^2$$

From Table A-3, select 2 No. 9 bars, $A_s = 2.00$ in$^2$.

Review for shear:

With no stirrups: $v_c = 2\sqrt{f'_c} = 126$ psi

concrete shear $V_c = v_c b_w d = 126 \times 8.7 \times 21 = 23$ kips

acting shear (from shear diagram) $= 21.7$ kips $< 23$ kips

(Section O.K. without stirrups).

Use pans 36 in. wide × 20 in. deep; stem width 7 in. at base, spacing 37 in.; reinforcement 2 No. 9 bars, 2-in. cover (minimum).

It would be possible in Example 6-5 to reduce the bar size by using two layers of bars, four bars total. The depth $d$ would then be reduced to about $19\frac{1}{2}$ in., and all calculations would have to be repeated for this new value of $d$.

Even with the long span and relatively heavy live load in Example 6-5, it can be seen that a moderately sized stem of 7 in. is adequate for the shear loads. It could easily be increased if necessary just by opening up the spacing of the pans. It may be concluded that relatively high shears can be sustained by joist systems with little penalty in increased stem size.

It should be apparent that one-way joist systems are one of the more adaptable floor systems in the industry. Where a supporting structure occurs at all four sides of a floor system in a reasonably square pattern, joists may be placed in both directions. Such two-way systems, called "waffle" systems, permit comparatively heavy loads to be carried on comparatively long spans with maximum economy of materials. Two-way joist systems are beyond the scope of this book, not due to unusual complexity but to lack of space.

## REVIEW QUESTIONS

1. A floor slab is 4 in. thick. If it is used as part of a tee-beam system, what maximum depth should be observed in selecting the depth of the stem? Why?
2. How can one accurately determine whether the neutral axis of a tee beam is actually in the slab or whether it falls in the stem?
3. Under what circumstances would one suspect that the neutral axis of a particular tee beam might fall in the stem of the tee?
4. For simple spans, how is the size of the tee-beam stem usually determined?
5. For continuous beams, how is the size of the tee-beam stem usually determined?
6. Tee beams are commonly spaced at about one-fourth of the span length. Why?
7. What width of beam is used to compute the minimum allowable steel ratio $\rho$? How is it used in regions of positive moment?
8. Visualizing the "internal couple" concept for moment in beams as shown in Fig. 6-6, what is the approximate length of the moment arm $jd$ under service levels of load? At ultimate load?

9. What are the most common widths of steel pans used to form one-way pan-joist systems?
10. What is the usual means to increase the shear capacity of a pan-joist system without resorting to the use of shear reinforcement?

## PROBLEMS

An interior span of a continuous tee beam has the shear and moment diagrams indicated below. The live load and the beam spacings are listed in the tabulation. The dead load of the stem must be estimated, then confirmed later when design is final. Using the strength method, select a suitable tee section for the given conditions, to include both positive and negative reinforcement and any shear reinforcement required. $f'_c = 4000$ psi, grade 60 steel.

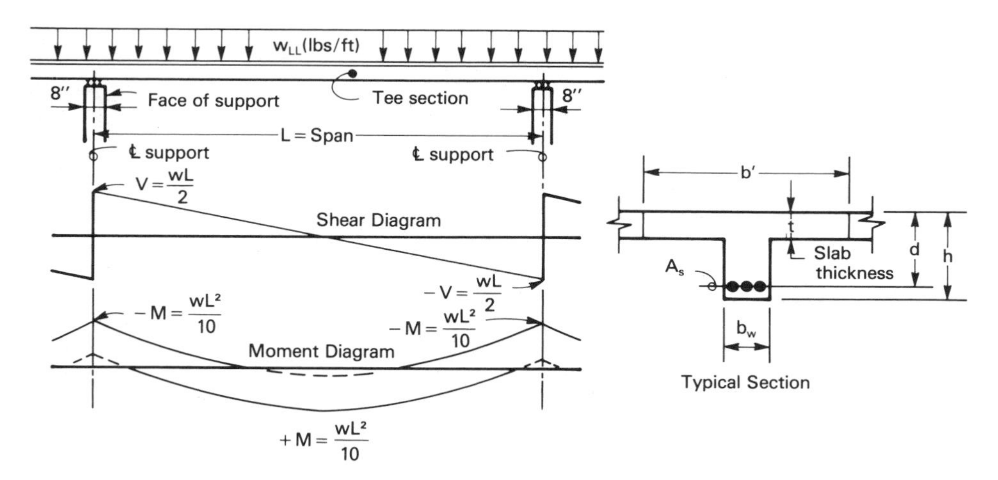


| Problem | Live load (psf) | Slab thickness (in.) | Tee spacing | span | Limits on $h$ (in.) |
|---|---|---|---|---|---|
| **1** | 250 | 4 | 4′-0″ | 18′-4″ | None |
| **2** | 210 | 4 | 4′-8″ | 21′-8″ | None |
| **3** | 170 | 4 | 6′-0″ | 25′-0″ | 19 |
| **4** | 170 | 6 | 8′-0″ | 30′-0″ | $22\frac{1}{2}$ |

Design a tee-beam floor system for a simple span with the conditions and live loads shown below. The dead load of the system must also be included. Select the flexural reinforcement and any shear reinforcement required. $f'_c = 3000$ psi, grade 60 steel.

| Problem | Live load (psf) | Slab thickness (in.) | Tee spacing | Clear span |
|---|---|---|---|---|
| **5** | 250 | 4 | 5′-4″ | 20′-0″ |
| **6** | 230 | 4 | 6′-0″ | 23′-4″ |
| **7** | 210 | 5 | 7′-4″ | 26′-8″ |
| **8** | 190 | 5 | 8′-0″ | 30′-0″ |

9. Confirm the required areas of tensile reinforcement in Problems 9 through 12 using the approximate formulas of Eqs. (6-2).
10. For the continuous tee beam of Example 6-2, determine the moment of inertia of the chosen tee section at midspan where moments are positive. Then, assuming the negative $A_s$ at the support to be equal to the positive $A_s$, determine the moment of inertia of the rectangular section near the supports where moments are negative. Compare the two values of moment of inertia to see how much the moment of inertia varies in going from the rectangular section to the tee section.
11. Design a pan-joist floor system for a simple clear span of 30 ft, for a live load of 125 psf. $f'_c = 3000$ psi, grade 60 steel.

# 7

# Bond and Anchorage

## GENERAL

The term *development length* is used frequently in subsequent discussions. The development length of a reinforcing bar is the length of bar that must be bonded to the concrete to develop the full strength of the bar. The end of a bar could thus be embedded in a concrete block to a depth equal to its development length and when the protruding end is loaded to failure, the bar itself will fail just before it pulls out of the concrete block.

The adhesion of concrete to its reinforcement is called *bond*. It is pointed out in Chapter 1 that bond strength is one of the more uncertain properties of concrete that must be used in design. To reduce the dependence on theoretical values when using this uncertain property, the development lengths of reinforcement under a wide variety of circumstances have been determined by extensive tests; requirements for embedment have been established from these test results and are prescribed by Code.

There are several special circumstances defined by the Code where the development lengths may be reduced somewhat. Those special circumstances that are encountered frequently enough to justify the time and effort to learn them are included in the subsequent sections. Other cases, infrequently applicable, are omitted to avoid a confusion of special cases. The results herein may therefore be slightly conservative at times.

The working stress method does not have a separate means for determination of development lengths. For the working stress method, the Code (B.4) simply references the strength method for development lengths and splices. The Code therefore has only one set of requirements for both the strength method and the working stress method.

In earlier years, Code requirements were based on an embedment length based on a certain number of diameters of the bar. For example, under certain circumstances, a development length may have been required to be 48 bar diameters. The computation of this embedment length, say 48 times $\frac{3}{4}$ or 36 in. for a No. 6 bar, was then repeated each time the information was needed.

Present Code requirements are prescribed by an empirical formula, to be used for computing the required embedment length under various circumstances. These lengths have been computed for various bar sizes and are tabulated in Tables A-11 through A-13. They are equally applicable whether the design is being done according to the strength method or the working stress method.

The following discussions are limited to bar sizes No. 11 or smaller for flexural reinforcement and to bar sizes No. 5 or smaller for stirrups and ties. Bar sizes larger than these are subject to additional Code requirements requiring additional design considerations and conditional checks. Further, these larger sizes would apply primarily to heavy construction, which is beyond the scope of this book.

In addition, consideration is limited only to deformed bars; smooth wire and mesh are not included. ACI 318-83 includes several additional special formulas and checks that apply to the use of smooth bars and mesh. As late as 1979, the British Code of Practice CP 110 handled the matter simply by limiting the anchorage load on plain bars to about 70% of that for deformed bars of the same size.

The strength reduction factor $\phi$ is not used in determination of development lengths. The empirical formulas adopted by ACI inherently include the effects of placement tolerances and other factors that might affect ultimate strength.

## CUTOFF POINTS

When determination of tensile reinforcement was made (Chapter 4), the amount of reinforcement was computed on the basis of maximum tensile force at a particular point. At some distance away from this maximum value of tension, where tensile force becomes reduced or even nonexistent, it is possible to cut off a part of the tensile reinforcement, retaining only that amount of reinforcement required for the particular area. Code places strict limits on the minimum amount of reinforcement that must be retained.

A typical moment envelope for a flexural member is shown in Fig. 7-1b. The moment envelope is defined as the outermost moment diagram at any point resulting from any one of the possible loads; any other load produces a lesser value of moment at that point. (Moment envelopes are discussed more fully in Chapter 9.) Note that the inflection point can shift a significant distance laterally under different loadings.

The point at which the bar must attain its full yield strength is called the *critical section*. The development length must of course lie entirely outside this point. For example, the critical section for bars mk c in Fig. 7-1 is at midspan, and for bars mk d is at face of support.

Typical cutoff points for flexural reinforcement are shown in Fig. 7-1b. In most cases, these cutoff lengths can be determined graphically simply by scaling the moment

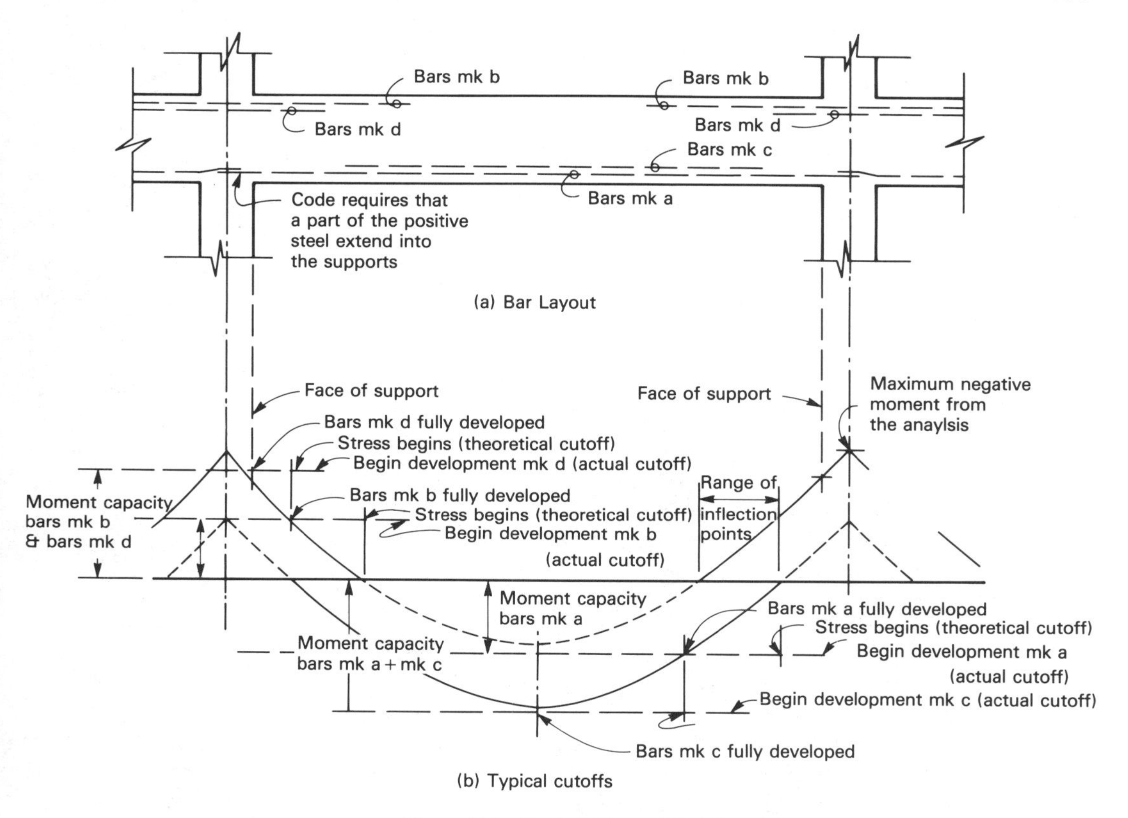


**Figure 7-1** Typical Flexural Reinforcement.

diagram. A higher degree of accuracy is difficult to defend in view of the empirical nature of the design.

It is required that the reinforcement that is to be cut must be extended past its theoretical cutoff point (the point where it is theoretically no longer needed) by a distance not less than $d$ nor less than $12d_b$. By this means, most of the strength of the bar is available where it is first needed and the remaining strength will be available further along the bar. At these theoretical cutoff points, however, the bars that have not been cut must carry the full moment and must be fully developed at such points.

It should be recognized that all cutoff points are measured from the theoretical point where the bar is needed, not from the actual cutoff points of adjacent bars. Such a case is shown for bars mk a in Fig. 7-1b, where the cutoff length is measured from the theoretical point of cutoff of bars mk c, not their actual point of cutoff. It should also be recognized that more than one criterion may apply; the cutoff point for bars mk b must satisfy the requirement of extending at least a distance $d$ or $12d_b$ beyond the point of inflection as well as extending at least their full development length beyond the theoretical point of cutoff of bars mk d. The development lengths are of course those of Table A-13, as discussed earlier. Cutoff points are included in the examples presented later in this chapter.

## DEVELOPMENT LENGTHS OF STRAIGHT BARS

Development length $L_d$ for deformed bars in tension is prescribed by Code (12.2.2), for bars not larger than No. 11 and having less than 12 in. of fresh concrete cast below them:

$$L_d \geq \frac{0.04 \times A_b f_y}{\sqrt{f'_c}} \qquad (7\text{-1a})$$

$$\text{but not less than } 0.0004 d_b f_y \qquad (7\text{-1b})$$

where $L_d$ = development length of the bar
$A_b$ = cross-sectional area of the bar
$d_b$ = diameter of the bar

and all other symbols are those used previously.

Horizontal bars having 12 in. or more of fresh concrete cast below them can lose a considerable amount of bond strength due to collections of air, water, and laitance underneath them. These impurities rise and collect under the bars during compaction of the concrete. Code (12.2.3.1) requires that an additional 40% development length be provided for such bars, called *top bars*. Development lengths for top bars are therefore 1.4 times the development lengths computed by Eqs. (7-1). In addition, the development length $L_d$ for deformed bars in tension may not be less than 12 in. except for lap splices and for web reinforcement, both of which are discussed in subsequent sections.

The development lengths of deformed bars in tension prescribed by Eqs. (7-1) have been computed and tabulated in Table A-13. Since the development length depends on three variables—concrete strength, steel strength, and bar size—the tables are quite voluminous. The tabulated values take into account all the foregoing requirements.

Similarly, Code (12.3.2) prescribes a development length for deformed bars in compression:

$$L_d \geq \frac{0.02 d_b f_y}{\sqrt{f'_c}} \qquad (7\text{-2a})$$

$$\text{but not less than } 0.0003 \times d_b f_y \qquad (7\text{-2b})$$

and not less than 8 in.

where all symbols are as defined previously.

The development lengths of deformed bars in compression prescribed by Eqs. (7-2) have also been computed and tabulated in Table A-13. It is observed that the classification "top bar" does not exist for bars in compression, only for bars in tension.

It is also observed from Table A-13 that the development length for bars in compression is considerably less than for bars in tension.

## SPLICE LENGTHS OF STRAIGHT BARS

Reinforcing bars may be spliced simply by lapping them a specified length within a concrete member. The efficiency of a lap splice is 100%; full strength is transferred. Code (12.14.2.1) does not permit bars larger than No. 11 to be lap spliced except in exceptional circumstances.

Reinforcing bars need not contact each other in a lap splice. It may be noted in Fig. 3-2 that the bars in the cantilevered retaining wall are not even at the same spacing as the bars projecting from the base, yet a full transfer of load is affected between the two sets of bars. In this case, 100% of the bars are spliced within the given location, although in other arrangements the splice may involve only 50% of the bars or even less.

Code (12.15.1) separates lap splices into three classes, depending on the amount of excess steel at a particular splice location and the percentage of bars being spliced at the location, whether 50%, 75%, or 100%. Table 7-1 defines the three classes of splices. It should be apparent from the requirements of Table 7-1 that class A splices rarely occur. Almost all splices fall within class B or class C, and even class B splices usually occur as a result of deliberately arranging to splice 50% of the bars at one location and the remainder at another location. In practice, class C splices are by far the most common splices. The lap lengths for the three classes are based on the development lengths $L_d$ given by Eqs. (7-1):

$$\text{Class A splice:} \quad \text{lap length} = 1.0L_d$$

$$\text{Class B splice:} \quad \text{lap length} = 1.3L_d$$

$$\text{Class C splice:} \quad \text{lap length} = 1.7L_d$$

but in no case may the lap length be less than 12 in.

As before, these splice lengths have been computed and tabulated for ready reference in Table A-13. It should be noted that the factors 1.0, 1.3, and 1.7 are applied before the minimum criteria applicable to Eqs. (7-1) are applied and that the effects of being "top bars" apply equally to splices. It may also be observed that splice lengths

**TABLE 7-1 TENSION LAP SPLICES**

| $\frac{A_s \text{ provided}}{A_s \text{ required}}$ | Percent $A_s$ spliced at the location | | |
|---|---|---|---|
| | 50% | 75% | 100% |
| ≥2.0 | Class A | Class A | Class B |
| <2.0 | Class B | Class C | Class C |

greater than 10 ft can occur in larger bar sizes, effectively prohibiting their use simply by their cost. Mechanical splices or welded splices in such cases become economically feasible; such alternative splices are discussed in a subsequent section.

Reinforcement may also be lap spliced in compression, although the largest use of compression splicing is undoubtedly in columns. The required lap length for compression splicing does not include a category "top bars," nor are there any class A, B, or C conditions. The compressive splice length is the same as the general development length given earlier by Eqs. (7-2); Code (12.16.1) places the following additional limitations on the minimum splice length in compression:

$$L_d \geq 0.0005 d_b f_y \qquad \text{but} \qquad L_d \geq 12 \text{ in.} \tag{7-3}$$

where all symbols are as used earlier.

Compressive splice lengths have been computed from the applicable Eqs. (7-2) and (7-3) and have been tabulated in Table A-13. It should be observed that the splice lengths are governed by Eq. (7-3) for the higher concrete strengths.

## DEVELOPMENT LENGTHS WITH STANDARD HOOKS

It was noted earlier that where high-strength steels are used with low-strength concretes, development lengths of straight bars can become prohibitively long. The amount of steel that is duplicated in these straight splices and anchorages can add considerably to the total tonnage of reinforcement, and except for the convenience of permitting the use of shorter reinforcing bars, such splices and anchorages provide no particular benefits to the structure. A more efficient method of anchoring and splicing steel for such cases is in the use of hooks, where the end of the reinforcement is bent into a standard shape that is known to improve its anchorage to the concrete.

The two types of hooks prescribed by Code (7.1) are shown in Fig. 7-2. Both types of hooks produce equal load transfer, both are used extensively throughout the industry, but the 90° hook is probably the more frequently used. As a general rule, the cost of bending the reinforcement into hooks is more than offset by the savings in materials. In some cases, however, there simply is not enough room to provide for straight development length, and the use of a hook becomes necessary regardless of cost.

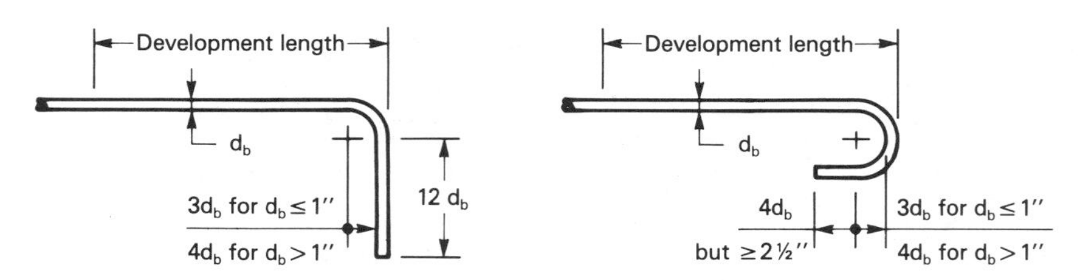


**Figure 7-2** Standard Tension Hooks.

The development lengths for hooked bars in tension are specified by Code (12.5.2). The development length where the concrete cover is at least that given in Chapter 2 is given by

$$L_d = \frac{0.02 d_b f_y}{\sqrt{f'_c}}$$

but not less than $8d_b$

and not less than 6 in. (7-4)

where all symbols are as used previously. Where concrete cover is increased, Code (12.5.3.2) permits a reduction in the development length, but such special cases are not included here.

Hooks are not effective in compression.

The development lengths for hooked bars in tension have been computed from Eq. (7-4) and have been tabulated in Table A-11. It should be observed that the development lengths are roughly half those given in Table A-13 for the larger-diameter straight bars. It should also be noted that the development length is given to the back of the hook; clear concrete cover is then computed from this point, simplifying the calculations somewhat.

A hook transfers a rather large concentration of load into the concrete in a relatively small distance. As a consequence, a rather confused stress pattern develops around hooks in which there may be locally high tensile stresses. Splitting of the concrete in the vicinity of hooks, particularly where several hooks occur together, is likely to occur when cover is small. For such cases, particular care must be taken when detailing the concrete to assure that the required cover will be maintained.

Some examples can now be presented utilizing the development lengths. The first example is a simple cantilever beam where the only reinforcement is that for the negative moment at the face of support. The second example is typical ''quickie'' problem in lifting a precast concrete member.

**Example 7-1**

*Strength method or working stress method*
*Determination of anchorage for reinforcement and*
*to select a workable anchorage*

*Given:*
Rectangular beam as shown
Grade 40 steel
$f'_c = 3000$ psi

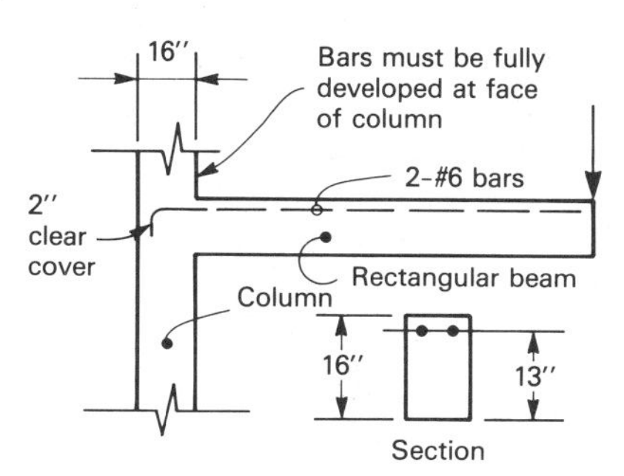

The required development length for a straight No. 6 bar is given in Table A-13. Since the bars shown in the sketch have more than 12 in. of fresh concrete below them, they are classed as top bars, and $L_d$ = 18 in. in tension. There is not enough room at the support to provide this development length. If a hook is used, the development length from Table A-11 is 11 in., which can be provided. Use hooked bars; at least 2 in. of cover, as shown; the embedment length becomes 14 in.

**Example 7-2**

*Strength method or working stress method*
*Determination of anchorage for embedded item*
*to select embedment for a lifting eye*

*Given:*
Precast tee as shown
Grade 40 steel
$f'_c$ = 3000 psi

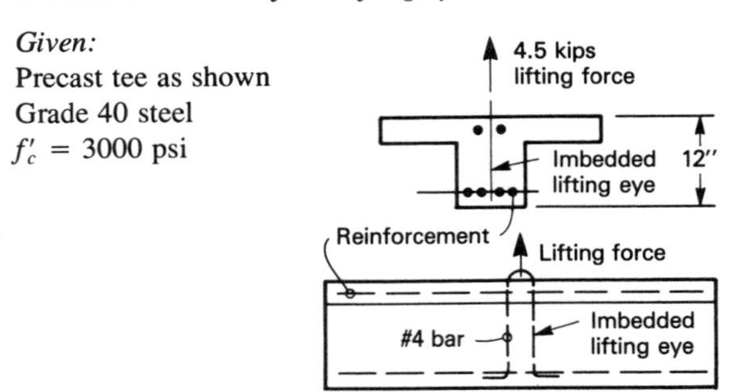


To develop full strength of the embedded No. 4 bars, a development length of 12 in. is required for a straight bar or 7 in. for a hooked bar. Since there is again not enough room for the straight bar, it must be hooked as shown. The capacity is then $2f_yA_s = 2 \times 40{,}000 \times 0.2 = 16$ kips.

In Example 7-2 it should be recognized that the solution did not include the actual load on the lifting eye. Nor do the tables include any provision for loads other than full capacity. A brief discussion of this omission is presented toward the end of this chapter.

## ABBREVIATED CRITERIA FOR CUTOFF POINTS

In addition to requirements for development of strength, Code also has requirements for embedment of reinforcement (and cutoff lengths) which must be met regardless of strength, stress levels, or load capacities. A few of these general requirements have been discussed elsewhere but are repeated below for ready reference. Other practices that are permitted but are not necessarily required are also included in the following list.

1. Tension reinforcement may be anchored by bending it across the web to provide a suitable anchorage length.
2. Tension reinforcement may be made continuous with reinforcement on the opposite face of the member.
3. Flexural reinforcement shall extend past the point where it is no longer needed by

a distance not less than $d$ nor less than $12d_b$, whichever is greater, except at supports of simple spans and at free ends of cantilevers.

4. Continuing reinforcement shall have an embedment length not less than the development length $L_d$ beyond the theoretical cutoff of adjacent noncontinuing reinforcement.
5. Flexural reinforcement shall not be terminated in a tension zone unless one of the following conditions is satisfied:
   (a) Shear at the cutoff point does not exceed two-thirds of that permitted, to include effects of any shear reinforcement.
   (b) Continuing reinforcement provides double the area required for flexure at the theoretical cutoff point and shear does not exceed three-fourths of that permitted.
6. At least one-third of the positive moment reinforcement in simple members and one-fourth of the positive moment reinforcement in continuous members shall extend along the same face of the member into the support; in beams, such reinforcement shall extend into the support not less than 6 in.
7. At simple supports and at points of inflection, positive moment tension reinforcement shall be limited to a diameter such that the development length $L_d$ satisfies the following equation:

$$L_d \le \frac{M_n}{\phi V_n} + L_a \tag{7-5}$$

where $M_n$ = nominal ultimate moment capacity of the section at the support with its reduced $A_s$
$V_n$ = nominal ultimate shear force on the section
$\phi$ = strength reduction factor for shear and at a support
$L_a$ = embedment beyond center of support and at a point of inflection
$L_a$ = depth $d$ or $12d_b$, whichever is greater

8. The value of $M_n/V_n$ in Eq. (7-5) may be increased 30% when the ends of the reinforcement are confined by a compressive reaction. Equation (7-5) need not be satisfied for reinforcement terminating beyond the centerline of simple supports by a standard hook, or by mechanical anchorage at least equivalent to a standard hook.
9. Negative moment reinforcement in a continuous, restrained or cantilever member shall be anchored in or through the supporting member by embedment length, hooks, or mechanical anchorage.
10. At least one-third of the total tension reinforcement provided for negative moment at a support of a continuous beam shall have an embedment length beyond the point of inflection not less than the depth $d$ nor less than $12d_b$ nor less than one-sixteenth the clear span, whichever is greater.
11. Bars spliced by noncontact lap splices in flexural members shall not be spaced transversely farther apart than one-fifth the required lap splice length nor more than 6 in.

The foregoing list of requirements may be worth learning if one is deeply involved in detailing concrete every working day. Otherwise, a simplified approach is needed that can quickly be relearned whenever needed. One such approach is to make steel cutoffs of up to one-half of the positive area of steel at one time, or all of the negative area of steel at one time. The foregoing list then reduces to the following rules.

1. In regions of negative moment:
   (a) Continue the entire area of flexural steel past the point of inflection by a distance $d$, $12d_b$, or span/16, whichever is farthest.
   (b) At integral columns, provide full development for all flexural steel at face of support by providing full development length in or through the column, hooking as necessary, keeping reinforcement diameters small enough that full development is achieved.
2. In regions of positive moment:
   (a) At interior supports, extend at least one-fourth of the area of flexural steel past the center of supports by a distance not less than 6 in., or past the face of support by a hook; extend the remaining flexural steel past the point of inflection by a distance not less than $d$ nor less than $12d_b$, whichever is greater.
   (b) At end columns or simple supports, extend all tension reinforcement past

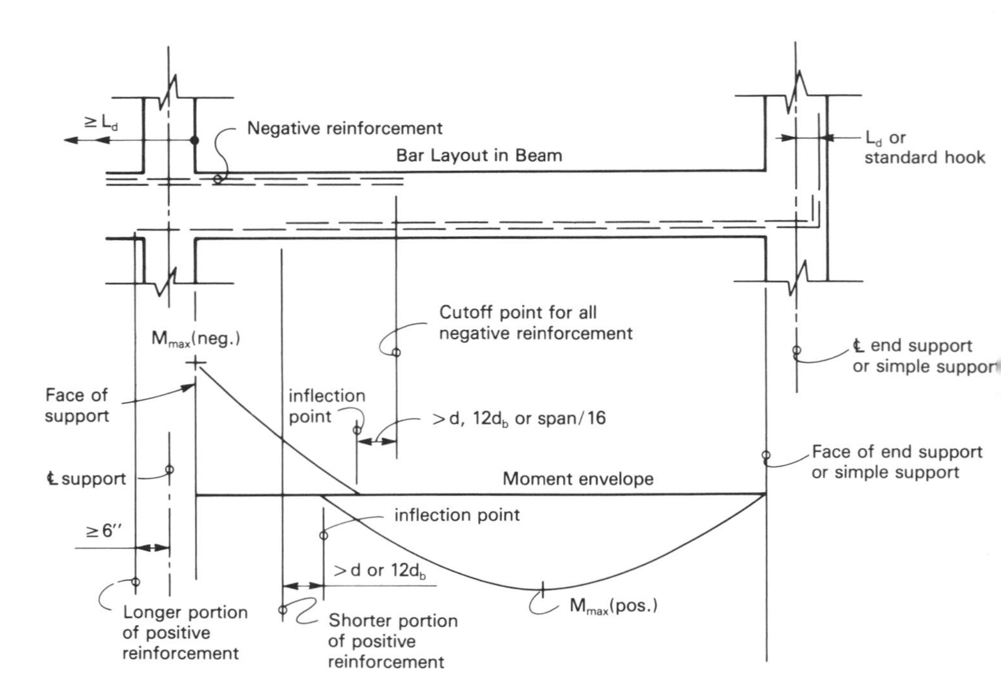


**Figure 7-3** Potential Cutoff Points.

the center of the support by a distance not less than $L_d$ or by a standard hook.

The foregoing rules are summarized in Fig. 7-3. Such an abbreviated set of rules is obviously conservative and will require more steel than if the full set of criteria were applied. In small projects, however, the savings in steel is at least partially offset by the additional labor hours required to fabricate and place the extra mark numbers. In larger projects, the savings would be significant and the more refined criteria become worthwhile.

For an example in the use of the foregoing criteria, the beam of Example 5-2 is used. The final reinforcement sizes and stirrup sizes have already been determined, but no attempt has been made to check the development of the reinforcement. For the example, the various distances on the moment diagram have been computed; in practice, they would probably be scaled.

**Example 7-3**

*Strength method*
*Reinforcement cutoff points*

Determine the cutoff points of the flexural reinforcement of Example 5-2. The beam and the moment diagram are shown below. Steel is grade 60, $f'_c$ is 3000 psi.

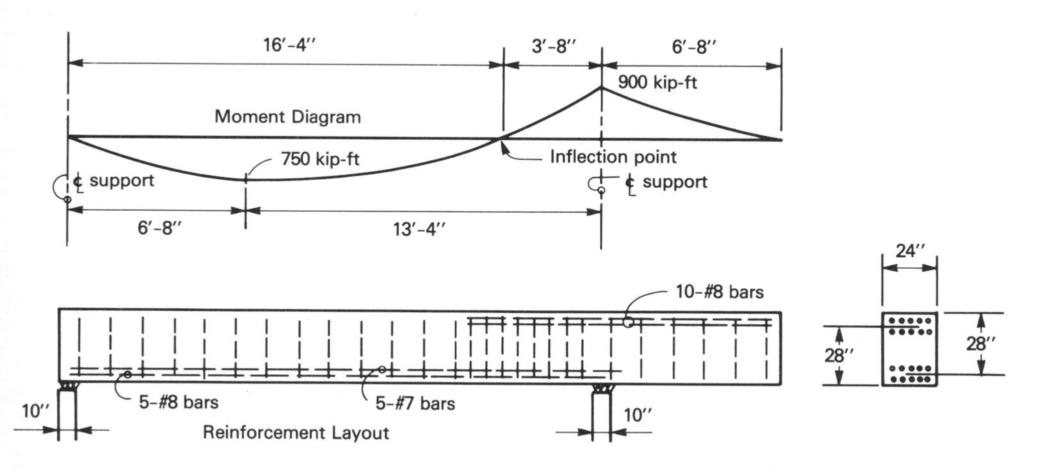


Determine the development lengths from Table A-13:

For No. 8 bars: $L_d = 34$ in. or 48 in. for top bars

For No. 7 bars: $L_d = 26$ in.

For negative reinforcement at the right support, extend the bars to the end of the beam and provide standard hooks (see the sketch).

For negative reinforcement to the left of the right support,

cutoff point: $> d$ (or 28 in.) past inflection point

$> 12d_b$ (or 12 in.) past inflection point

$> \frac{\text{span}}{16}$ (or 15 in.) past inflection

Use a cutoff point 28 in. past the inflection point, or 6 ft. 0 in. from the centerline of support (see the sketch).

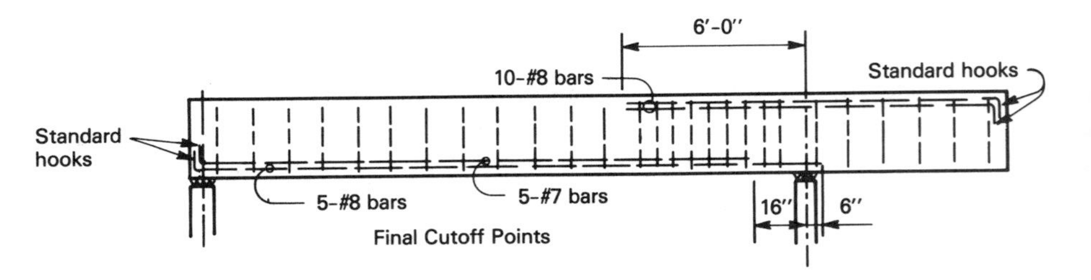


For positive reinforcement at the left support, extend all bars past centerline of support and provide standard hooks (see the sketch). At the right support, extend the bottom layer of bars 6 in. past the centerline of the support (see the sketch). For the remainder of positive bars at the right support,

cutoff point: $> d$ (or 28 in.) past the inflection point

$> 12d_b$ (or $10\frac{1}{2}$ in.) past inflection point

Use a cutoff point 28 in. past the inflection point or 16 in. from centerline of support (see the sketch).

## DEVELOPMENT OF STIRRUPS AND TIES

Development lengths for stirrups and ties are of course comparable to those of flexural reinforcement. The problem is complicated by the fact that the available space to provide the development length is quite limited. Although Code (12.13.2.2) permits development of stirrups by straight extensions, space limitations are usually so restrictive that the ends are almost always hooked.

A typical stirrup is shown in Fig. 7-4, with a general case of loading as shown. The load in the stirrup occurs as a result of diagonal cracking in the tensile zone, requiring that the resisting force be developed at the other end of the stirrup in the compression zone. This effect is so marked that Code (12.3.2) requires the strength of the stirrup to be fully developed within a distance $d/2$ from the compression side of the beam.

By their nature, stirrups are usually made of the smaller bar sizes, spaced rather closely together. Since the development lengths of smaller bar sizes are smaller, this feature automatically alleviates some of the problem in finding room to develop the bar.

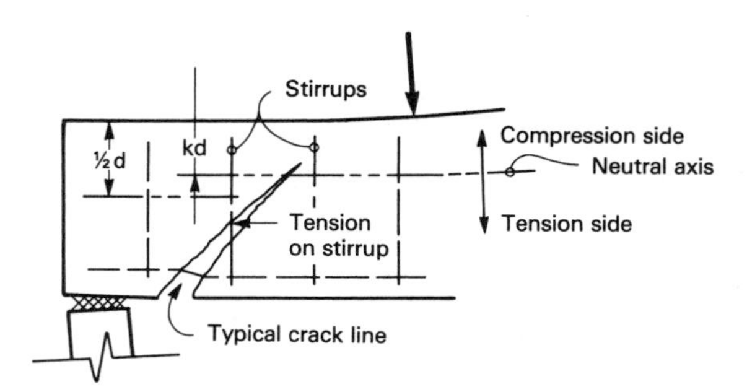


**Figure 7-4** Loading on Stirrups.

Code (7.2.2) also recognizes that smaller bars can be bent to tighter radii and has established separate criteria for hooks in stirrups and ties.

The criteria for hooks in stirrups and ties are shown in Fig. 7-5 for bar sizes No. 5 and smaller. In addition to the standard 90° hook, a hook of 135° (rather than 180°) is permitted for stirrups. In the smaller bar sizes, a bend of 135° has essentially the same capacity as that of a 180° bend, but is more easily handled in bending machines.

The following development lengths for stirrups are prescribed by Code (12.13.2) as a portion of the straight development lengths $L_d$ given earlier:

1. For straight bars, the development length shall be not less than $L_d$, nor less than $24d_b$, nor less than 12 in., located within a distance $d/2$ from the compression face.
2. With a 90° hook, the development length shall be not less than $0.5L_d + 3d_b$, located within a distance $d/2$ from the compression face.
3. With a 135° hook in grade 50 or 60 steel, the development length shall be not less than $0.33L_d + 3d_b$, located within a distance $d/2$ from the compression face.
4. With a 135° hook in grade 40 steel, the full strength of bars No. 5 and smaller are considered to be developed by the hook, where the hook is bent around longitudinal reinforcement on the compressive side.

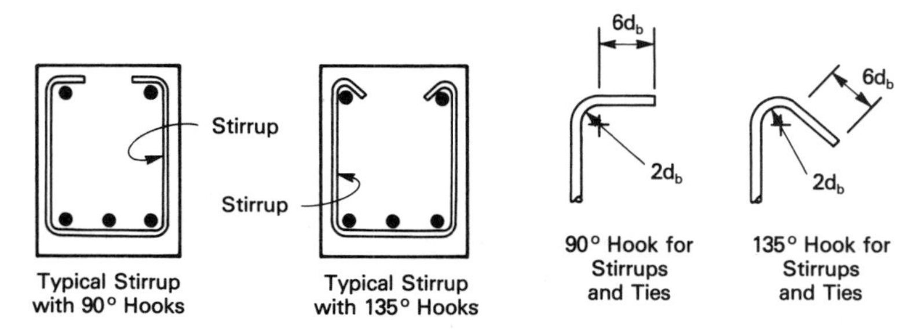


**Figure 7-5** Typical Stirrups and Ties.

The development lengths prescribed by the foregoing criteria have been computed and tabulated in Table A-12. For ready reference, a sketch of the bend criteria is included. For concrete strengths greater than 3000 psi, the concrete strength is not a factor in determining the minimum development lengths, hence the table is quite small.

One feature of the criteria for stirrups overshadows most of the others: a 135° hook in grade 40 steel provides full development of the stirrup. Even when grade 60 steel is to be used, the calculations are commonly made as if the steel were grade 40 and the stirrups are designed with 135° hooks. There is then no problem with providing anchorage for the stirrups; the difference in cost between grades 40 and 60 steels for such limited quantities is negligible.

For an example in checking development lengths for stirrups, consider the stirrups selected for the beam of Example 5-1. The stirrups were selected as $\frac{3}{8}$-in. diameter, with hooked ends, to be placed in a beam having an effective depth of 22 in. This selection can now be checked. From Table A-12, the development length for the hooked end of this stirrup is shown as 7 in., which must lie completely within a distance of $d/2$, or 11 in., from the compression face of the beam. It is now seen that this choice is satisfactory.

Alternatively, had a stirrup with straight ends been selected, the development length of the straight ends are found from Table A-12 to be 12 in. This length cannot be accommodated in the 11 in. available, so this selection was not acceptable.

## SPECIAL PROVISIONS AND LIMITATIONS

The development lengths given in the preceding sections will apply to the vast majority of applications in routine building construction. Code also provides for reductions in these development lengths for a rather large number of special cases (which are beyond the scope of this book). In sufficiently repetitive applications, these reductions can afford worthwhile savings in steel and one may wish to examine the provisions of the Code in detail for such special cases. The general cases presented earlier are always on the "safe" side, although they may not always be on the most economical side.

Typical circumstances that would allow reductions in development lengths occur when repetitive bars are spaced more than 6 in. apart with more than 3 in. cover to the nearest face. Another case occurs when the minimum side cover over a hook exceeds $2\frac{1}{2}$ in. with end cover in excess of 2 in. In addition, there are several special circumstances in providing anchorage for stirrups which one might investigate if the number and size of the stirrups justify the time. Again, the study of these items is beyond the scope of an introductory text such as this, but the reader should be aware of their existence.

Additionally, the scope of the foregoing sections does not include the use of wire mesh or wire fabric, neither for flexural reinforcement nor for shear reinforcement. Such applications are becoming more common and one may wish to study further such applications in larger projects. Development lengths for these materials are also prescribed by Code.

One of the more frequently applied reductions is based on the amount of excess reinforcement that may be present at the point where the splice is to be made. Such excess is rather common, since the areas of steel are selected for peak loadings and are often continued through areas of lesser loadings. In general, Code permits a reduction in development lengths proportional to the required $A_s$ in the area, that is, by the ratio ($A_s$ required)/($A_s$ provided). Specific details may be found in the Chapter 12 of the Code.

Where restrictions on space are particularly severe or where development lengths become excessively long, it may become worthwhile to provide mechanical anchorages, welded or otherwise attached to the reinforcement. Code (12.6) permits such commercially made connectors to be used where there is sufficient full-scale test data available to show that the anchorage is, in fact, adequate. Code (12.14.3.4) requires that such anchorages develop 125% of the yield strength of the bar rather than the full ultimate strength.

Similarly, Code (12.14.3.3) permits welding of reinforcement to anchorages or to other reinforcement as a means of splicing. As with mechanical anchorages, such welding must develop 125% of the yield strength of the bar rather than the full ultimate strength.

''Bundling'' of bars, or placing several longitudinal bars in contact with each other to save space, is permitted by Code (7.6.6), although such practices are severely restricted by several Code provisions. The development lengths of bundled sets of bars must be extended considerably. Consequently, Code (12.4) provisions for development lengths should be studied closely before the decision is made to use bundled bars.

## ELASTIC ANALYSIS

The elastic equations for bond stress and development length do not appear directly in the 1983 ACI Code. All development lengths are simply stated as empirical formulas, usually nonhomogeneous, based on test data without regard for theoretical development. Although the elastic analysis is not overtly recognized, it can offer valuable insight into the workings of bond and development length. A highly abbreviated discussion of the elastic analysis of bond is presented in the following paragraphs.

A typical segment of a beam is removed and is shown with its internal resultants of load in Fig. 7-6. Note that the segment is shown as $\Delta$ inches thick, with all signs in the positive sense. The tensile reinforcement is removed and shown at larger scale, with the average bond stress $u$ indicated around the bar.

Moments are summed about point A, yielding

$$V\Delta + Tjd - T_2jd = 0 \qquad \text{hence} \qquad T_2 - T = \frac{V\Delta}{jd} \tag{7-6}$$

Forces acting on the reinforcement are also summed, yielding

$$T_2 - T = u\left(\sum \pi d_b\right)\Delta \tag{7-7}$$

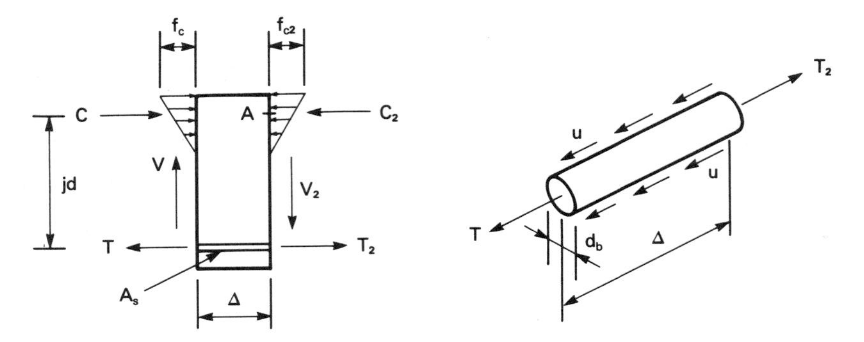


**Figure 7-6** Bond Stress.

where $\Sigma\ \pi d_b$ is the perimeter, or the sum of perimeters, of the reinforcement. $\Sigma\ \pi d_b$ was conventionally designated in older ACI Codes as $\Sigma o$.

Equation (7-7) is substituted into Eq. (7-6) and the symbol $\Sigma o$ is used to indicate sum of perimeters; the result is solved for $u$:

$$u = \frac{V}{\sum ojd} \tag{7-8}$$

where $V$ = shear on the section
$u$ = average bond stress around the bar
$jd$ = distance between tensile and compressive resultants.

It should be obvious from Eq. (7-7) that the bond stress is highest where $T_2 - T$ is highest, or where the moment is changing rapidly. Such points are shown in strength-of-materials work to be located where shears are highest, as verified by Eq. (7-8); these points are usually near the supports or near concentrated loads. It may be concluded that high bond stress will occur around the flexural reinforcement wherever high values of shear occur, most commonly at the beam supports and at other points of concentrated loads.

Overall anchorage is now examined; a typical flexural anchorage is shown in Fig. 7-7. The embedded length $L_e$ must develop an average bond stress $u$ if the moment at face of support is to be sustained. Moments are summed about the resultant of compressive force $C$ at point $A$:

$$M = Tjd \tag{7-9}$$

where $M$ is the static moment on the section. Forces are now summed along the length of embedment,

$$T = u \sum oL_e \tag{7-10}$$

where $\Sigma o$ is the sum of perimeters, as defined previously.

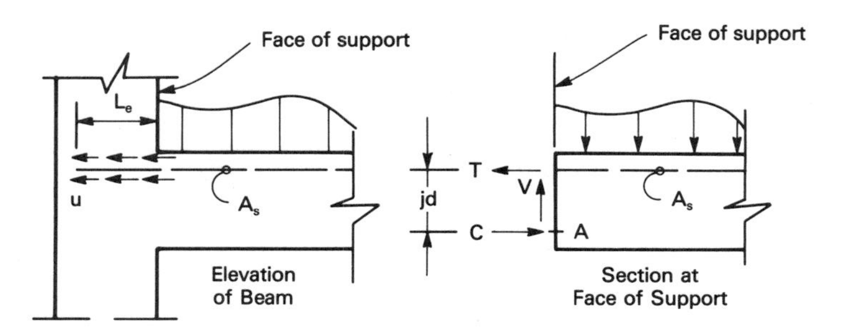


**Figure 7-7** Anchorage Length.

Equation (7-10) is substituted into Eq. (7-9) to eliminate $T$, and the result solved for the required embedment length $L_e$:

$$L_e = \frac{M}{u \sum ojd} \tag{7-11}$$

Equations (7-8) and (7-11) relate the bond stress $u$ and required embedment length $L_e$ to the shear and moment acting on the section. If an allowable bond stress $u$ were known, the necessary embedment lengths and bar perimeters could be computed. In past years, the bond stress $u$ was in fact specified by Code and the other unknowns were computed from Eqs. (7-8) and (7-11); in more recent years, the Code has dropped that approach in favor of the more accurate approach presented earlier in this chapter.

Circumstances do arise, however, where the older approach using an allowable stress can be very useful. For example, consider the following problem, which occurs frequently where pad eyes are embedded in the foundations, to be used in subsequent steel erection.

A grade 40 deformed No. 6 reinforcing bar is embedded a distance of 6 in. in a concrete footing. Determine the safe working load that may be applied as a pullout load. ($f'_c$ = 3000 psi. The allowable bond stress $u$ may be taken as 350 psi.)

For pullout load $P$, the allowable load is obviously the allowable bond stress $u$ times the embedded contact area (circumference times length):

$$P = u(\pi \times d_b \times L_e) = 350 \times \pi \times 0.75 \times 6$$

$$= 4950 \text{ lb}$$

If, instead of using the allowable bond stress, the current Code provisions must be used, the only value that can be computed is the development length at full yield, $L_d$, computed as 13 in. (Table A-13). The load $P_y$ in the bar at yield is given by $f_yA_b$, or 17.6 kips. If it is assumed that the pullout load is proportional to the depth of embedment, the reduced pullout load $P_n$ is given by

$$P_n = P_y\left(\frac{6}{13}\right) = 17.6 \times \frac{6}{13}$$

$$= 8.1 \text{ kips}$$

Further, if a factor of safety (F.S.) of 1.67 is assumed to working levels, the allowable load $P$ is then

$$P = \frac{8.1}{\text{F.S.}} = 4.85 \text{ kips}$$

This value of 4.85 computed under the foregoing assumptions compares well with the value of 4.95 kips computed using older criteria. The reliability of the older criterion can be defended further by recognizing that it was used quite successfully for many years before the newer and more accurate criterion was developed. Judiciously applied, the older criterion can still be used with satisfactory results.

The following formulas for computing usable bond stress for deformed bars in tension are extracted from the 1963 ACI Code. The stresses are in psi and apply to bar sizes not larger than No. 11.

| | Ultimate stress | Working stress |
|---|---|---|
| Top bars: | $u = \dfrac{6.7\sqrt{f'_c}}{d_b}$ but $< 560$ | $u = \dfrac{3.4\sqrt{f'_c}}{d_b}$ but $< 350$ |
| Other bars: | $u = \dfrac{9.5\sqrt{f'_c}}{d_b}$ but $< 800$ | $u = \dfrac{4.8\sqrt{f'_c}}{d_b}$ but $< 500$ |

For plain bars, the usable bond stresses may be taken as one-half of these values but not more than 160 psi at working stresses nor 250 psi at ultimate stress.

For deformed bars in compression, the following stresses apply.

| | Ultimate stress | Working stress |
|---|---|---|
| All bars: | $u = 13\dfrac{\sqrt{f'_c}}{d_b}$ but $< 800$ | $u = 6.5\dfrac{\sqrt{f'_c}}{d_b}$ but $< 400$ |

The foregoing method is now outdated and should not be used where the more recent approach may be applied. The more recent approach, however, can be seen to be somewhat more limited in its range of applications. The older criterion can offer a reassuring check in unusual applications or when the newer criterion is only vaguely applicable.

## REVIEW QUESTIONS

1. What is meant by the "development length" of a reinforcing bar?
2. What is the "bond strength" of concrete, and what are some of the factors that can affect it adversely?
3. How does the bond of concrete to smooth reinforcing bars compare to its bond to deformed bars?
4. In view of the answer to Question 3, about what percentage of the total load on a deformed bar might be attributed to the mechanical bearing of its deformations against the adjacent concrete?
5. How have development lengths of reinforcing bars been determined in recent years?
6. What is the primary advantage in using hooks?
7. What is the primary point of caution in using hooks?
8. When only a part of the required development length of a bar can be embedded in concrete, how can its allowable working load be computed using the strength method? Using the working stress method?
9. On flexural reinforcement in a simple span, where can bond stresses be expected to be highest? Lowest?
10. Why is it almost always necessary to provide hooks for anchoring stirrups?
11. What particular advantages does the 135° hook provide when detailing ties and stirrups?
12. Where in the cross section of the beam does the development of stirrups occur?
13. What is "bundling" of reinforcement?
14. When mechanical anchorages or welding is used, how much of the strength of the reinforcement must be developed?

## PROBLEMS

1. A concrete tee-beam roof having a simple span is carried on 16-in.-thick masonry walls. The tee beams are reinforced with No. 5 reinforcing bars. What is the required end anchorage for those reinforcing bars that must be extended into the support? $f'_c = 3000$ psi, grade 40 steel.
2. For the bars of Problem 1, what would be the end anchorage requirement if the steel were grade 60?
3. In the cantilever beam of Example 7-1, 3 No. 5 bars would provide as much steel area as the 2 No. 6 bars shown. What would be the anchorage requirement for the No. 5 bars if they were to be used instead?

For the following rectangular beams, determine the required areas of positive and negative reinforcement and the cutoff points (or anchorage requirements) for both. Use the abbreviated criteria for cutoffs and anchorages. Assume that the given dead load includes an allowance for the weight of the rectangular beam. It is not necessary to include a design for any required shear reinforcement. $f'_c = 4000$ psi, grade 60 steel.

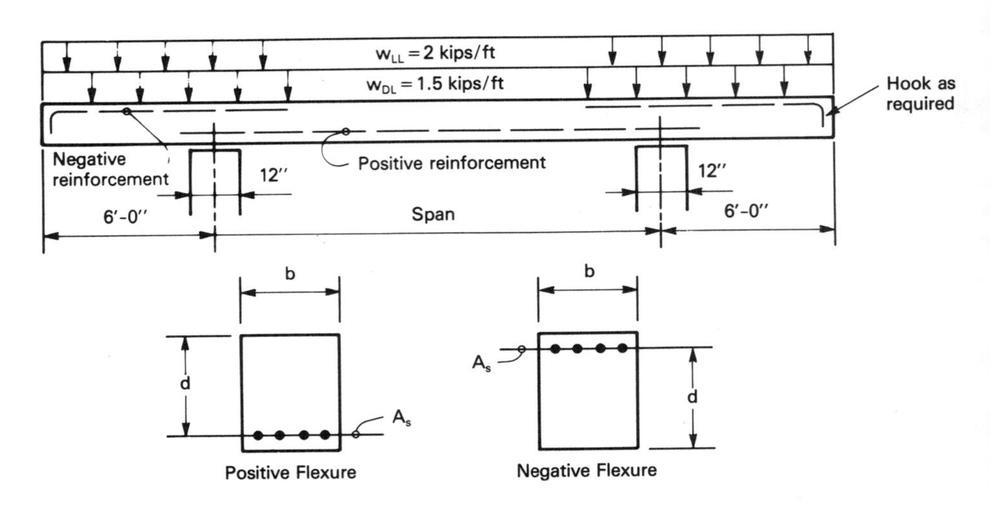


| Problem | Span (ft) |
|---|---|
| **4** | 14 |
| **5** | 16 |
| **6** | 18 |
| **7** | 20 |
| **8** | 22 |

# 8

# Reinforced Concrete Columns

## GENERAL

Code (B.6.1) requires that all columns be designed using the strength method of design. Code (B.6.2) does not recognize a working stress method for the design of columns, except to specify that a factor of safety of 2.5 shall be applied to the ultimate load $P_n$ to obtain the allowable working load $P_w$. The allowable load $P_w$ in the working stress method is thus obtained by dividing the nominal ultimate load $P_n$ by 2.5, or

$$P_w = \frac{P_n}{2.5} \quad \text{(working stress method)} \tag{8-1}$$

It has been observed in earlier discussions that an average load factor of 1.52 can be expected when the strength method is used. The strength reduction factor $\phi$ is applied after this load factor is applied; the strength reduction factor for tied columns is 0.7. The nominal ultimate load $P_n$ can therefore be approximated as a multiplier times the working load:

$$P_n = \frac{\text{average load factor}}{\phi} P_w = \frac{1.52}{0.7} P_w$$

or

$$P_w = \frac{P_n}{2.17} \quad \text{(strength method)} \tag{8-2}$$

When Eqs. (8-1) and (8-2) are compared, it is seen that the factor of safety in column design using the strength method can be expected to be appreciably less than for the working stress method (about 15% less). This reduction in factor of safety for column design is consistent with a similar reduction in the factor of safety for beam design already noted in Chapter 4. It may be concluded that the overall factor of safety in the strength method is appreciably less than in the working stress method; closer attention to detail is therefore warranted when using the strength method.

It is again emphasized that Code (B.6) requires all column design in either method to be based on ultimate load. Stress is never a consideration, neither at day-to-day service levels nor at some maximum allowable working level. For those who may be interested in knowing the approximate stresses at service levels, however, the derivations presented in the following sections include a means to find them.

## CONFIGURATIONS AND PRACTICES

As indicated in Fig. 8-1, many shapes may be used for columns, such as square, rectangular, hollow, circular, Y-shaped, L-shaped, and so on. By far the most common shape is the rectangular shape, which includes the square as a special case. Only the rectangular shape is treated in the succeeding sections; other shapes may be treated similarly.

In designing a column, it is always necessary to make an initial selection for the gross dimensions of the column section, $b$ and $t$, as shown in Fig. 8-1. The reinforcement is then determined for this particular gross size. If the amount of reinforcement required for this gross size is considered to be unacceptable, a new gross size must be selected and the process repeated until an acceptable result is found.

In making the first guess at the size of a column, the rules of thumb given in Chapter 3 can be helpful. Quite often, though, when a group of columns are being designed, the gross size of the column carrying the heaviest load is established first, and all other columns are made the same size. Then only the reinforcement of the other columns need be varied to suit the actual loads.

The rule of thumb for the size of a square column is repeated from Chapter 3:

column width = 12 in. plus 1 in. per story above

Where the column is to be rectangular, it should have roughly the same gross area as this square column, or slightly more. Where significant moments are present, the accuracy of this rule of thumb becomes even worse than usual.

A low concrete building (five stories or less) having a column module of 18 to 20 ft and concrete floors and roof can be expected to have a column loading of 30 to 40 tons per floor. This load is the working load, not the ultimate load. A nominal column load for small buildings will be seen to be a useful index when guessing sizes of columns.

Older codes specified a minimum dimension of 10 in. for columns, but more recent codes have discontinued this minimum limit. The column size is now controlled by the designer and may be set to match the masonry or concrete walls used elsewhere in the

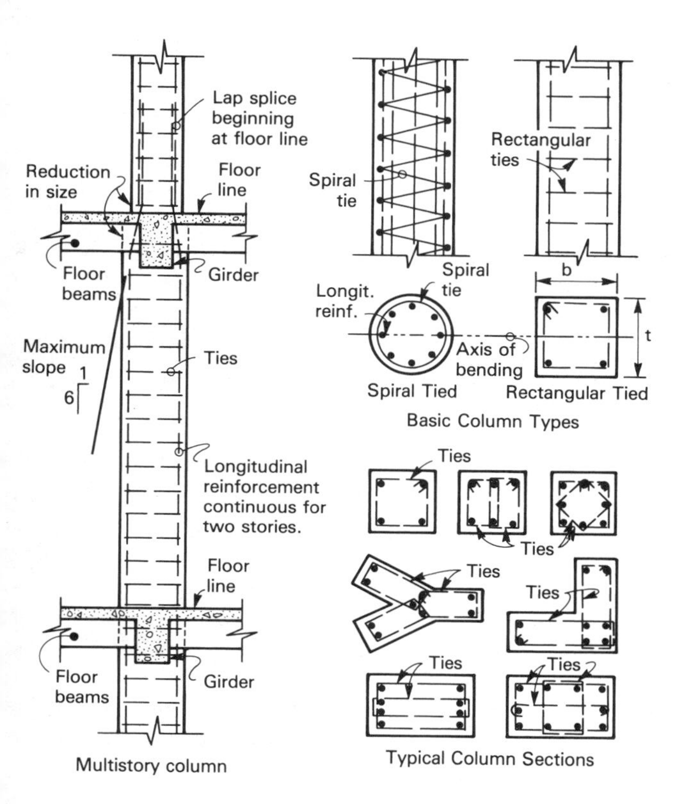


**Figure 8-1** Column Details.

design. In normal practice, however, the difficulty in casting columns 10 in. square (or less) severely discourages their use. A minimum dimension of 12 in. is commonly observed.

For one- and two-story buildings, the loads are so low that a minimum-sized column with minimum longitudinal steel is the usual design. Even these sizes usually provide a capacity much larger than needed. Example 8-5 lists the capacities of minimum-sized columns for various grades of steel and concrete.

In a typical concrete building, the strength of the concrete specified for columns is usually higher than for the beams and slabs. The primary reason is the susceptibility to creep under their high sustained loads. Similarly, the longitudinal reinforcement is almost always the higher-strength steels wherever they are available.

In general, the longitudinal reinforcement in columns is spliced each two stories. Such frequent splicing requires large amounts of additional steel, but the reinforcing bars will not cantilever out of the forms far enough to permit going more than two stories. The splice is almost always placed just above floor level.

Code (7.10.5) requires that the longitudinal bars be tied to prevent their buckling outward under load. Design requirements for the ties themselves are also given by Code:

1. Ties shall be at least No. 3 for longitudinal bar sizes No. 10 or smaller and at least No. 4 for larger bars and bundled bars.
2. Vertical spacing shall not exceed 16 longitudinal bar diameters or 48 tie-bar diameters or the least dimensions of the column.
3. Ties shall be arranged such that every corner bar and alternate longitudinal bars shall have lateral support provided by the corner of a tie having an included angle of not more than 135°; no bar shall be farther than 6 in. clear on each side along the tie from such a laterally supported bar.

Typical tie arrangements are shown in Fig. 8-1 for some of the more common configurations.

Although Code (10.9.1) allows the steel areas in columns to be as high as 8% of the gross concrete area *bt*, the congestion in such heavily reinforced columns severely limits their use. The congestion is particularly severe where the vertical column bars must pass by the horizontal bars in the two intersecting girders that an interior column usually supports. As a practical matter, the steel areas are kept less than 3% of the gross concrete area *bt*; steel areas above 4% are rare.

To decrease congestion, it is common practice to use the larger bar sizes for longitudinal reinforcement. A larger steel area is thus provided by fewer bars. The use of the larger bar sizes in columns usually does not create problems with bond since the shears can be expected to be quite low.

Two of the more common arrangements of longitudinal reinforcement are shown in Tables A-8 through A-10 of the Appendix. Where moments are low, the longitudinal steel is usually distributed uniformly around the column. As the magnitude of the moment increases, it obviously becomes more efficient to concentrate the steel at the flexure faces.

Longitudinal reinforcement is almost always arranged symmetrically about the axes of bending. Even where wind loads are not being resisted by the columns and load reversals are thus minimized, the column reinforcement is still arranged symmetrically. There is thus no chance that the column can be constructed backwards.

## BEHAVIOR UNDER LOAD

Columns can fail structurally by either of two separate modes: buckling or crushing. Long columns fail by buckling and short columns fail by crushing. However, many concrete columns are neither long nor short, but are somewhere in an intermediate range. For such columns, failure occurs through an indistinct mixture of the two modes.

The "critical" buckling stress of long columns is given by the well-known Euler column formula:

$$\frac{P_c}{A} = \frac{\pi^2 E}{(L/r)^2} \tag{8-3}$$

where $P_c/A$ = stress just as buckling impends
$E$ = modulus of elasticity
$L$ = column length
$r$ = radius of gyration, $\sqrt{I/A}$

The physical conditions (boundary conditions) assumed for the derivation of the Euler column formula are shown in Fig. 8-2. Under axial load, the column fails by lateral displacement, but it will return to its original straight configuration when the load is removed. Note that the length $L$ is measured between the hinge points and that the moment of inertia $I$ is oriented about the axis of bending. Note also that the end conditions are hinges (or points of zero moment, such as inflection points).

It is important to be aware that the end moments are zero in the derivation. The existence of end moments, even small ones, can seriously reduce the critical load $P_c$. The formula is valid, however, between any two consecutive points of zero moment.

Equally important in the derivation, it is assumed that the column is perfectly straight; there can be no accidental offset in its alignment that would produce in initial moment. It it also assumed that the load is always concentric on the section; there can be no moments introduced due to eccentric loading. It is further assumed that the moment of inertia $I$ is constant throughout the length of the column; there can be no variations in the dimensions, nor can there be irregularities in reinforcement or in splices.

By this point in the presentation, it should be obvious that very few concrete columns could meet the conditions required by the Euler formula. Even so, the basic parameters appearing in the Euler formula remain valid indicators for all column performance. The way in which these parameters enter into the semiempirical design of concrete columns is discussed in succeeding sections.

One such parameter that will appear later is the ratio $L/r$, which has a special significance in column design. Called the *slenderness ratio*, it affords an indication of

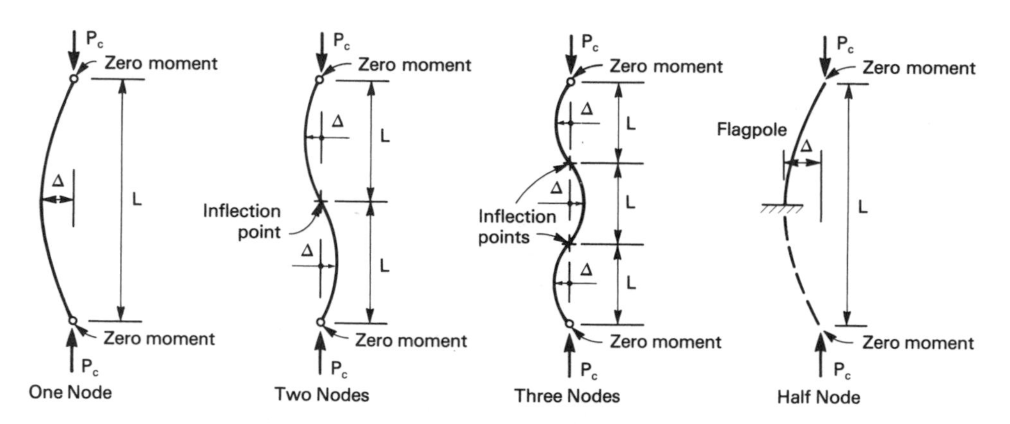


**Figure 8-2** Euler Column Conditions.

the susceptibility of a column to buckling. The higher the slenderness ratio is, the higher the susceptibility to buckling becomes and the lower the buckling load is.

At the other extreme from long columns are short concrete pedestals. When the ratio of height to least lateral dimension is 3 or less, the member is not subject to Code criteria for columns. The design is performed by direct $P/A + Mc/I$ design procedures.

## ACI COLUMN FORMULA

For the vast majority of columns that are not long enough to be subject to design by the Euler formula nor short enough to be designed as a pedestal, ACI has developed a simple approach. For rectangular tied columns, Code (10.3.5.2) requires that the nominal eccentric load $P_n$ shall not exceed 80% of the concentric load at full plasticity, $P_0$, where

$$P_0 = 0.85f'_c(A_g - A_{st}) + f_yA_{st} \tag{8-4}$$

In this formula, herein called the ACI column formula,

$$A_g = \text{gross area of concrete, } bt$$

$$A_{st} = \text{area of longitudinal reinforcement}$$

$$0.85f'_c = \text{idealized yield stress of concrete}$$

$$f_y = \text{yield stress of reinforcement}$$

It is emphasized that the load $P_0$ given by Eq. (8-4) is a concentric load. Also, it is the absolute maximum load that the cross section could sustain when it is in a fully plastic state of deformation. There is no extra capacity for moment.

When the nominal axial load $P_n$ is less than 80% of this peak axial load $P_0$ at full plasticity, there is some excess capacity available to take moment. This moment, $M_n$, is viewed by Code (10.3.6) as "the maximum moment that can accompany the axial load." The section is thus designed *first* for axial load; moment can then be added to take up any excess capacity.

The nominal ultimate axial load $P_n$ and the nominal ultimate moment $M_n$ are computed as always:

$$P_n = \frac{P_u}{\phi} \quad \text{and} \quad M_n = \frac{M_u}{\phi} \tag{8-5a, b}$$

where $P_u$ and $M_u$ are the factored ultimate loads and $\phi$ is the strength reduction factor defined and used previously. For columns, $\phi = 0.7$, but as in beams, the strength reduction factor $\phi$ may be varied where $P_u$ is less than $0.10f'_cbt$. The same relationship used earlier applies:

$$\phi = 0.9 - 0.2\frac{P_u}{0.10f'_cbt} \tag{4-36}$$

This allowable increase in $\phi$ applies only where axial load is very small and moment is very large; under such circumstances, the member begins to act as a beam subject to axial load rather than a column subject to flexure.

It is important to recognize that the ACI column formula has no provisions for length, radius of gyration, or buckling. Buckling criteria and effects of length are treated by separate limits and requirements, imposed and computed independently from the column formula. The column formula provides only the maximum capacity of a section under concentric load, without regard to any other limitations.

## BUCKLING CRITERIA

The parameter used by Code (10.11.4) to classify buckling criteria in columns is the slenderness ratio, taken from the Euler column formula:

$$\text{slenderness ratio} = \frac{KL_u}{r}$$

where $K$ = numerical factor dependent on end conditions of the column and on the type of overall lateral-load-carrying system
$L_u$ = unsupported length of the column, taken as the clear distance between attached or supporting members at top and bottom
$r$ = radius of gyration of the column section, taken as 0.3 times the least dimension for rectangular sections

The value of $K$ is given by Code (10.11.2.1) as 1.0 for all buildings or structures braced against sidesway; these are the only structures considered in this text. For these structures, the columns are not used to resist any part of the lateral loads. Another common system, although not considered in this text, is that of an unbraced frame where lateral loads are resisted entirely by flexure on the columns; for such frames the value of $K$ may be much higher.

Although the value of $K$ is known to be 1.0 herein, the symbol $K$ will still be shown wherever the slenderness ratio is referenced. Its inclusion will indicate where the effects of end conditions and sidesway must be considered.

For columns braced against sidesway, Code (10.11.4.1) permits the effects of slenderness to be neglected when

$$\frac{KL_u}{r} < 34 - 12\frac{M_1}{M_2} \tag{8-6}$$

where $M_1$, $M_2$ = factored end moments from the analysis for vertical loads (no lateral loads included)
$M_1$ = smaller moment, having a positive sign if the column is bending in single curvature and a negative sign if in double curvature
$M_2$ = larger moment, always having a positive sign

The least value of the slenderness ratio is seen to occur when $M_1 = M_2$, for which the slenderness ratio is 22.

All columns considered in this book fall within this intermediate category, where slenderness effects may be neglected. For the other category, where slenderness effects must be considered, Code (10.11.5) prescribes a means to magnify the design moments. The result of such moment magnification is to cause an overdesign of an intermediate column such that it will carry a lighter load (the actual design load) under the more severe conditions.

A final point concerns the length $L_u$. The Euler formula, from which the slenderness ratio is drawn, uses the length between hinge points as its buckling length, while the ACI criteria specifies an "unsupported" length without regard to end conditions. ACI has thus chosen a simple but conservative value for length, recognizing the myriads of combinations of loads, inflection points and end rotations that could affect the buckling length of a structural column.

## ULTIMATE STRENGTH ANALYSIS

In column analysis, a very useful dimension is the eccentricity of load, $e_n$, where

$$e_n = \frac{M_n}{P_n} \tag{8-7}$$

This eccentricity of the load $P_n$ is shown in Fig. 8-3 and is used frequently throughout the subsequent discussions and analyses. The eccentricity is measured from the centerline of the section, not from the neutral axis.

In applying the ACI column formula to the analysis of columns, there are other conditions prescribed by ACI that must be met:

1. Longitudinal reinforcement may not be less than 1% nor more than 8% of the gross cross-sectional area of concrete, $A_g$.
2. Columns shall be designed for the maximum moment $M_n$ that can accompany the

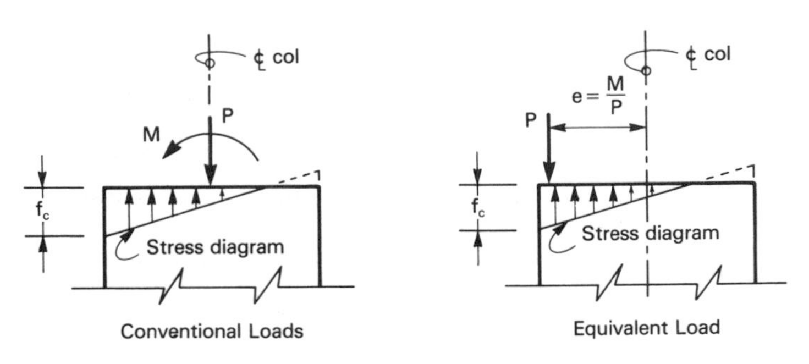


**Figure 8-3** Eccentricity of Load.

axial load $P_n$. (Where moment varies along the length of the column, the largest value of moment is used.)

3. Strains shall be assumed to vary linearly across the section.
4. Maximum strain in the concrete shall be assumed to be 0.003.
5. Tensile stress in steel when in the elastic range shall be taken as the modulus $E_s$ times the strain; when in the plastic range the stress shall be taken at the specified yield.
6. Tensile strength of concrete is zero.
7. Stress variation in concrete may be assumed to be any shape that provides results in substantial agreement with tests.
8. When $P_n$ is less than the axial load at balanced conditions, $P_b$, or less than $0.10f'_cA_g/\phi$, the ratio of reinforcement $\rho$ provided shall not exceed 75% of the balanced ratio $\rho_b$ that would be required for flexure only.
9. Where there is no computed moment on a column, the column shall be designed for an eccentricity not less than $0.1t$ for the column load $P_n$ (not a Code requirement but a recommended practice).

All the foregoing conditions are included in the subsequent analyses. Condition 2, concerning the design for moments, recognizes that the allowable moment on a column is interrelated with the axial force; as axial load is decreased, moment can increase (up to a point). Condition 9 was a Code requirement prior to 1971; it recognizes that there cannot be such a thing as a perfectly concentric load or a perfectly straight column.

A typical column section is shown in Fig. 8-4a. The total area of steel is designated as $A_{st}$ and its steel ratio as $\rho$. That part of the steel distributed to the flanges is $A'_s$ and its steel ratio is $\rho'$. That part of the steel distributed to the web is $A''_s$ and its steel ratio is $\rho''$.

In the equivalent section of Fig. 8-4b, the web steel $A''_s$ is replaced by an equivalent imaginary strip of steel having a finite width, as shown. All other symbols are the same as used previously.

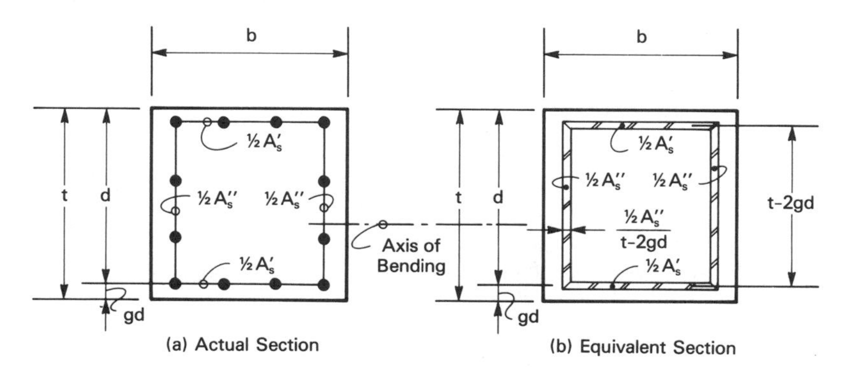


**Figure 8-4** Column Section.

An elevation of the column at the section is shown in Fig. 8-5. The line $OA$ represents the case of fully plastic deformation with a strain of 0.003. The yield strain of steel is shown as $0.003s$ and of concrete as $0.003c$. By definition, the yield strain of concrete is $0.85f'_c/E_c$. The two values of strain are equated and the result solved for $c$, yielding

$$c = 333 \frac{0.85f'_c}{E_c} \tag{8-8a}$$

Similarly for steel,

$$s = 333 \frac{f_y}{E_s} \tag{8-8b}$$

The section is now allowed to rotate up to line $OB$. At any angle of rotation within the sector 1, all steel and all concrete are in full plastic yield. Stresses are therefore unchanged regardless what angle of rotation is imposed; the section simply cannot develop any resistance to moment.

Anywhere within sector 1, the concentric axial force $P_{n1}$ required to maintain the state of strain is seen to be the same as the concentric load given by the ACI column formula:

$$P_{n1} = P_0 = 0.85f'_c(A_g - A_{st}) + f_y A_{st} \tag{8-9a}$$

Equivalently, in terms of steel ratios,

$$P_{n1} = \left[0.85f'_c(1 + g - \rho' - \rho'') + f_y(\rho' + \rho'')\right] \frac{bt}{1 + g} \tag{8-9b}$$

where $A_{st} = A'_s + A''_s$, $\rho = A_{st}/bd$, $\rho' = A'_s/bd$, $\rho'' = A''_s/bd$ and $d = t/(1 + g)$.

The second subscript for $P_{n1}$ denotes the sector number, a practice that will be followed hereafter to denote the sector. The superscripts (′) and (″) denote flange steel and web steel, respectively.

Again referring to Fig. 8-5, for the section to rotate beyond the line $OB$ into sector 2 requires that the steel on the right side emerge from compression yield. The force in that steel is thus reduced; the change produces a moment about the centerline of the section. In terms of the variable $k$, the stress in the steel in sector 2 is found by ratios to be

$$f_{s2} = 0.003E_s \frac{1 - k}{k} \tag{8-10}$$

where the sign is negative when stress is being reduced.

The change in force $\Delta P'_{s2}$ in sector 2 is the change in stress times the steel area, where the steel area is $\frac{1}{2}\rho' bd$:

$$\Delta P'_{s2} = \tfrac{1}{2}(f_y + f_{s2})\,\rho' b \frac{t}{1 + g} \tag{8-11}$$

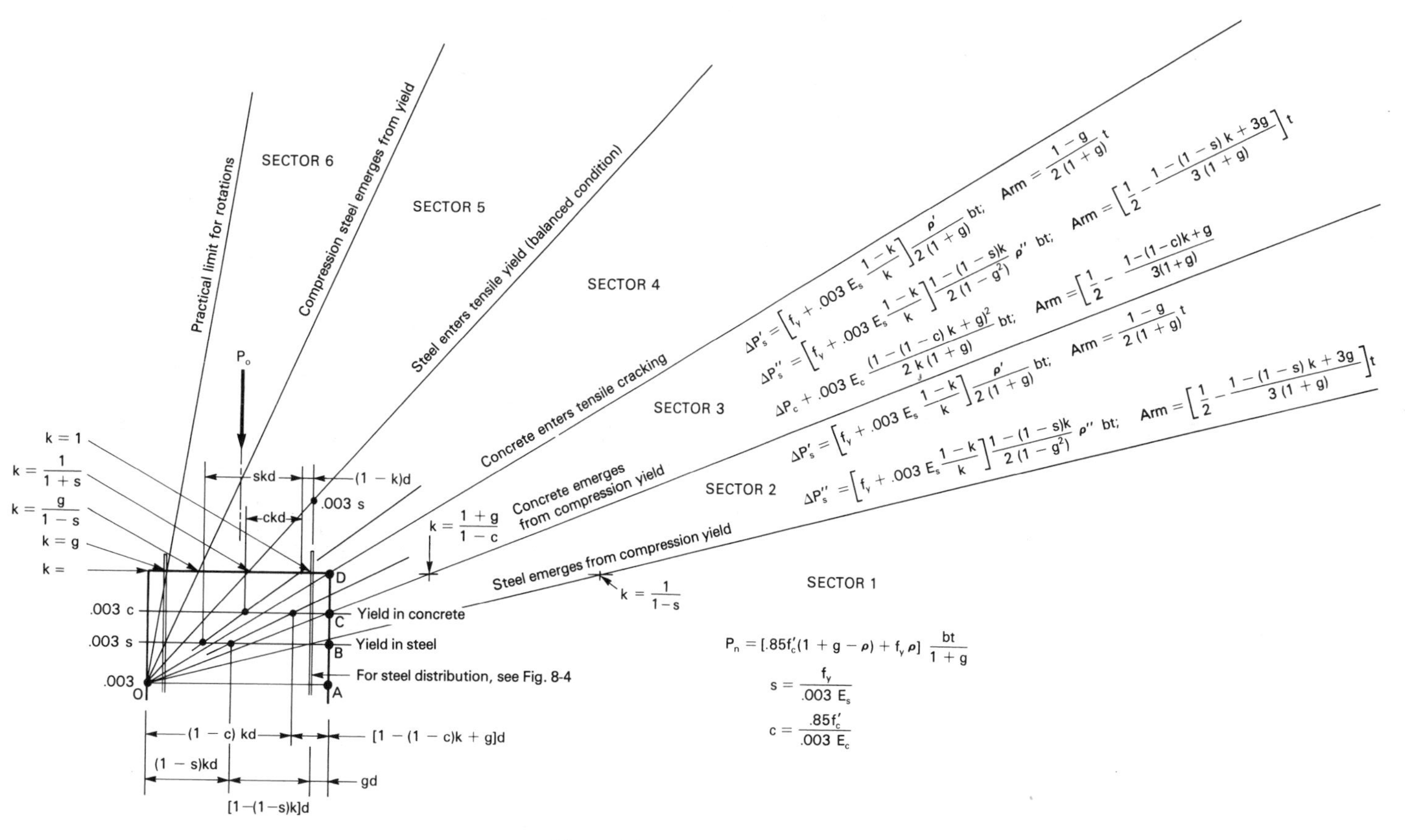


**Figure 8-5** Column Rotations.

**TABLE 8-1 SUMMARY OF COLUMN EQUATIONS**

<table>
<tr><th colspan="3">SECTOR</th><th>6</th><th>5</th><th>4</th><th>3</th><th>2</th></tr>
<tr><td rowspan="4">FLANGE STEEL</td><td rowspan="2">Fixed Side</td><td>$\Delta P'_s$</td><td>$0.003E_s \frac{g-k(1-s)}{2k(1+g)} \rho' bt$</td><td></td><td></td><td></td><td></td></tr>
<tr><td>Arm</td><td>$\frac{1-g}{2(1+g)} t$</td><td></td><td></td><td></td><td></td></tr>
<tr><td rowspan="2">Elev-ated side</td><td>$\Delta P'_s$</td><td colspan="2">$f_y \frac{\rho'}{(1+g)} bt$</td><td colspan="3">$\left[f_y + 0.003E_s \frac{1-k}{k}\right] \frac{\rho'}{2(1+g)} bt$</td></tr>
<tr><td>Arm</td><td colspan="2">$\frac{1-g}{2(1+g)} t$</td><td colspan="3">$\frac{1-g}{2(1+g)} t$</td></tr>
<tr><td rowspan="4">WEB STEEL</td><td rowspan="2">Tri-angle shape</td><td>$\Delta P''_s$</td><td colspan="2">$f_y \frac{2sk}{1-g^2} \rho'' bt$</td><td colspan="3">$\left[f_y + 0.003E_s \frac{1-k}{k}\right] \frac{1-(1-s)k}{2(1-g^2)} \rho'' bt$</td></tr>
<tr><td>Arm</td><td colspan="2">$\frac{2k(3+s)-3(1+g)}{6(1+g)} t$</td><td colspan="3">$\frac{1-3g+2k(1-s)}{6(1+g)} t$</td></tr>
<tr><td rowspan="2">Rect-angle shape</td><td>$\Delta P''_s$</td><td colspan="2">$2f_y \frac{1-(1+s)k}{1-g^2} \rho'' bt$</td><td></td><td></td><td></td></tr>
<tr><td>Arm</td><td colspan="2">$\frac{(1+s)k-g}{2(1+g)} t$</td><td></td><td></td><td></td></tr>
<tr><td rowspan="4">CONCRETE</td><td rowspan="2">Tri-angle shape</td><td>$\Delta P_c$</td><td colspan="3">$0.003E_c \frac{c^2k}{2(1+g)} bt$</td><td>$0.003E_c \frac{(1-(1-c)k+g)^2}{2k(1+g)} bt$</td><td></td></tr>
<tr><td>Arm</td><td colspan="3">$\frac{2k(3-c)-3(1+g)}{6(1+g)} t$</td><td>$\frac{1+g+2k(1-c)}{6(1+g)} t$</td><td></td></tr>
<tr><td rowspan="2">Rect-angle shape</td><td>$\Delta P_c$</td><td colspan="3">$0.003E_c c \frac{1+g-k}{1+g} bt$</td><td></td><td></td></tr>
<tr><td>Arm</td><td colspan="3">$\frac{k}{2(1+g)} t$</td><td></td><td></td></tr>
</table>

The moment arm of the force $\Delta P'_{s2}$ to centerline of column is given by

$$a'_{s2} = \frac{t}{2} - gd = \frac{1}{2}\left(\frac{1-g}{1+g}\right)t \tag{8-12}$$

Also in sector 2, there is a change in force on the steel in the web, $A''_s$. The maximum change in stress on the voided triangular stress block is seen to be the same as that derived for $f_{s2}$:

$$f_{s2} = 0.003E_s \frac{1-k}{k} \tag{8-13}$$

The total force is the area of the voided triangular stress block:

$$\Delta P''_{s2} = \frac{f_y + f_{s2}}{2} \rho'' \frac{1-(1-s)k}{1-g} b \frac{t}{1+g} \tag{8-14}$$

The moment arm for this force to the centerline of the column is

$$a''_{s2} = \left[\frac{1}{2} - \frac{1 - (1 - s)k + 3g}{3(1 + g)}\right] t \tag{8-15}$$

In sector 2, the final force on the section is then

$$P_{n2} = P_0 - \Delta P'_{s2} - \Delta P''_{s2} \tag{8-16}$$

and the final moment is

$$M_{n2} = \Delta P'_{s2} a'_{s2} + \Delta P''_{s2} a''_{s2} \tag{8-17}$$

The eccentricity ratio $e_{n2}/t$ is given by

$$\frac{e_{n2}}{t} = \frac{M_{n2}}{P_{n2}} \tag{8-18}$$

If these equations for $P_n$ and $M_n$ in sector 2 are written out completely, they become quite cumbersome. Even so, they can be solved readily by a small computer and the results tabulated for easy reference. Before introducing the final design tables, however, further development of these highly discontinuous equations is warranted.

Continuing with the development, as the rotation is increased beyond the line $OC$ into sector 3, a new discontinuity is introduced. A part of the concrete emerges from yield and enters the elastic range. It is noted that Eqs. (8-10) through (8-14) previously derived for sector 2 remain valid in sector 3, changing only their subscript from 2 to 3.

Proceeding as before, the stress in the concrete in sector 3 is found by ratios:

$$f_{c3} = 0.003E_c \frac{1 + g - k}{k} \tag{8-19}$$

where the sign is negative when the stress is being reduced. The voided area of the concrete stress block which has emerged from yield is found to be

$$A_{c3} = [1 - (1 - c)k + g] \frac{bt}{1 + g} \tag{8-20}$$

The change in force on the concrete is then the volume of the voided stress block:

$$\Delta P_{c3} = \tfrac{1}{2}(0.003cE_c + f_{c3}) A_{c3} \tag{8-21}$$

The moment arm of this force is

$$a_{c3} = \frac{t}{2} - \frac{1 - (1 - c)k + g}{3(1 + g)} t \tag{8-22}$$

The final force on the section is then

$$P_{n3} = P_0 - \Delta P'_{s3} - \Delta P''_{s3} - \Delta P_{c3} \tag{8-23}$$

The final moment is

$$M_{n3} = \Delta P'_{s3} a'_{s3} + \Delta P''_{s3} a''_{s3} + \Delta P_{c3} a_{c3} \tag{8-24}$$

The eccentricity ratio $e_{n3}/t$ is given by

$$\frac{e_{n3}}{t} = \frac{M_{n3}}{P_{n3}} \tag{8-25}$$

The foregoing procedure is repeated for each sector, using the equivalent section shown in Fig. 8-4 for all derivations. A new sector begins at each discontinuity; a discontinuity occurs whenever materials enter or emerge from yield and when concrete cracks in tension. A summary of the resulting equations is presented in reduced form in Table 8-1, (see p. 163).

When the concrete cracks in tension (sector 4 and beyond), the voided part of the stress block is no longer a triangle but becomes a rectangle plus a triangle. These two shapes are listed separately for the concrete in Table 8-1. The same shapes occur in the web steel when the steel enters tensile yield.

The expressions resulting from the foregoing analysis (Table 8-1) are unwieldly and complicated. Fortunately, the computer solves complicated expressions as readily as simple ones, with the end result of the computer output being the column tables, Tables A-8 through A-10. In developing these design tables, the values of $f'_c$, $f_y$, $A'_s$, and $A''_s$ were selected, then the value of $k$ was varied to produce the desired values of $e_n/t$; the ratio $R_n$ was the final computation.

The ratio $R_n$ defined in the column design tables is something of an artificial quantity. It is simply a convenient parameter suggested by the 1971 ACI design manual. It is dimensionless, so the values given in the design tables of the Appendix are valid either for Imperial (English) units or for SI units.

It should be observed that the lowest value of $e_n/t$ in the design tables is 0.1. As stated at the beginning of this section, the assumption of a minimum eccentricity of $0.1t$ is considered to be good practice. Eccentricities therefore begin at $e_n/t = 0.10$.

## EXAMPLES IN THE STRENGTH METHOD

The following examples illustrate the use of the column design tables, Tables A-8 through A-10. Comments are included within the examples to clarify assumptions or identify arbitrary choices. The first example is the design of a simple column to take only axial load, without moment.

**Example 8-1**

*Strength method*
*Design of a column subject only to axial load*
*No limitations on dimensions or reinforcement*

*Given:*
Apartment building, five floors
At a first-floor column, $L_u$ = 9 ft 0 in.
$P_{DL}$ = 229 kips, $P_{LL}$ = 201 kips
Grade 60 Steel, $f'_c$ = 4000 psi

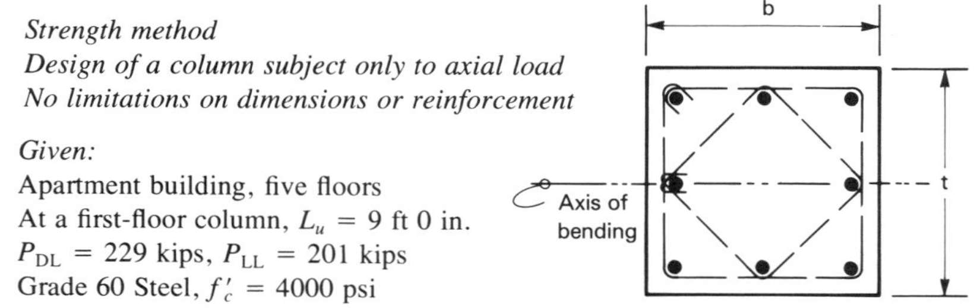

Compute the nominal ultimate load $P_n$:

$$P_n = \frac{1.4P_{DL} + 1.7P_{LL}}{\phi} = \frac{1.4 \times 229 + 1.7 \times 201}{0.7}$$

$$= 946 \text{ kips}$$

Choose an overall size and gross area (square column): Use the rule of thumb: $b = t = 12$ in. + 1 in. × stories above = 12 in. + 1 in. × 4. Try $t = 16$ in., $b = 16$ in.

Check the buckling criteria:

$$\frac{KL_u}{r} < 34 - 12\frac{M_1}{M_2} \qquad \frac{1 \times 9 \times 12}{0.3 \times 16} = 22.5 < 34 \quad \text{(O.K.)}$$

Compute $R_n$ and $e_n/t$:

$$R_n = \frac{P_n/bt}{0.85f'_c} = \frac{946{,}000/16 \times 16}{0.85 \times 4000} = 1.09$$

$$\frac{e_n}{t} = \text{minimum} = 0.1$$

From Table A-9, with steel uniformly distributed; use $A_s = 2.4\% = 0.024 \times 16 \times 16 = 6.14$ in$^2$.

From Table A-3, select 8-No. 8 bars.
Select the ties: For No. 8 bars, use tie size No. 3.

Spacing: < 16 bar diameters or < 16 in.
< 48 tie diameters or < 18 in.
< least dimension or < 16 in.

Use a column 16 in. × 16 in.; 8-No. 8 longitudinal bars, No. 3 ties at 16 in. o.c., as shown in the sketch above.

### Example 8-2

*Strength method*
*Design of a column subject only to axial load*
*One dimension restricted*

*Given:*
Same column as Example 8-1
One dimension limited to 12 in.

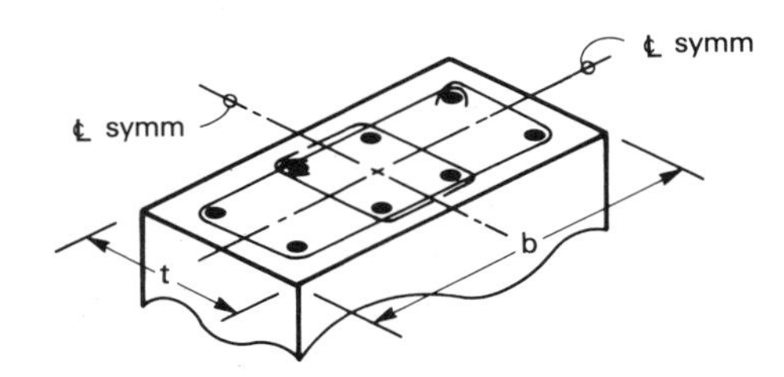

Compute the nominal ultimate load $P_n$:

$$P_n = 946 \text{ kips} \qquad \text{(same as Example 8-1)}$$

Choose an overall size and gross area. Use the rule of thumb for a square column: 16 in. × 16 in.

$$\text{gross area} = 16 \times 16 = 256 \text{ in}^2$$

With one side 12 in., try 12 in. × 22 in., $A_g = 264$ in$^2$.

Check the buckling criteria:

$$\frac{KL_u}{r} < 34 - 12\frac{M_1}{M_2} \qquad \frac{1 \times 9 \times 12}{0.3 \times 12} = 30 < 34 \quad \text{(O.K.)}$$

Compute $R_n$ and $e_n/t$:

$$R_n = \frac{P_n/bt}{0.85 f'_c} = \frac{946{,}000/12 \times 22}{0.85 \times 4000} = 1.054$$

$$\frac{e_n}{t} = \text{minimum} = 0.1$$

From Table A-9, with steel on two faces only; use 2% steel, $A_s = 0.02 \times 12 \times 22 = 5.28$ in$^2$.

From Table A-3, select 4 No. 8 and 4 No. 7 bars.

Select the ties: for No. 8 bars, use tie size No. 3.

Spacing: < 16 bar diameters or < 14 in. (for No. 7 bars)
< 48 tie diameters or < 18 in.
< least dimension or < 12 in.

Use 12 × 22-in. column, 4 No. 8 and 4 No. 7 bars, No. 3 ties at 12 in. o.c., No. 8 bars at the four corners.

**Example 8-3**

*Strength Method*
*Design of a column that is subject to flexure*
*Configuration limited to square sections*

*Given:*
Same criteria as Example 8-1
At top of column, add $M_{DL} = 65$ kip-ft, $M_{LL} = 48$ kip-ft
At bottom of column, add $M_{DL} = 33$ kip-ft, $M_{LL} = 24$ kip-ft

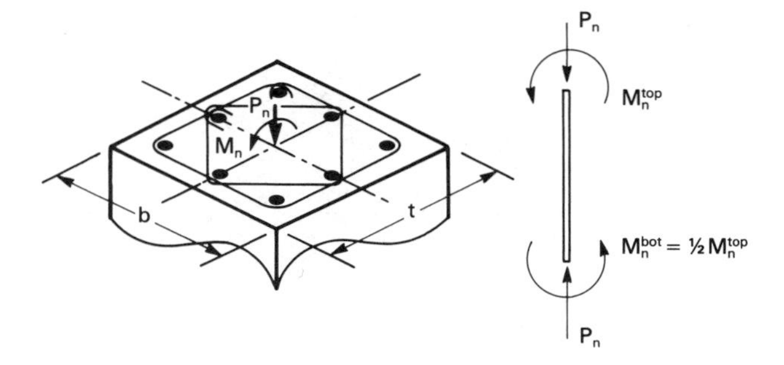


Compute the nominal ultimate loads $M_n$ and $P_n$:

$$P_n = 946 \text{ kips (from Example 8-1)}$$

$$\text{At top:} \quad M_n = \frac{1.4M_{\mathrm{DL}} + 1.7M_{\mathrm{LL}}}{\phi} = \frac{1.4 \times 65 + 1.7 \times 48}{0.7}$$

$$= 247 \text{ kip-ft}$$

$$\text{At bottom;} \quad M_n = 124 \text{ kip-ft}$$

Choose an overall size: $b = t = 16$ in. (same as Example 8-1).

Check the buckling criteria:

$$\frac{KL_u}{r} < 34 - 12\frac{M_1}{M_2} \qquad \frac{1 \times 9 \times 12}{0.3 \times 16} \overset{?}{<} 34 - 12 \times \frac{-124}{247}$$

$$22.5 < 40 \quad \text{(O.K.)}$$

Compute $R_n$ and $e_n/t$:

$$R_n = \frac{P_n/bt}{0.85f'_c} = \frac{946{,}000/16 \times 16}{0.85 \times 4000} = 1.09$$

$$\frac{e_n}{t} = \frac{M_n/P_n}{t} = \frac{247{,}000 \times 12/946{,}000}{16} = 0.196$$

From Table A-9, with steel uniformly distributed, select 4.8% steel (too high).

Since this column section is considered to be unacceptable, the trial size must be increased and the entire procedure repeated for this new trial size. An increment of 4 in. will be added, producing a trial section of 20-in square. For the new gross size, compute $R_n$ and $e_n/t$:

$$R_n = \frac{P_n/bt}{0.85f'_c} = \frac{946{,}000/20 \times 20}{0.85 \times 4000} = 0.70$$

$$\frac{e_n}{t} = \frac{M_n/P_n}{t} = \frac{247{,}000 \times 12/946{,}000}{20} = 0.16$$

From Table A-9, select $\rho < 1\%$ steel (too low).

Again the trial size is unacceptable. For the next trial, select 18-in square, halfway between the first and second trial sizes. For the new gross size, compute $R_n$ and $e_n/t$:

$$R_n = \frac{P_n/bt}{0.85 f'_c} = \frac{946{,}000/18 \times 18}{0.85 \times 4000} = 0.86$$

$$\frac{e_n}{t} = \frac{M_n/P_n}{t} = \frac{247{,}000 \times 12/946{,}000}{18} = 0.174$$

From Table A-9, select $\rho = 2.0\%$ (O.K.—use 2%).

$$A_s = 0.02 \times 18 \times 18 = 6.48 \text{ in}^2$$

From Table A-3, select 4-No. 9 and 4-No. 8 longitudinal bars.

Select the ties: for No. 9 longitudinal bars, use No. 3 ties.

Spacing: $<$ 16 bar diameters or $<$ 16 in. (least)

$<$ 48 tie diameters or $<$ 18 in.

$<$ least dimension or $<$ 18 in.

Use a column 18 in. × 18 in., 4-No. 9 and 4-No. 8, tie size No. 3 at 16 in. o.c., No. 9 bars at the four corners.

The final column section is that shown at the beginning of the example.

**Example 8-4**

*Strength Method*
*Design of a column that is subject to flexure*
*Depth of section limited on the weak axis of bending to a maximum of 11 in.*
*Given:*
*Same criteria as Example 8-1*
*At top of column, add* $M_{DL}$ = 50 kip-ft, $M_{LL}$ = 30 kip-ft
At bottom, add hinge, $M_{DL}$ = 0, $M_{LL}$ = 0

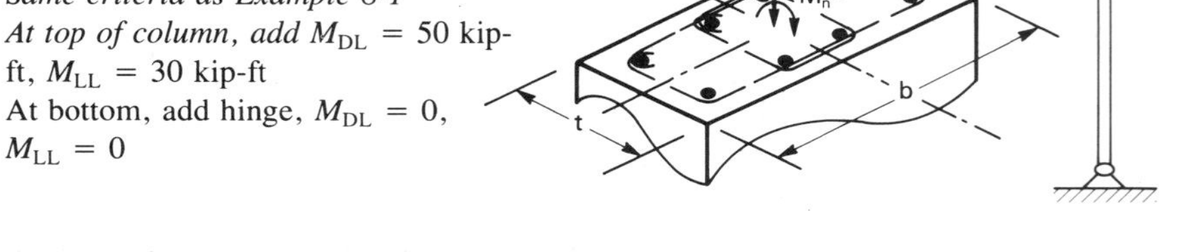


Compute the nominal ultimate loads $P_n$ and $M_n$:

$$P_n = 946 \text{ kips} \qquad \text{(see Example 8-1)}$$

$$M_n = \frac{1.4 M_{DL} + 1.7 M_{LL}}{\phi} = \frac{1.4 \times 50 + 1.7 \times 30}{0.7}$$

$$= 173 \text{ kip-ft}$$

Choose an overall size and gross area for a square column. From the rule of thumb, try 16 × 16 in., $A_g = 256 \text{ in}^2$.

Try a column size 11 in. × 24 in., $A_g = 264 \text{ in}^2$.

Check the buckling criteria:

$$\frac{KL_u}{r} < 34 - 12\frac{M_1}{M_2} \qquad \frac{1 \times 108}{0.3 \times 11} \overset{?}{<} 34 - 12 \times \frac{0}{173}$$

$$33 < 34 \quad \text{(O.K.)}$$

Compute $R_n$ and $e_n/t$:

$$R_n = \frac{P_n/bt}{0.85 f'_c} = \frac{946{,}000/11 \times 24}{0.85 \times 4000} = 1.054$$

$$\frac{e_n}{t} = \frac{M_n/P_n}{t} = \frac{173{,}000 \times 12/946{,}000}{11} = 0.20$$

From Table A-9, steel on two faces, read $\rho$ of almost 4%. (no good—try a larger section with lower $\rho$.)

Try a section 11 in. × 30 in. For the new trial size, compute $R_n$ and $e_n/t$:

$$R_n = \frac{P_n/bt}{0.85 f'_c} = \frac{946{,}000/11 \times 30}{0.85 \times 4000} = 0.84$$

$$\frac{e_n}{t} = \frac{M_n/P_n}{t} = \frac{173{,}000 \times 12/946{,}000}{11} = 0.20$$

From Table A-9, with steel at two faces, read $\rho$ of 2%.

Compute the steel area:

$$A_s = \rho bd = 0.02 \times 11 \times 30 = 6.60 \text{ in}^2$$

From Table A-3, select 4-No. 9 and 4-No. 8 bars.

Select the ties: for No. 9 bars, use tie size No. 3.

Spacing: < 16 bar diameters or < 16 in. (for No. 8 bars)

< 48 tie diameters or < 18 in.

< least dimension or < 11 in.

Use a column 11 in. × 30 in., 4-No. 9 and 4-No. 8 bars, No. 3 ties at 11 in. o.c.

The final section is shown in the sketch at the beginning of the example.

A summary of the first four examples is tabulated below.

| Example | $P_n$ (kips) | $M_n$ (kip-ft) | $b$ (in.) | $t$ (in.) | $e/t$ | $A_s$ (in$^2$) |
|---|---|---|---|---|---|---|
| 8-1 | 946 | 0 | 16 | 16 | 0.10 | 6.14 |
| 8-2 | 946 | 0 | 22 | 12 | 0.10 | 5.28 |
| 8-3 | 946 | 247 | 18 | 18 | 0.17 | 6.48 |
| 8-4 | 946 | 173 | 30 | 11 | 0.20 | 6.60 |

In these four examples, the axial load was taken at the same value each time and other factors were then varied. Comparing Example 8-3 with Example 8-1, it is noted that adding a significant amount of moment had only a small effect on the size of the column and its reinforcement. Comparing Example 8-2 with Example 8-4, it is seen that rather drastic changes in one dimension does not change the total area $b \times t$ by a large amount, and even adding a large moment in the shallow direction changes the area of steel only 20%. It is concluded that *for braced frames*, the axial load dominates to such an extent that other factors become relatively minor.

When axial loads become small, as in one- or two-story buildings, very often the minimum-sized column sections must be used. The next example lists the allowable loads on minimum-sized columns.

**Example 8-5**

*Strength method*
*Design of columns subject only to axial load*
*Minimum dimensions and minimum reinforcement*

Determine the nominal axial load $P_n$ for column sections fabricated using $f'_c$ of 3000, 4000, and 5000 psi, and for grades 40, 50, and 60 steels. Tabulate the results for $e_n/t = 0.1$ for minimum or no moment. Use $A_s$ of 1%, steel uniformly distributed. Read $R_n$ from the Appendix tables and compute $P_n = 0.85 f'_c btR_n$.

| | | | $P_n$ (kips) | |
|---|---|---|---|---|
| $f_y$ (ksi) | $f'_c$ (psi) | $R_n$ | 10 in. × 10 in. $A_g = 100$ in$^2$ | 12 in. × 12 in. $A_g = 144$ in$^2$ |
| 40 | 3000 | 0.91 | 232 | 334 |
| 40 | 4000 | 0.87 | 296 | 426 |
| 40 | 5000 | 0.85 | 361 | 520 |
| 50 | 3000 | 0.93 | 237 | 341 |
| 50 | 4000 | 0.89 | 303 | 436 |
| 50 | 5000 | 0.87 | 370 | 532 |
| 60 | 3000 | 0.96 | 245 | 353 |
| 60 | 4000 | 0.91 | 309 | 446 |
| 60 | 5000 | 0.89 | 378 | 545 |

Without incurring load reductions due to length:

Maximum length of a 10-in. column: 100 in. = 8 ft 4 in.

Maximum length of a 12-in. column: 120 in. = 10 ft 0 in.

Typical loads per story can be expected to be:

$$P_w \text{ per story} = 30 \text{ tons} \quad \text{or} \quad 60 \text{ kips}$$

$$P_n \text{ per story} = P_w \times 2.5 = 150 \text{ kips}$$

From an examination of the tabulation, it is seen that the allowable load on a minimum-sized column should be adequate for a two- or three-story building, even where nominal moments are present. For lengths longer than the indicated maximum lengths, Code requires significant reductions in the allowable loads.

The reverse of the design problem can occur occasionally, when the size of the column and its reinforcement are known, and it is one of the loads that is to be determined. Since the loads $M_n$ and $P_n$ are interrelated, one of the loads must be known, else there are an infinite number of combinations that could be solutions. In most cases it is the axial load $P_n$ that is known and it is desired to find how much moment $M_n$ the section can take in addition to the axial load; the next example is such a case.

**Example 8-6**

*Strength method*
*Investigation of a section to find the working moment, given the axial load*

Determine the allowable moment at working levels for the section shown in the following sketch, where the axial load is given.

*Given:*
Section as shown
$P_{DL} = 86$ kips
$P_{LL} = 70$ kips
Grade 60 steel
$f'_c = 4000$ psi

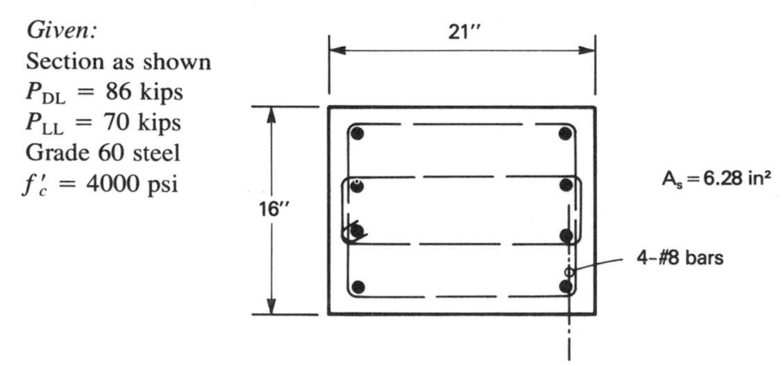


Compute the nominal ultimate load $P_n$:

$$P_n = \frac{1.4P_{DL} + 1.7P_{LL}}{\phi} = \frac{1.4 \times 86 + 1.7 \times 70}{0.7}$$

$$= 342 \text{ kips}$$

Compute $R_n$ and the steel ratio $\rho$:

$$R_n = \frac{P_n/bt}{0.85f'_c} = \frac{342{,}00/16 \times 21}{0.85 \times 4000} = 0.30$$

$$\rho = \frac{A_s}{bt} = \frac{6.28}{16 \times 21} = 0.019$$

From Table A-9, read $e_n/t = 0.78$.

Compute $M_n$ from the $e_n/t$ ratio:

$$\frac{e_n}{t} = \frac{M_n/P_n}{t} \qquad 0.78 = \frac{M_n/342{,}000}{21}$$

Solve for $M_n$: $M_n = 5602$ kip-in. $= 467$ kip-ft.

Use a factor of safety of 2.5 to obtain $M_w$:

$$\text{working moment } M_w = \frac{467 \text{ kip-ft}}{2.5} = 187 \text{ kip-ft}$$

## ELASTIC ANALYSIS

In the elastic range, the analysis of a column section is much like that of a beam section except that an axial force has been added and there may be additional reinforcement in the middle of the section. A typical column section is shown in Fig. 8-6, with the load components $P_w$ and $M_w$ representing the axial force and moment at working levels. Similar to the strength method, the web steel is replaced by an imaginary strip of steel having a finite width as shown.

A discontinuity is introduced when the value of $k$ is equal to $1 + g$. For higher values of $k$, all concrete is effective; for lower values of $k$, the concrete in tension is assumed to be cracked and ineffective. Two sets of equations are therefore required to bridge the discontinuity.

For $k$ greater than $(1 + g)$, the stress diagrams of Fig. 8-6 are valid. The sum of forces yields an equation for the ratio $R_w$:

$$R_w = \frac{P_w/bt}{f_c} = \frac{2k - 1 - g}{2k\,(1 + g)}\left[1 + g + (2n - 1)\,(\rho' + \rho'')\right] \qquad (8\text{-}26)$$

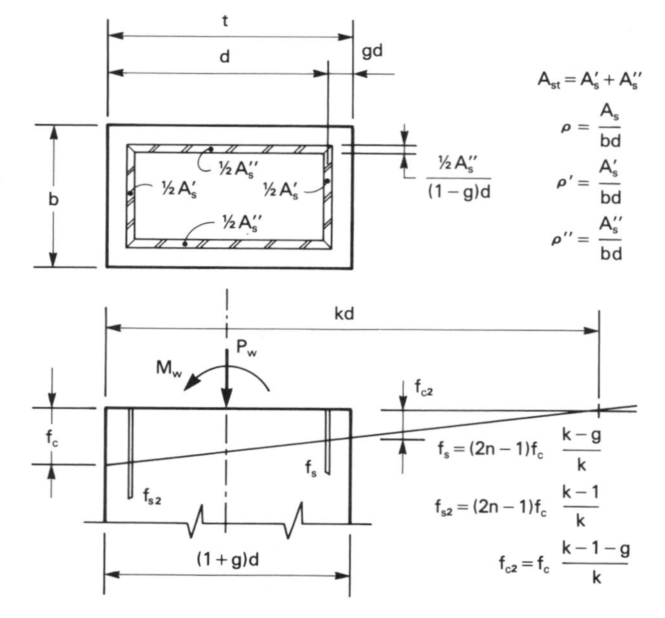


**Figure 8-6** Elastic Column Stresses, k > (1 + g)

The sum of moments about the centerline yields a similar equation in moments:

$$R_m = \frac{M_w/bt^2}{f_c} = \frac{1}{12k}\left[1 + g + (2n - 1)(3\rho' + \rho'')\frac{(1 - g)^2}{(1 + g)^2}\right] \tag{8-27}$$

Dividing Eq. (8-26) by Eq. (8-27) yields a solution for the eccentricity ratio:

$$\frac{e_w}{t} = \frac{R_m}{R_w} \tag{8-28}$$

For $k$ less than $(1 + g)$, the state of stress is shown in Fig. 8-7. For those stresses, the sum of horizontal forces yields the solution for $R_w$:

$$R_w = \frac{P_w/bt}{f_c} = \frac{1}{2k(1 + g)}\left[k^2 + (2n - 1)\rho'(k - g) + (2n - 1)\rho''\frac{(k - g)^2}{1 - g} - n\rho''\frac{(1 - k)^2}{1 - g} - n\rho'(1 - k)\right] \tag{8-29}$$

The sum of moments about centerline yields a similar equation in moments:

$$R_m = \frac{M_w/bt^2}{f_c} = \frac{1}{4k(1 + g)^2}\left[\frac{k^2}{3}(3 + 3g - 2k) + n\rho'(1 - k)(1 - g) + (2n - 1)\rho'(k - g)(1 - g) + (2n - 1)\rho''\frac{(k - g)^2}{3(1 - g)}(3 - g - 2k) + n\rho''\frac{(1 - k)^2}{3(1 - g)}(1 - 3g + 2k)\right] \tag{8-30}$$

For this range of $k$, the eccentricity ratio is again given by

$$\frac{e_w}{t} = \frac{R_m}{R_w} \tag{8-31}$$

The actual numerical values of the stress ratio $R_w$ are tabulated in the column tables of the Appendix, Tables A-8 through A-10, at the lower portion of each table. In developing the table, the value of $k$ was varied to produce the desired value of $e_w/t$ [Eq. (8-28) or (8-31)]; then the value of $R_w$ corresponding to that $e_w/t$ was listed [Eq. (8-26) or (8-29)]. The tabulation is seen to be independent of $f_y$, the yield stress of steel; it applies to all grades of steel.

The tables are set up in terms of two parameters, $e_w/t$ and $R_w$. The stress ratio $R_w$ is analogous to the stress ratio $R_n$ introduced earlier in the strength method; $R_w$, of course, uses working load and working stress, whereas $R_n$ uses nominal ultimate load and yield stress. For easy reference, both $R_n$ and $R_w$ are defined at the top left corner of each table.

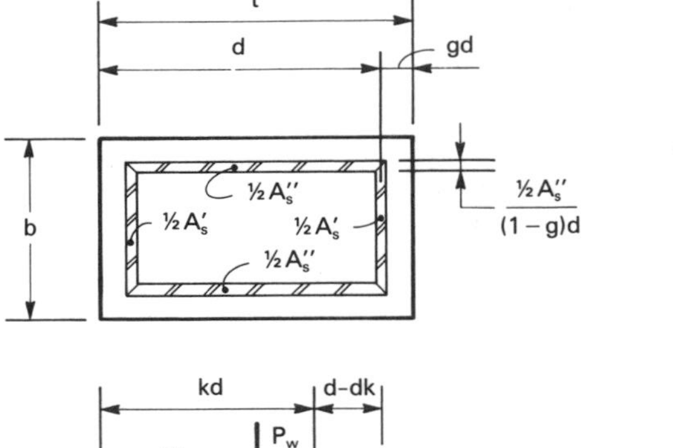


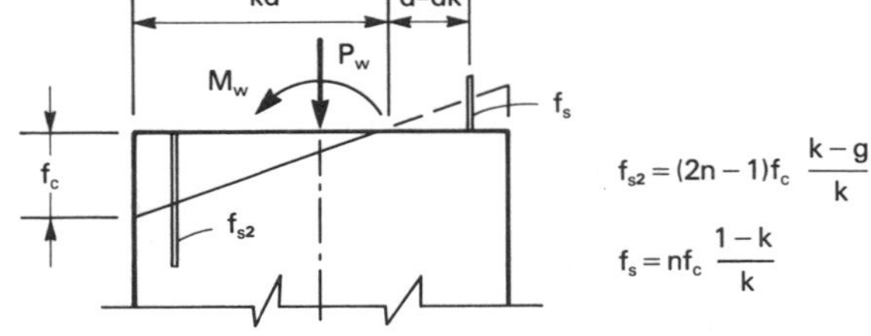


**Figure 8-7** Elastic Column Stresses, k < (1 + g)

The stress ratio $R_w$ affords a means to determine the compressive stress in the concrete during service conditions. The stress $f_c$ is the maximum working stress in the concrete and includes both axial compression plus flexure. When $R_w$, $P_w$ and the overall dimensions $b$ and $t$ are known, the service stress can be computed.

When investigating the service stress as it affects brittle finishes or coverings, it should be borne in mind that creep and shrinkage in concrete affect strain but not stress. When the elastic stress is known, the related elastic strain is found simply by dividing elastic stress by the modulus of elasticity:

$$\text{elastic } \epsilon_c = \frac{f_c}{E_c} \tag{8-32}$$

The total strain must then be computed to include both elastic strain and inelastic creep and shrinkage. As discussed in Chapter 3, the inelastic strain may be as much as the elastic strain; hence the total strain may be as much as twice the elastic strain,

$$\text{total } \epsilon_c \leq 2\,\frac{f_c}{E_c} \tag{8-33}$$

The additional strain can have severe effects on finishes and coatings.

## EXAMPLES IN THE ELASTIC RANGE

The Appendix tables include provisions for the analysis of columns in the elastic range. An indication of the general performance of a column during its day-to-day working life can therefore be obtained. The procedure is directly parallel to that for the plastic range, as illustrated in the following example.

**Example 8-7**

*Elastic analysis of columns*
*Investigation of service stresses*
*Column subject to axial load only*

For the column of Example 8-1, determine the stress in concrete under working loads.

*Given:*
$P_{DL} = 320$ kips
$P_{LL} = 200$ kips
Grade 60 steel
$f'_c = 4000$ psi

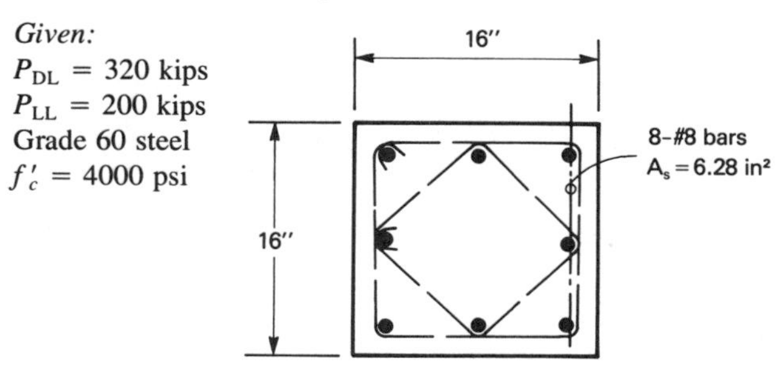


Compute the working load $P_w$:

$$P_w = P_{DL} + P_{LL} = 230 + 200 = 430 \text{ kips}$$

Compute the steel ratio $\rho$:

$$\rho = \frac{A_s}{bt} = \frac{6.28}{16 \times 16} = 0.025,$$

From Table A-9, using $e_w/t = 0.1$ to account for any accidental eccentricities, read $R_w = 0.89$.

Compute the service stress:

$$R_w = \frac{P_w/bt}{f_c} \qquad 0.89 = \frac{430{,}000/16 \times 16}{f_c}$$

Solve for $f_c$: $f_c = 1887$ psi.

The service stress in concrete is 1887 psi, or 47% of $f'_c$.

It should be recognized that the computed stress $f_c$ in Example 8-7 includes the effects of long-term creep and shrinkage. Until the long-term effects actually happen, the computed value may be considerably in error. Investigation of service stresses can serve a very practical purpose, as demonstrated in the next example.

**Example 8-8**

*Elastic analysis of columns*
*Investigation of service stresses*
*Column subject to axial load plus flexure*

The column shown below is to be clad with sculptured marble; the marble is to be rigidly attached to the concrete. Determine the highest stress that could occur in the marble at normal service levels. The modulus of elasticity of the marble is $6 \times 10^6$ psi.

*Given:*
Section as shown
$P_{DL} = 230$ kips
$P_{LL} = 200$ kips
$M_{DL} = 66$ kip-ft
$M_{LL} = 47$ kip-ft
Grade 60 steel
$f'_c = 4000$ psi
$E_c = 57{,}000\sqrt{f'_c} = 3{,}600{,}000$ psi

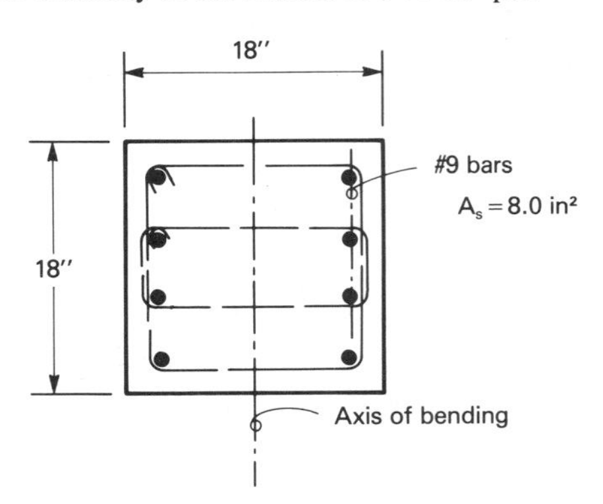


Compute the working loads $P_w$ and $M_w$:

$$P_w = P_{DL} + P_{LL} = 230 + 200 = 430 \text{ kips}$$

$$M_w = M_{DL} + M_{LL} = 66 + 47 = 113 \text{ kip-ft}$$

Compute $\rho$ and $e_w/t$:

$$\rho = \frac{A_s}{bt} = \frac{8.00}{18 \times 18} = 0.0247, \text{ say } 2.5\%$$

$$\frac{e_w}{t} = \frac{M_w/P_w}{t} = \frac{113{,}000 \times 12/430{,}000}{18} = 0.18, \text{ say } 0.2$$

From Table A-9, read $R_w = 0.69$.

Compute $f_c$:

$$R_w = \frac{P_w/bt}{f_c} \qquad 0.69 = \frac{430{,}000/18 \times 18}{f_c}$$

Solve for $f_c$, $f_c = 1923$ psi (maximum).

Compute the elastic strain in the concrete:

$$\epsilon_c = \frac{f_c}{E_c} = \frac{1923}{3.6 \times 10^6} = 0.000534 \text{ in./in.}$$

The column tables of the Appendix include the effects of creep and shrinkage in the concrete. The computed elastic strain of 0.000534 in./in. cannot occur, therefore, until an additional inelastic strain of 0.000534 in./in. has also occurred. The total strain at the surface of the concrete is thus the sum of the two,

$$\epsilon_c = 0.000534 + 0.000534 = 0.00107 \text{ in./in.}$$

The stress in the marble can now be computed from this total strain. Since the marble is rigidly attached to the concrete, the total strain in the marble $\epsilon_m$ will be equal to the total strain in the concrete; the stress can therefore be found from Hooke's law.

$$\epsilon_m = \epsilon_c = 0.00107 \text{ in./in.}$$

$$f_m = E_m \epsilon_m = 6 \times 10^6 \times 0.00107 = 6420 \text{ psi}$$

The maximum stress that can occur in the marble is 6420 psi.

It should be recognized that the computed stress in the marble in Example 8-8 is quite high, probably approaching the ultimate strength of the marble. Not all of it is likely to occur, however, since a large part of the inelastic strain in the concrete will occur during construction. As construction progresses, the dead load increases and much of the creep that is going to occur will occur progressively during that time. Attaching the marble would, of course, be one of the last items on the construction schedule.

Nonetheless, the elastic strain by itself induces a stress in the marble of 3210 psi (half the total), which is still a significant stress. A great deal of attention will have to be paid to the attachments, or the marble may separate from its concrete backing. Such failures do occur.

## DISCUSSION OF THE STRENGTH ANALYSIS

An important assumption was made in the section on strength analysis that is not included in the conditions listed at the beginning of that section: It was tacitly assumed that when the concrete emerges from yield and enters its elastic range (Fig. 8-5), it follows the idealized stress–strain curve of Fig. 1-4. The assumption is significant, since part of the concrete must behave elastically whenever moment occurs on the section. The accuracy of the ultimate load analysis is therefore dependent to some degree on assumptions regarding the elastic behavior of concrete.

With regard to these assumptions about elastic behavior, the Code permits any stress–strain relationship to be used that yields results in substantial agreement with test results. The idealized curve of Fig. 1-4 is seen to fulfill this requirement since the results of the foregoing analysis are in substantial agreement with published results.

As stated earlier, Code requires that the maximum strain in the concrete at failure load shall be 0.003 in./in. The Code makes no further requirements regarding any other strains, such as shrinkage strains or creep strains. In the foregoing analysis, the strain of 0.003 was assumed to include all contributions to strain, whether elastic or inelastic.

As a separate item, the strength analysis yields the values of the two interrelated loads, $P_n$ and $M_n$, throughout much of their range of values. If these values were plotted, the resulting graph would be an "interaction diagram," showing graphically how the two loads are interrelated. ACI publishes a comprehensive (and expensive) set of such interaction diagrams covering the most common arrangements of steel and types of concrete.

The tables of the Appendix, although limited, will provide values of $e_n/t$ and $P_n/bt$ adequate for the design of small projects. If the project is so large that the tables are inadequate, the project is certainly beyond the intentions of this book. For larger projects, a great deal more study of columns is required than can be presented in an introductory volume such as this.

## DISCUSSION OF THE ELASTIC ANALYSIS

The accuracy of the elastic stresses computed using the column tables of the Appendix is of course subject to question. For example, it was assumed that stress in compressive steel was doubled due to shrinkage and creep. If the actual shrinkage and creep ever reach the assumed values, the stress in the compression steel will double, as assumed. Until that happens, if ever, the computed value of $f_c$ will always be in error, but always on the "safe" side.

As a second point, the accuracy of the idealized stress–strain curve used in the analysis was discussed in the preceding section. Although its accuracy may be somewhat uncertain, it was concluded there that it was at least accurate enough that it does not detract from reasonable agreement with test results. Its accuracy is not diminished by being used again in the elastic analysis.

Due to these and other uncertainties and approximations, the accuracy of the elastic analysis is somewhat unreliable. The use of the elastic tables as a means to design columns is therefore not recommended. It is intended that the elastic tables of the Appendix should be used only to obtain a reasonable indication of stresses and strains at working levels; there is no other method sanctioned by the Code to investigate these stresses and strains.

## REVIEW QUESTIONS

1. In the working stress method for design of columns, how is the allowable axial working load $P_w$ obtained?
2. Why should the factor of safety be higher for columns than for beams?
3. What purpose do lateral ties serve in a reinforced concrete column?
4. What is the minimum required steel area for longitudinal reinforcement in a column?
5. What is the maximum allowable steel area for longitudinal reinforcement in a column?
6. Why is the steel area in columns usually held to less than 4% of the gross concrete area $bt$?
7. What is the radius of gyration, and how is it computed?
8. What is the slenderness ratio, and how is it used?
9. Describe the state of stress in a column section when it is loaded by the load $P_0$ as defined by the ACI column formula. What is the magnitude of the strain in the concrete under this load?
10. In the strength method, at what axial load does a member change from a column carrying a flexural load to a beam carrying an axial load?

11. How are end moments on a column accounted for when using the ACI column formula?
12. How is the effect of slenderness of a column accounted for when using the ACI column formula?
13. The ACI column formula is expressed only in terms of axial load. How is a moment accommodated?
14. Why aren't columns designed elastically?

## PROBLEMS

Select a column section and its reinforcement for the conditions indicated below. $f'_c = 4000$ psi, grade 50 steel.

| Problem | $P_{DL}$ (kips) | $P_{LL}$ (kips) | $M_{DL}$ (kip-ft) | $M_{LL}$ (kip-ft) | $L_u$ ft | Limits on $t$ (in.) |
|---|---|---|---|---|---|---|
| **1** | 170 | 160 | 0 | 0 | 10′-0″ | None |
| **2** | 190 | 270 | 0 | 0 | 13′-4″ | 16 |
| **3** | 215 | 180 | 85 | 75 | 10′-0″ | None |
| **4** | 280 | 225 | 110 | 75 | 13′-4″ | 16 |

Select a column section and its reinforcement for the conditions indicated below. $f'_c = 5000$ psi, grade 40 steel.

| Problem | $P_{DL}$ (kips) | $P_{LL}$ (kips) | $M_{DL}$ (kip-ft) | $M_{LL}$ (kip-ft) | $L_u$ ft | Limits on $t$ (in.) |
|---|---|---|---|---|---|---|
| **5** | 135 | 205 | 90 | 60 | 13′-4″ | None |
| **6** | 225 | 205 | 90 | 60 | 16′-8″ | None |
| **7** | 315 | 225 | 220 | 120 | 20′-0″ | None |
| **8** | 405 | 225 | 220 | 120 | 16′-8″ | None |

Select a column section and its reinforcement for the conditions indicated below. $f'_c = 3000$ psi, grade 60 steel.

| Problem | $P_{DL}$ (kips) | $P_{LL}$ (kips) | $M_{DL}$ (kip-ft) | $M_{LL}$ (kip-ft) | $L_u$ ft | Limits on $t$ (in.) |
|---|---|---|---|---|---|---|
| **9** | 90 | 180 | 232 | 214 | 10′-8″ | None |
| **10** | 90 | 160 | 111 | 133 | 10′-8″ | 16 |
| **11** | 135 | 135 | 111 | 140 | 10′-8″ | None |
| **12** | 135 | 180 | 160 | 133 | 10′-8″ | 16 |

13. Derive the expressions given in Table 8-1 for $\Delta P_c$ and its moment arm in sector 4.
14. Determine the actual service stress in the columns of Problems 2, 6, and 10.
15. Determine the maximum service strain that could occur anywhere in the concrete of the columns in Problems 1, 5, and 9.

# 9

# Continuity

## GENERAL

In Chapters 4 through 8 the design of concrete members was limited to that of selected parts of the total solution, dealing with only one aspect of the design at any one time. No attempt was made to include other effects or even to see what influence such effects might have on the problem in question. Although such an approach is useful in presenting a particular topic, a comprehensive look at the design must eventually be taken.

Such a comprehensive look at two typical design problems is presented in this chapter, to include the inherent interrelationships between flexure, shear, and bond. The fragmented solutions of the foregoing chapters are combined into a complete design, first for a relatively simple beam-and-slab roof and second for a structural frame in concrete. Only the structural design is considered; no attempt is made to present the architectural features.

Shear and moment envelopes have been used superficially in earlier chapters. They are used in greater detail in the comprehensive examples of this chapter; a review of these envelopes is included in the next section.

Included also in this chapter is an approximate method of structural analysis of continuous slabs, beams, and frames. This approximate analysis of certain types of regular modularized beams and frames is widely used throughout the industry, not only for concrete structures but for continuous beams and frames of any material. It is particularly appropriate for concrete, however, since so many concrete structures are exactly the type of regular modularized structures that are suited to this type of ''blanket'' analysis.

## SHEAR AND MOMENT ENVELOPES

The concept of shear and moment envelopes is used extensively in the design of concrete structures. The moment envelope, when accurately drawn, may be scaled to determine the cutoff points for longitudinal reinforcement. The shear envelope may be similarly scaled to determine the limiting points where stirrups are required.

For the sake of simplicity in the following presentation, only uniform loads will be considered, although the same concepts apply regardless of the type of load. Also, only working levels of dead load and live load will be considered throughout the structural analysis. The shears and moments due to dead loads are kept separate from those due to live loads, since they must be used separately later in the design of columns and again in the design of the foundations.

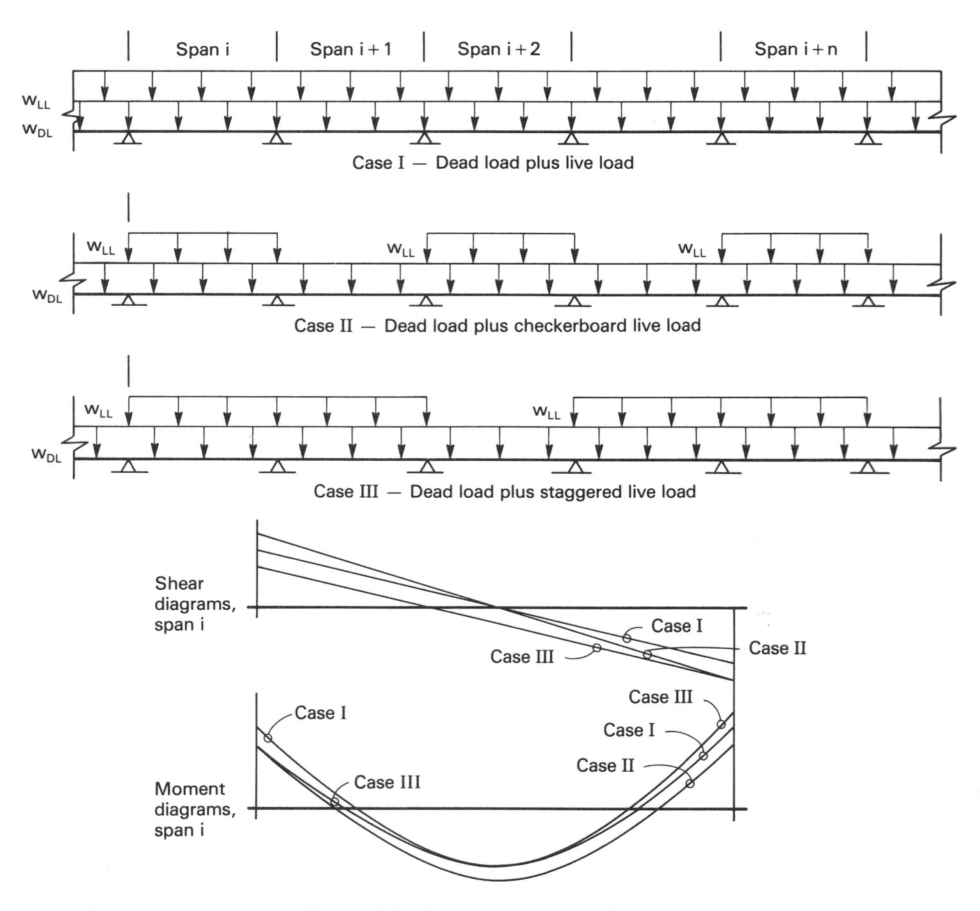


**Figure 9-1** Shear and Moment Diagrams.

As indicated in Fig. 9-1, a beam may be subject to various combinations of load, any one of which may occur in the life of the structure. Each combination of load produces its own shear and moment diagrams, as indicated by the shear and moment diagrams of Fig. 9-1. Depending on the circumstances, there could be numerous such combinations. It would indeed be a tedious pursuit to check every load case through every point on the beam to assure that the highest value of load has been accounted for. Rather, the shear and moment diagrams are drawn reasonably accurately to scale, and the outermost values of these diagrams at any point form the "envelope" of shears or moments which the beam must sustain. It should be noted that the envelopes for Fig. 9-1 are different for every span.

It is well to note at this point that in concrete design, the moment diagram is drawn on the "tension side" of the beam, opposite to the sign convention of basic strength of materials. The reason should be obvious: The reinforcement goes on the side where the moment diagram falls. The cutoff points for reinforcement determined in Chapter 7 have already demonstrated the usefulness of the convention. In addition, the moment diagram for this convention corresponds more closely to the elastic curve of the beam, providing a suggestion of the curvatures.

For timber or steel beams, a moment envelope is rarely required since the strength of the beam is constant throughout the span and is equal for both positive and negative moments. Once the peak value is known, either for positive or negative moment, the beam can be sized for that value and it is then known to be adequate for all other points in the span. In concrete, however, where strength of a beam varies across the span and where strength under positive moment may not be equal to strength under negative moment, the capacity of the beam at every point of the span must be assured.

If the proper load cases are considered, the moment envelope of Fig. 9-1 will contain the maximum negative moment at the supports and the maximum positive moment at midspan. Given only these numerical values at a later time, the envelopes for uniform loading can then be reconstructed using very simple sketching techniques. The following observations concerning shear and moment diagrams have been found useful in making such sketches of both shear and moment envelopes. Although approximate, the resulting envelopes are accurate enough for final design of concrete members.

The shear and moment diagrams for a simply supported beam under uniform loading are shown in Fig. 9-2. The maximum value of moment is $wL^2/8$ and occurs at

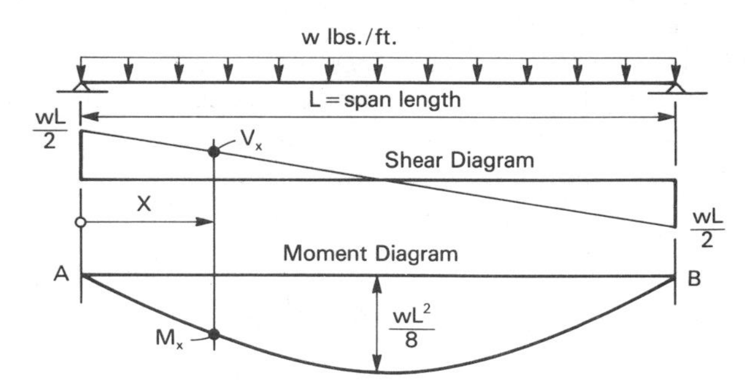


**Figure 9-2** Simply Supported Beam.

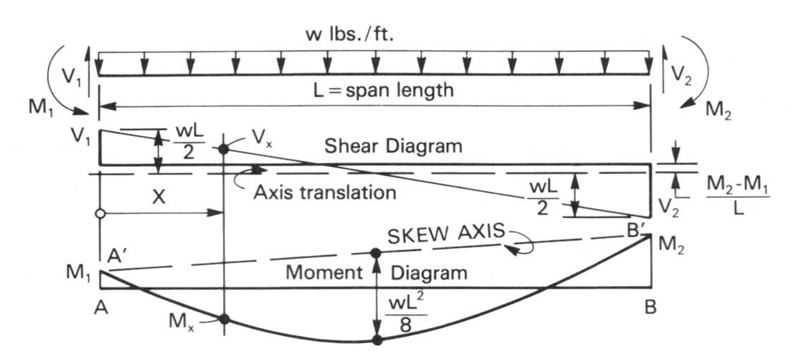


**Figure 9-3** One Span of a Continuous Beam.

midspan. The moment at any point $x$ is found by summing moments to the left of the point $x$:

$$M_x = \tfrac{1}{2}wLx - \tfrac{1}{2}wx^2 = \tfrac{1}{2}wx(L - x). \tag{9-1}$$

Similarly for continuous beams, the shear and moment diagrams for one span of a continuous beam are shown in Fig. 9-3. As before, the value of moment at any point $x$ is found by summing moments to the left of point $x$, assuming $M_2 > M_1$,

$$M_x = -\left[(M_2 - M_1)\frac{x}{L} + M_1\right] + \frac{1}{2}wx(L - x) \tag{9-2}$$

The second term in Eq. (9-2) is seen to be identical to the simple span moment of Eq. (9-1). The first term is seen to be the equation of a straight line drawn from $M_1$ to $M_2$ at the supports, shown as $A'B'$ in Fig. 9-3. The final moment diagram for this interior span of a continuous beam is thus the moment diagram of a simple beam drawn from the axis $A'B'$ rather than the zero axis $AB$. Even though the horizontal axis $A'B'$ is skewed, all moments are still measured vertically.

It should also be recognized that the shear diagram of Fig. 9-3 is the shear diagram of a simple span, translated up or down by the distance $(M_2 - M_1)/L$. Note that for shear diagrams the horizontal axis is only translated; it is not skewed.

The foregoing observations will be used extensively in sketching the shear and moment envelopes in subsequent discussions. The following examples demonstrate the procedure that will be used when only key values are known.

**Example 9-1**

*Shear and moment envelopes*
*Key values given (negative moments only) for sketching the shear and moment diagrams*

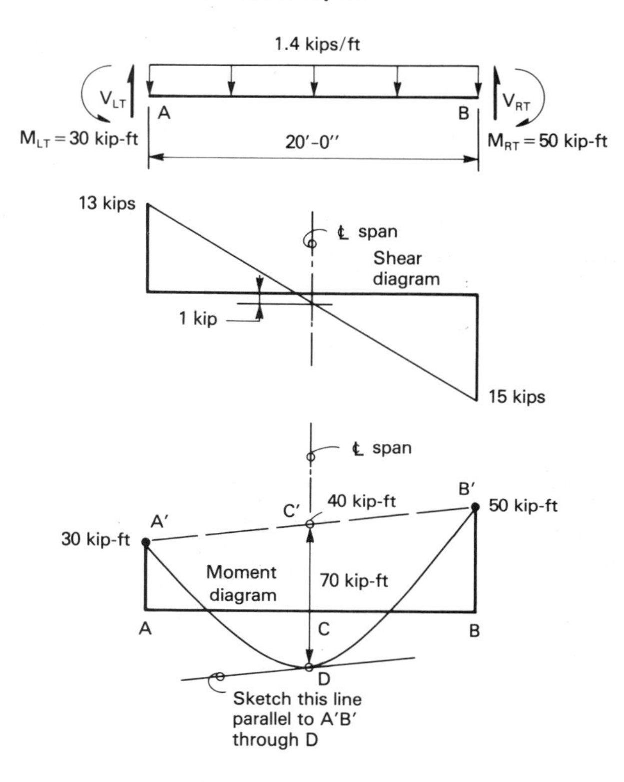


Compute end shears $V_{LT}$ and $V_{RT}$:

$$\Sigma M \text{ about } B = 0$$

$$-30 + V_{LT} \times 20 - \frac{1.4 \times 20^2}{2} + 50 = 0$$

$$V_{LT} = \frac{310 - 50}{20} = 13 \text{ kips}$$

$$V_{RT} = 1.4 \times 20 - 13 = 15 \text{ kips}$$

*Shear diagram:* Plot $V_{LT}$ and $V_{RT}$ and draw shear diagram through these points as shown.

*Moment diagram:* Compute the moment for a simple span:

$$M = \frac{wL^2}{8} = \frac{1.4 \times 20^2}{8}$$

$$= 70 \text{ kip-ft}$$

Plot $M_{LT}$ and $M_{RT}$ as shown.

Draw skew axis $A'B'$ through $M_{LT}$ and $M_{RT}$.

Draw the centerline of span and compute the moment at the centerline of span (at $C'$).

Plot $wL^2/8$ on the centerline, line $C'D$.

Sketch the moment diagram through $A'DB'$, tangent to a line through $D$ and parallel to $A'B'$.

Example 9-1 presents only half the moment envelope. To sketch the other half, it is necessary to know the maximum positive moment that can occur at midspan. The next example illustrates this case.

**Example 9-2**

*Shear and moment envelopes*
*Key values given (negative and positive moment) for sketching the moment envelope for Example 9-1*

For Example 9-1, the maximum positive moment at midspan is now given as 40 kip-ft. Sketch the corresponding moment diagram, assuming that the moments at the supports are in the same ratio as those of Example 9-1.

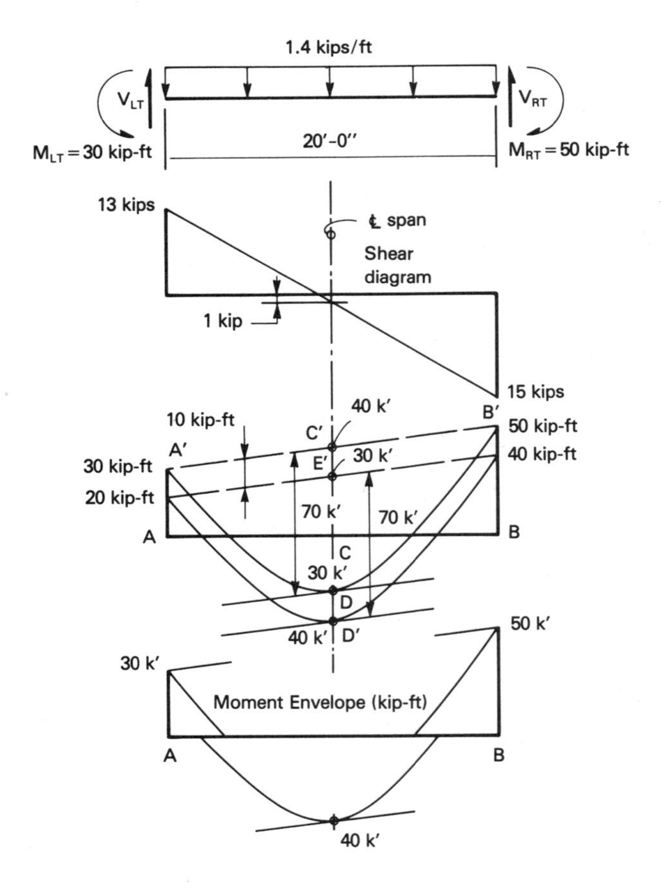

*Positive moment envelope:* Along centerline of span plot point $D'$, 40 kip-ft on positive side of the axis $AB$.

Move upward along centerline of span 70 kip-ft and plot $E'$ as shown.

Draw an axis through $E'$ parallel to $A'B'$.

Sketch a duplicate moment diagram identical to the first but displaced downward 10 kip-ft through point $D'$.

Delete all lines except the outermost lines, leaving only the moment envelope as shown.

The moment envelope of Example 9-2 is seen to be a complete moment envelope, constructed using only the key values of negative moment at the supports and the positive moment at midspan. Note that the maximum vertical ordinate that can occur in either case is the simple span moment of 70 kip-ft. Note also that the shear diagram is valid for either case and is the design shear diagram for this case.

The point to be realized is that the final diagrams for shear and moment are obtained simply by shifting the simple beam shear and moment diagrams up or down and, in the case of the moment diagram, skewing the horizontal axis.

## EXACT MAXIMUM VALUES OF SHEAR AND MOMENT

For braced frames, the maximum values of shear and moment that can occur on any given span can be developed rationally [1]. The structural analysis for such solutions follows conventional elastic methods of analysis and is therefore valid at working levels. Since the solution deals only with braced frames (braced against sidesway) the only loads to be considered are the vertical loads, dead and live.

A typical frame from a low multistory building is shown in Fig. 9-4a, loaded by

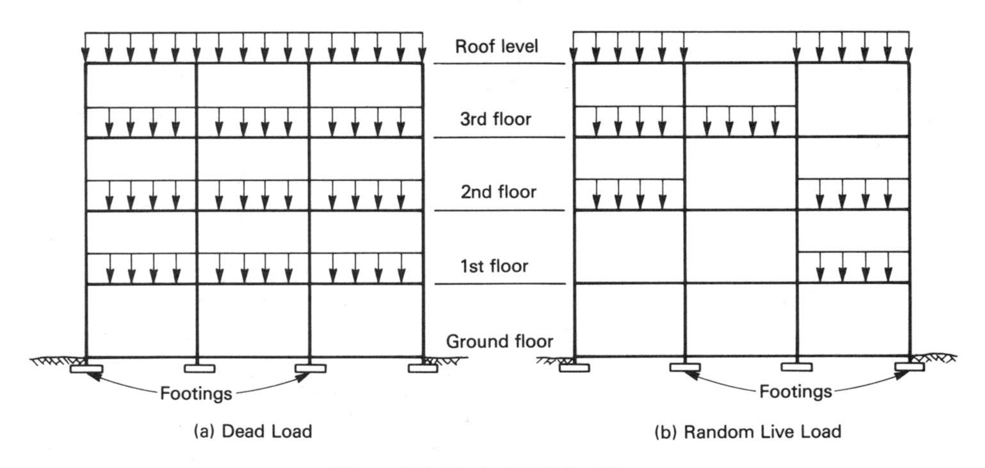


**Figure 9-4** Typical Building Frame.

its own dead weight. The same frame is shown in Fig. 9-4b with some of its spans loaded by the live load. It is easy to visualize numerous patterns for distributing the live load that might be combined to produce maximum moment at some particular point. The problem, of course, is to find the final configuration of loads that will produce maximum moment at a particular point. The problem is simplified by recognizing that only the live load may be shifted around; the dead load is fixed and, once computed, can be added to any one of the trial arrangements of live load. It is further simplified by recognizing that the transfer of moment or shear from floor to floor is minimal; the Code, in fact, permits each floor level to be removed from the building and to be designed independent of the adjacent floors. The problem is further simplified by recognizing that the stiffness (moment of inertia divided by span length) of the large girders is much higher than that of the columns. As far as any single floor is concerned, Code (8.8.3) permits the assumption to be made that the columns are, in effect, fixed at the adjacent floors.

With these assumptions, a typical floor level is taken from a building and shown in Fig. 9-5 with various patterns of load. The live-load patterns shown in Fig. 9-5 are

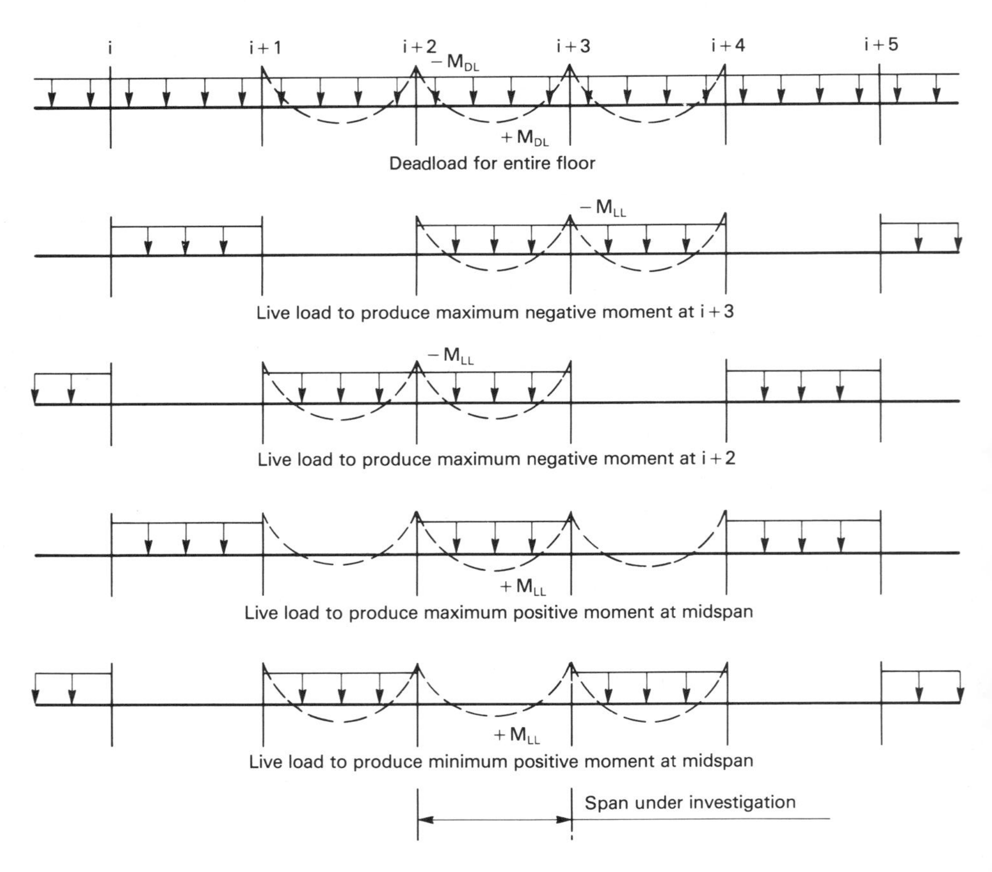


**Figure 9-5** Floor Loading Patterns.

those that produce the maximum positive and negative moments for these simplified cases; these patterns have been shown rationally to produce the maximum values of moment (Wang and Salmon, Reinforced Concrete Design, 3RD Ed. Harper and Row, 1979). The final design values are obtained by summing the values produced by dead load to one of the cases of live load. A summary of these maximum values is presented in a subsequent section.

The shear diagrams are not included in the foregoing analysis. It has been found that shear varies very little with the various load patterns and at most can be increased only by about 15%, and even then only at the face of the first interior support of the end span. The summary given later includes this 15% increase.

The foregoing discussions have made frequent reference to moments or shears "at

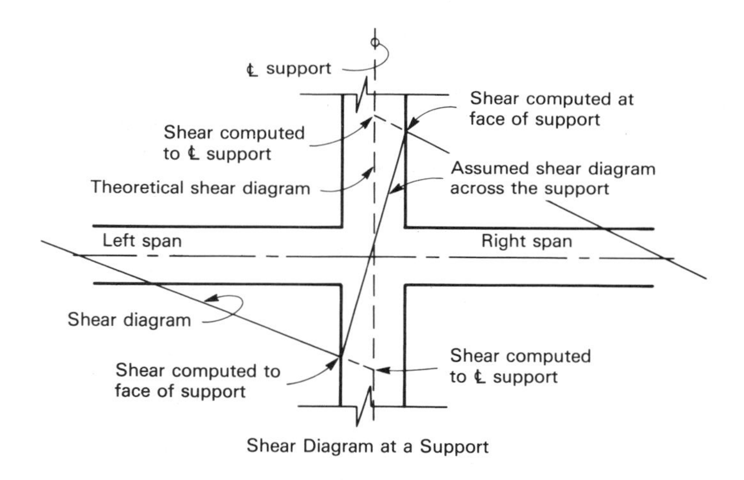


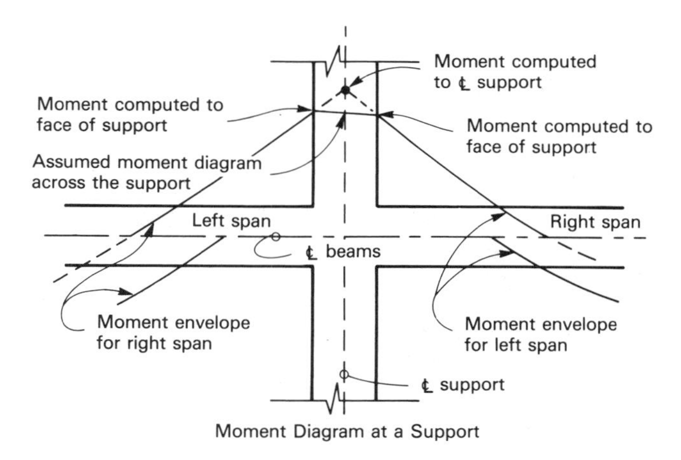


**Figure 9-6** Shear and Moment Across a Support.

a support,'' as if all supports were clean simple supports with no complications. Such supports are rare in concrete structures. In most cases, concrete beams are cast integrally with their supports; an interaction between the beam and its supports must occur.

Where a concrete beam is cast integrally with its supporting column, the pattern of flexure stress for that part of the beam within the lines of the column is, of course, disrupted. The shear stress and moment stress across the integral bearing (or of column) is not at all like that of a beam on a simple support. From full-scale tests, the stresses have been found to be much reduced; Code (8.3.3) recognizes this fact by taking the design values of both moment and shear at the face of the support. The assumed distribution is shown in Fig. 9-6.

## SUMMARY OF DESIGN VALUES FOR SHEAR AND MOMENT IN BEAMS

From an analysis similar to the foregoing, Code (8.3) has adopted the values given in Table 9-1 for shear and moment on beams. Termed the *ACI Approximate Method*, the use of these coefficients is acceptable for use in final design of continuous beams and frames. These values are valid where the following conditions are met:

1. There are at least two spans.
2. Spans are roughly equal, with the longer of any two adjacent spans not more than 20% longer than the shorter span.

**TABLE 9-1 ACI COEFFICIENTS[a]**

| | |
|---|---|
| Positive moment | |
| End spans | |
| Discontinuous end simply supported | $wL_n^2/11$ |
| Discontinuous end integral with support | $wL_n^2/14$ |
| Interior spans | $wL_n^2/16$ |
| Negative moment at exterior face of first interior support | |
| Two spans | $wL_n^2/9$ |
| More than two spans | $wL_n^2/10$ |
| Negative moment at other faces of interior supports | $wL_n^2/11$ |
| Negative moment at face of supports for slabs with spans not exceeding 10 ft and beams where ratio of sum of column stiffnesses ($\Sigma\, I/L_u$) to beam stiffnesses ($\Sigma\, I/L_n$) exceeds eight at each end of the span | $wL_n^2/12$ |
| Negative moment at interior face of exterior support for beams or slabs built integrally with supports | |
| Where support is a beam or girder | $wL_n^2/24$ |
| Where support is a column | $wL_n^2/16$ |
| Shear in end members at face of first interior support | $1.15wL_n/2$ |
| Shear at face of all other supports | $wL_n/2$ |

[a]For positive moment, $L_n$ is the clear span. For negative moment, $L_n$ is the average of the two adjacent clear spans. For shear, $L_n$ is the clear span. $w$ is the load per unit length.

3. Loads are uniformly distributed.
4. Live load is not more than three times the dead load.
5. Members are prismatic.

## LOADS AND MOMENTS ON COLUMNS

For columns, the slight increase in shears at the end spans need not be included in the column loads. At any floor level, the vertical axial load delivered to the column may be taken as the sum of all loads within an area halfway to the next support in any direction. This "tributary area" applies equally to dead load and to live load, although some reduction in live load (due to probability) is usually permitted by the building codes.

Even where no lateral loads are taken by the columns, as in braced frames, some moment is imparted to the columns by the beams framing into them. As indicated in Fig. 9-6, any difference in end moments between two beams at the same level must be taken by the column. Although small, this moment and its effect on the capacity of the column must be accounted for. In combination with the axial load, this moment and axial force constitute the design loads for the columns in a braced frame, as shown in Fig. 9-7.

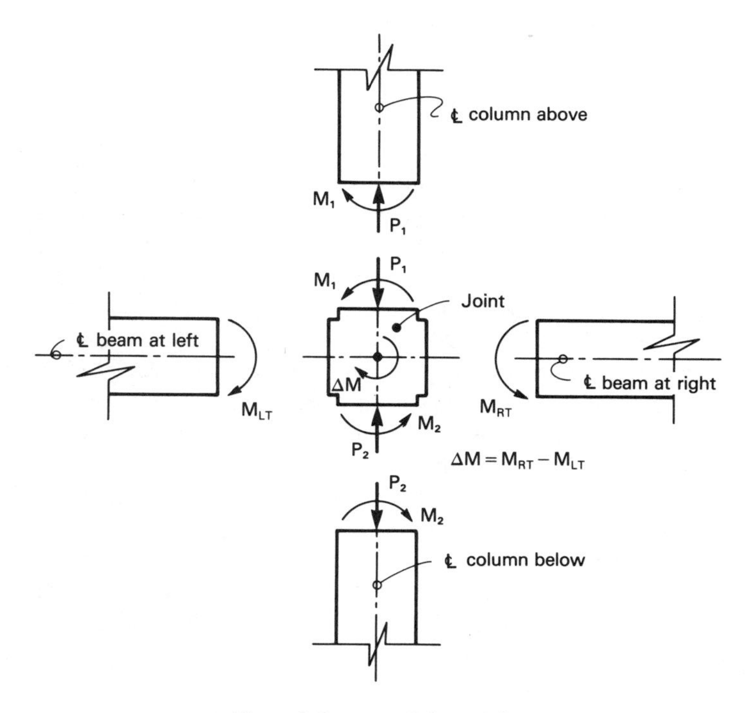


**Figure 9-7** Beam-Column Joint.

The moment that may be applied to a column by the beams framing into it is obviously the difference between the end moment of the beam on the left of the column and the end moment of the beam on the right. The magnitude of this moment is specified by Code (8.8.1) as the maximum moment that may occur due to loading in the adjacent spans only; that is, the effects of loads several bays away is not required. For the sake of simplicity, in this book the design moment is taken as the maximum value of moment already computed using the ACI coefficients (loads in distant spans are therefore included). For exterior columns, the moment on the outward side of the column is of course zero.

Code (8.8.1) recognizes two load cases for the design of columns. The first is that of maximum axial load concurrent with maximum moment. The second case is that of maximum moment combined with a reduced axial load (usually DL + 50%LL), recognizing that the full live load will not likely be applied 100% at every floor at all times. This second case produces a higher value of $e/t$, which at times can be more severe than the case of full live load.

## SUMMARY OF DESIGN LOADS ON COLUMNS

Code does not tabulate the loads for columns as neatly and succinctly as it does for beams. Code requires only that two cases be considered: that producing the highest absolute magnitudes of axial loads and moments, and that producing the highest value of $e/t$, where

$$\frac{e}{t} = \frac{M/P}{t} \tag{9-3}$$

The magnitude of the maximum moment $M$ in Eq. (9-3) is prescribed by Code; this moment was discussed in the preceding section. Determination of the lowest axial load $P$ that will produce maximum $e/t$ is left to the designer.

There is always some question concerning the lowest axial load $P$ that may exist on a column. It may be argued that the lowest possible load is that of dead load only, and this value should be used in all criteria involving buckling. It may also be argued that every building has some portion of the live load present at all times and the lowest axial load should include some realistic portion, say 50%, of the live load.

For this book the lowest axial load on a column is taken as dead load plus 50% of live load. This choice is influenced somewhat by the choice of the more conservative value of moment at full live load, as discussed earlier. In all cases, however, it must be recognized that lower values might conceivably occur and must be used wherever they are likely to occur.

It should also be recognized that municipal building codes usually permit some reduction in live load at each floor to compensate for probability of loading. For taller buildings, it is highly unlikely that every floor will be loaded to maximum live load at any one time. The municipal buildings codes therefore include an allowable "reduced

**TABLE 9-2 COLUMN LOADS**[a]

Maximum moment acting on an interior column:
Maximum moment computed for the beam at one side of the column less the minimum moment computed for the beam on the other side at the same level (both moments computed by ACI coefficients).
Maximum moment acting on an exterior column:
Maximum moment computed for the beam at one side of the column (ACI coefficients) plus any effects of torsion on the perimeter ell beam at the same level.
Maximum axial load acting on a column:
Maximum value of dead load plus live load acting on all tributary areas supported by the column. (Live load may have been reduced in accordance with municipal building code to account for probability of loading.)
Minimum axial load acting on a column:
Value of dead load plus 50% of live load acting on all tributary areas supported by the column (or, at the discretion of the designer, dead load only).

[a]The total maximum moment acting on the column is distributed between the column above and the column below in proportion to their stiffnesses $I/L_u$, where $I$ is given by $bh^3/12$. Far ends of columns are considered fixed at adjacent floor levels. Tributary area may be taken as the area enclosed within half the distance to the next vertical support in any direction.

live load'' to account for this improbability. The live load in the discussions above is this ''reduced live load'' where it applies.

With the foregoing conditions, the design loads for columns may be summarized as shown in Table 9-2.

## EXAMPLES

The following example illustrates the design of a complete cast-in-place roof system for a typical one-story shop building. A tee-beam system has been selected as shown with bridging at midspan only. Lateral loads are taken by the bearing walls, leaving only the vertical loads to be taken by the roof. Snow load is much less than the specified load and is not a consideration.

**Example 9-3**

*Design of a tee-beam roof system*

*Given:*
Box structure, layout as shown
Walls are concrete masonry units, 8 × 8 × 16 in.
Design roof live load is 40 psf
Rated capacity of monorail is 2.5 tons live load
Dead weight of monorail hoist is 500 lb
Impact load for monorail is 33% of live load
$f'_c = 3000$ psi
Grade 60 steel

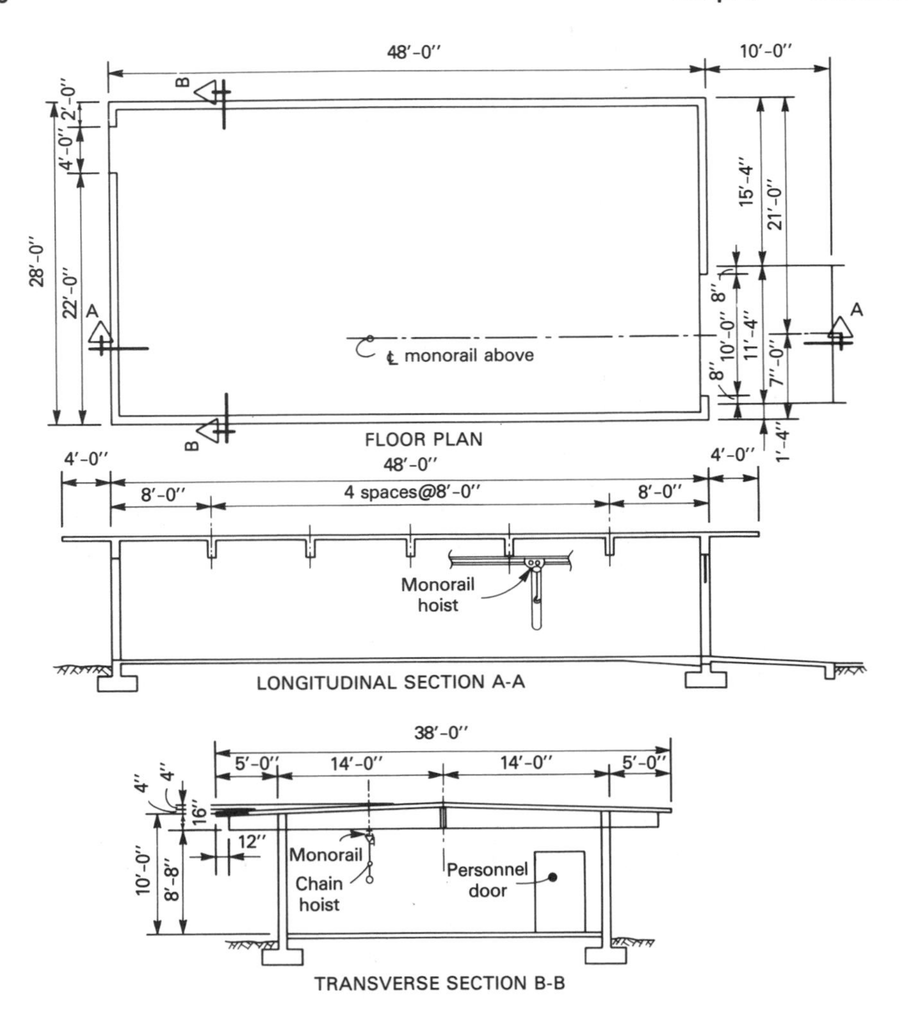


**Figure E9-3(a)** Layout and Configuration.

A tee-beam roof system is the trial system as shown in the layout and configuration sketch. A built-up roof covering is chosen, having a dead load of 10 psf. A dead load allowance of 5 psf is adopted for the underside of the roof, to allow for insulation, sprinklers, ducts, and future modifications. A slope of $2\frac{1}{2}\%$ is adopted for roof drainage.

*Design of roof slab:* A tentative spacing of 8 ft 0 in. is chosen for the tees, as indicated in the configuration sketch. For this spacing, a corresponding slab thickness of 4 in. has been assumed. Minimum thickness for control of deflections (from Table 3-1) is $L/28$ or 3.4 in., so the thickness chosen is adequate for deflection criteria. It is now necessary to assure that this slab thickness is, in fact, adequate for strength. Such a check follows.

Dead load of roof slab (plf):

| *Cantilever span* | | *Interior span* | |
|---|---|---|---|
| Roofing | 10 | Roofing | 10 |
| Slab | 50 | Slab | 50 |
| | | Misc. | 5 |
| $w_{DL}$ = | 60 plf | $w_{DL}$ = | 65 plf |

A sketch of half the beam follows, with loads indicated for the appropriate span.

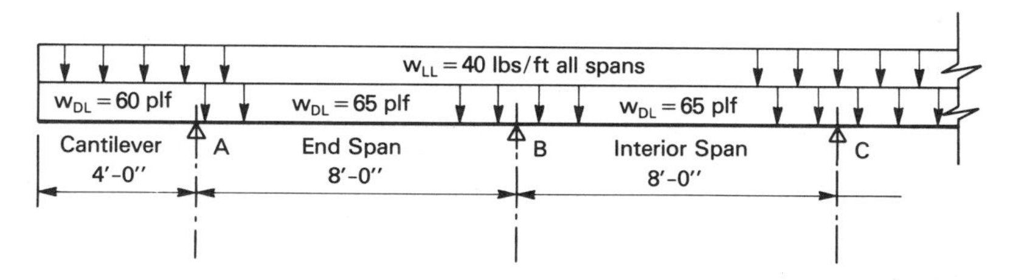


**Figure E9-3(b)** Roof Slab Loading.

Shear and moment diagrams are now plotted for the indicated loads using the ACI coefficients. A side calculation shows that the negative cantilever moment at $A$ is at least as large as the negative moment at the other supports; hence the end span will be treated in the analysis as another interior span. For the shear diagram, the loads must be factored and then divided by the undercapacity factor $\phi = 0.85$ for shear. At the exterior face of support $A$, where $V = wL$,

$$V_n = \frac{1.4 \times 60 \times 4 + 1.7 \times 40 \times 4}{0.85} = 715 \text{ lb}$$

At all other faces of all supports, $V = wL/2$,

$$V_n = \frac{1.4 \times 65 \times \frac{1}{2} \times 8 + 1.7 \times 40 \times \frac{1}{2} \times 8}{0.85} = 748 \text{ lb}$$

The final shear diagram follows.

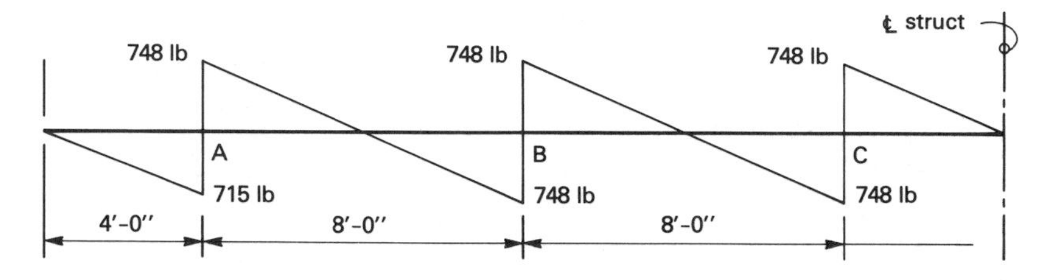


**Figure E9-3(c)** Roof Load Shear Diagram.

For the moment diagram, ACI coefficients may be used for all negative moments at the supports except at the end supports. At the end support, the actual cantilever moment is used.

At support $A$, $M = wL^2/2$ (actual cantilever moment):

$$M_n = \frac{1.4 \times 60 \times 4^2/2 + 1.7 \times 40 \times 4^2/2}{0.9}$$

$$= 1350 \text{ lb-ft/ft}$$

At all other supports, $M = wL^2/10$ (ACI):

$$M_n = \frac{1.4 \times 65 \times 8^2/10 + 1.7 \times 40 \times 8^2/10}{0.9}$$

$$= 1{,}130 \text{ lb-ft/ft}$$

For positive moment at midspan, $M = wL^2/16$ (ACI):

$$M_n = \frac{1.4 \times 65 \times 8^2/16 + 1.7 \times 40 \times 8^2/16}{0.9}$$

$$= 707 \text{ lb-ft/ft}$$

The final moment envelope follows.

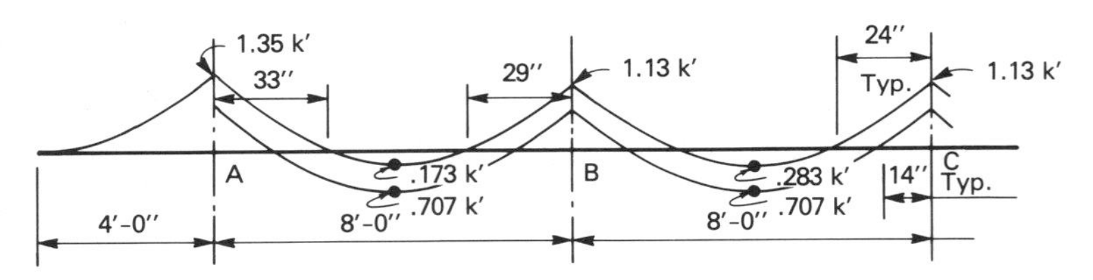


**Figure E9-3(d)** Roof Load Moment Diagram.

From the moment diagram it is seen that the maximum moments occur at the supports and are nearly equal. The same thickness of slab, 4 in., will therefore be used throughout. The value of $d$ thus provided can be computed by deducting the required concrete cover of $\frac{3}{4}$ in. and half the bar diameter from this thickness. Assume the bar diameter to be less than $\frac{1}{2}$ in.; hence

$$\text{actual } d = t - \tfrac{1}{2} \text{ bar diameter} - \text{cover}$$

$$= 4 - 0.25 - 0.75 = 3 \text{ in.}$$

With $b$ and $d$ known, the section modulus and steel ratio can be found. Apply the flexure formula:

$$\frac{M_n}{0.85 f'_c} = S_c \qquad \frac{1350 \times 12}{0.85 \times 3000} = \text{coeff.} \times bd^2 = \text{coeff.} \times 12 \times 3^2$$

Solve for coeff.: coeff. $= 0.059$.

From Table A-5, use $\rho = 0.0035$, coeff. $= 0.079$.

Note that for this value of $d$, the steel ratio is at its minimum value by Code, a rather common occurrence in lightly loaded roof slabs. The required steel area is then

$$A_s = \rho_{\min} bd = 0.0035 \times 12 \times 3 = 0.126 \text{ in}^2/\text{ft}$$

From Table A-2, select the steel arrangement. Use No. 3 bars at 10 in. o.c. Since all other moments in the slab, both positive and negative, are smaller than the moment used above, the minimum ratio of steel, 0.0035, will be required throughout. Use No. 3 bars at 10 in. o.c. for all other positive and negative moments.

The cutoff lengths for the slab steel may be determined graphically by laying out the reinforcement on the moment diagram and scaling it. The result follows.

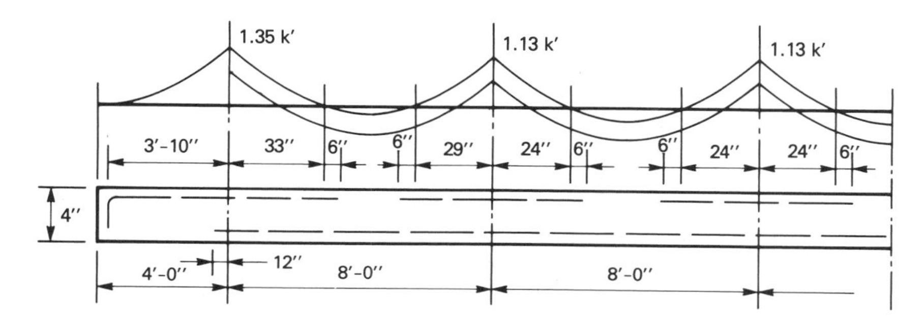


**Figure E9-3(e)** Slab Reinforcement Cutoff Points.

For the development length in the slab, negative reinforcement is extended past the inflection point by a distance $d$ (3 in.) or $12d_b$ ($4\frac{1}{2}$ in.), or span/16 (6 in.), whichever is greater. The positive reinforcement is extended past the centerline of the end supports by $L_d$; little benefit is gained by cutting the positive steel, so it is kept continuous.

Temperature reinforcement is required in the other direction, transverse to the flexural steel. Code requires a steel area of $0.0018bt$ for temperature reinforcement where grade 60 steel is used, with a spacing not to exceed 18 in. For the slab of this example, the required steel area is then

$$A_s = 0.0018bt = 0.0018 \times 12 \times 4 = 0.0864 \text{ in}^2/\text{ft}$$

From Table A-2, select No. 3 bars spaced at 16 in. on center. (A spacing of 16 in. permits a regular layout on the 8 ft 0 in. span). The final steel layout follows.

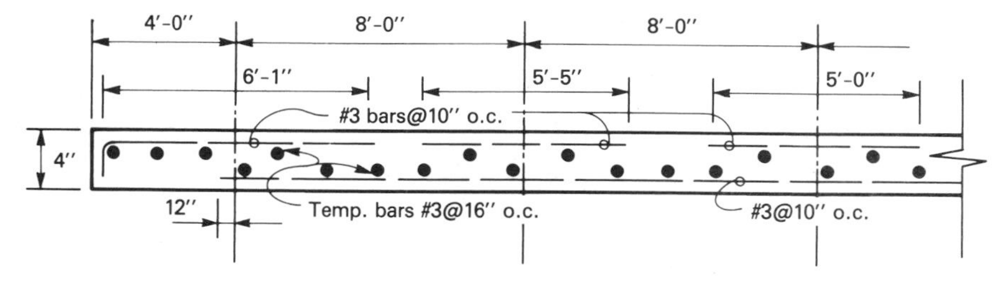


**Figure E9-3(f)** Final Steel Layout in Slab.

With the final design for the slab now completed, the tee sections can be designed. Since the tees are simply supported, the ACI coefficients cannot be applied. The shear and moment envelopes must be drawn specifically for these tee beams. A sketch of the governing dimensions follows.

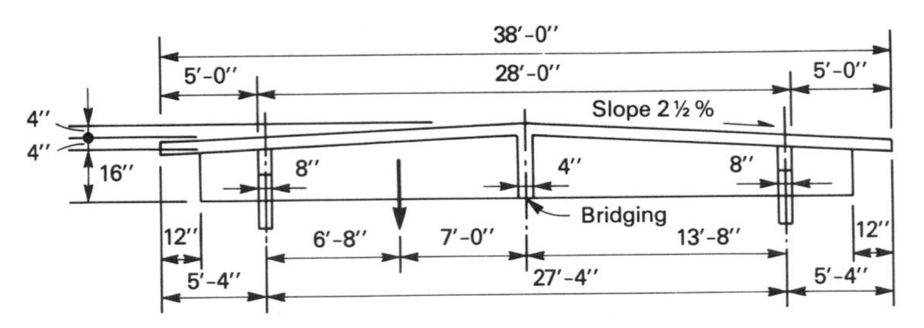


**Figure E9-3(g)** Tee Beam Dimensions.

As frequently happens, the size of the stem of the tee was set to match an outside influence of some sort. In this case it was the module of the concrete masonry units. To match this module, the stem was set at 8 in. × 16 in. at the supports. The filler blocks of the masonry walls will then fill the void space between the roof slab and the masonry wall without having to be cut. The slope of the roof was set at a convenient value consistent with the climate where the building is to be located; in this case the slope was set at $2\frac{1}{2}\%$.

For deflection control, the minimum overall depth of the section must be greater than $L/16$ or 20 in. (see Table 3-1). At the support, the overall depth is seen to be 20 in. and at midspan it is 24 in.; the depth is everywhere at least equal to the required minimum. Deflections, therefore, will not be a problem and require no further consideration. The dead load of the stem is calculated by averaging the stem weight over the full length of the tee.

$$\text{stem weight} = \text{avg. height} \times \text{width} \times \text{length} \times \text{conc. weight}$$

$$= 1.5 \times 0.67 \times 36 \times 145 = 5.22 \text{ kips}$$

$$\text{weight per foot} = \frac{5.22}{38} = 137 \text{ lb/ft}$$

The monorail live load is 2.5 tons (taken as 5000 lb), with a 33% impact load, for a total live load of 6.67 kips. The dead load of the hoist and 8 ft of rail is taken at 500 lb.

The dead loads (plf) acting on the tee beam are:

$$\begin{aligned}
\text{roofing DL} &= 10 \text{ psf} \times 8 \text{ ft 0 in.} && = 80 \\
\text{slab DL} &= 0.33 \times 8 \text{ ft 0 in.} \times 145 \text{ lb/ft} && = 387 \\
\text{interior DL} &= 5 \text{ psf} \times 8 && = 40 \\
\text{stem DL} &= && = \underline{137} \\
&\text{At overhang:} \quad w_{DL} && = 604 \text{ plf} \\
&\text{At interior:} \quad w_{LL} && = 644 \text{ plf}
\end{aligned}$$

The load diagram, shear diagram, and moment diagram for dead load are shown below. Two cases of load were used to obtain the envelopes, one where the monorail is directly under the beam and the other where the monorail is so far away that it has no influence. For computing critical shear at a distance $d$ from face of support, a value of 18 in. was assumed for $d$.

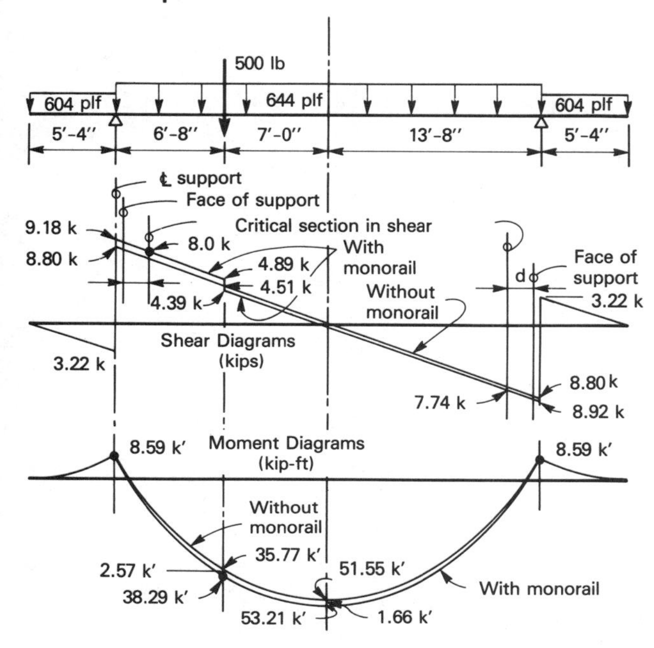


**Figure E9-3(h)** Dead Load Shear and Moment Diagrams.

The load, shear, and moment diagrams for live load are similarly computed and plotted. For live load, there is an extra case of loading that affects the maximum positive moment, the case where there is no live load on the overhang. The diagrams follow.

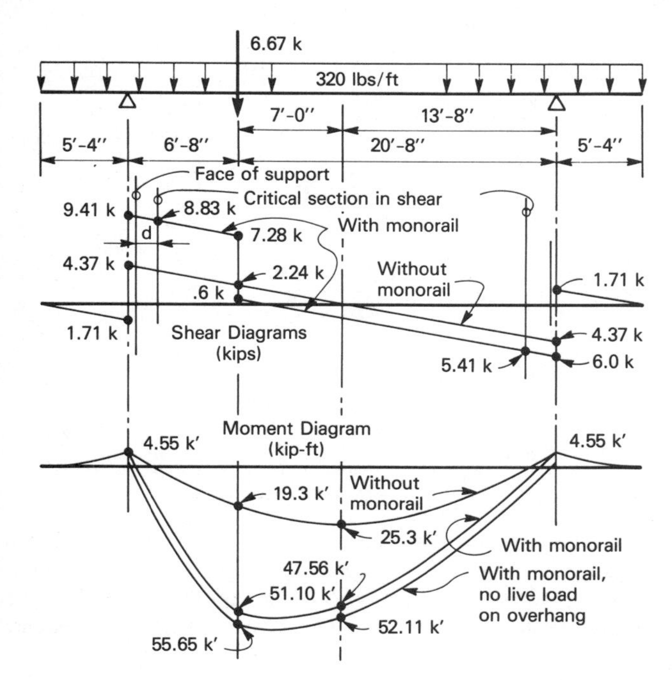


**Figure E9-3(i)** Live Load Shear and Moment Diagrams.

The composite shear and moment envelopes are shown below for $V_n$ and $M_n$, where

$$V_n = \frac{1.4V_{\mathrm{DL}} + 1.7V_{\mathrm{LL}}}{0.85}$$

$$M_n = \frac{1.4M_{\mathrm{DL}} + 1.8M_{\mathrm{LL}}}{0.90}$$

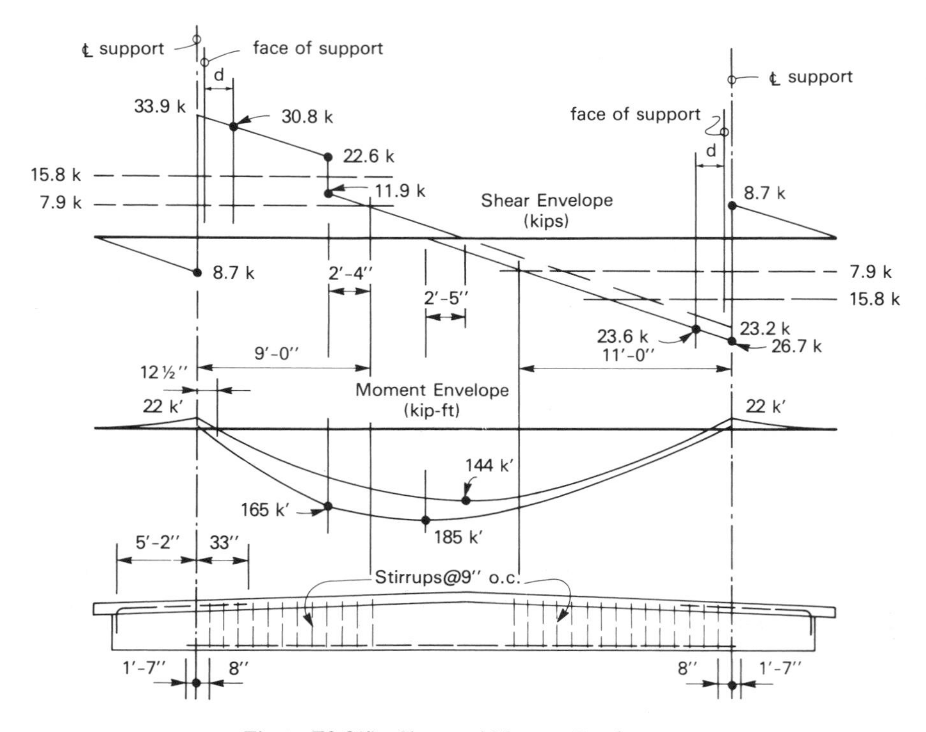


**Figure E9-3(j)** Shear and Moment Envelopes.

The capacity of the tee stem in shear is shown on the shear envelope, where

$$V_c = 2\sqrt{f'_c}b_w d = 110 \times 8 \text{ in} \times 18 \text{ in} = 15.84 \text{ kips}$$

With stirrups at $d/2$: $V_n = 3V_c = 47.52$ kips

With stirrups at $d/4$: $V_n = 5V_c = 79.20$ kips

The required locations for stirrups are shown on the shear envelope. Stirrups at a spacing of $d/2$ are required from the point where shear exceeds $\frac{1}{2}V_c$ or 7.92 kips. The coverage extends so close to midspan that stirrups are effectively required through the entire span. The actual locations of stirrups are shown on the diagrammatic sketch below the shear and moment envelopes.

The required cross-sectional area of a stirrup is given by the equation

$$V_s = V_n - V_c = \frac{A_v f_y d}{s}$$

By using $f_y$ for grade 40 steel in this equation, even though grade 60 steel is to be used, the anchorage for all stirrups can be provided by a 135° hook. This practice for anchorage of stirrups is used here. Substitute into the foregoing equation, where $s = d/2$, $f_y = 40{,}000$ psi:

$$30{,}800 - 15{,}840 = \frac{A_v \times 40{,}000 \times 18}{9}$$

Solve for $A_v$: $A_v = 0.187 \text{ in}^2$ (total area of two legs).

From Table A-3, choose No. 3 stirrups. Area of two legs is $2 \times 0.11 = 0.22 \text{ in}^2$. The configuration of the stirrups is shown on the final sketch. The design of shear reinforcement is thus complete; the design of the flexural reinforcement is taken next.

For negative moment at the overhang, the value of $d$ is taken at 18 in. as before. The section modulus is then

$$\frac{M_n}{0.85 f'_c} = S_c \qquad \frac{22{,}000 \times 12}{0.85 \times 3000} = \text{coeff.} \times 8 \times 18 \times 18$$

Solve for coeff.: coeff. = 0.040.

From Table A-5, choose $\rho_{\min} = 0.0035$ (minimum $\rho$).

Again, the area of steel is controlled by the minimum ratio $\rho_{\min}$. At the support,

$$A_s = \rho_{\min} b_w d = 0.0035 \times 8 \times 18 = 0.504 \text{ in}^2$$

Use 2 No. 5 bars throughout the cantilever, with standard 90° hooks at the outer ends.

For the development lengths of the negative reinforcement at the interior span, the bars must extend past the innermost inflection point by a distance $d$ (18 in.), or a distance $12d_b$ ($7\frac{1}{2}$ in.), or a distance span/16 (20 in.), whichever is greater. The development length of 20 in. is included on the final sketch.

To find the required steel ratio for the positive steel, the section modulus for the section must be found. The width of the compressive area $b'$ is shown in the following sketch. Its width, as prescribed by Code, is also computed below.

$b'$: $< 8t \times 2 + b = 72$ in.

$< \dfrac{\text{span}}{4} = 82$ in.

$<$ tee spacing $= 96$ in.

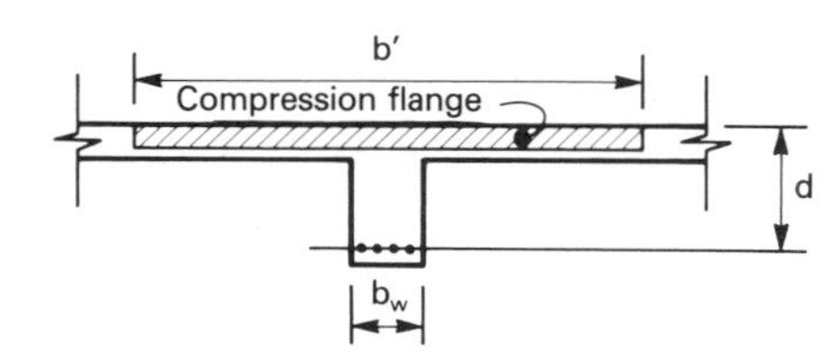

The coefficient of the section modulus can now be computed:

$$\frac{M_n}{0.85 f'_c} = S_c \qquad \frac{185{,}000 \times 12}{0.85 \times 3000} = \text{coeff.} \times 72 \times 22 \times 22$$

Solve for coeff.: coeff. = 0.025.

From Table A-5, select $\rho = 0.0010$ (based on $b'$).

Check for the location of the neutral axis. At service levels,

$$k_w = 0.128$$

$$k_w d = 0.128 \times 22 = 2.82 \text{ in.} < 4 \text{ in.}$$

The neutral axis is less than 4 in. below top of slab, so the neutral axis falls in the slab at service levels.

Required $A_s = \rho b'd = 0.0010 \times 72 \times 22 = 1.58 \text{ in}^2$.

*Check:*

$$\text{minimum } A_s = \rho_{\min} b_w d = 0.0035 \times 8 \times 22$$

$$= 0.616 < 1.58 \text{ in}^2 \quad \text{(O.K.)}$$

Use 4 No. 6 diameter bars for positive reinforcement.

For anchorage, a visual examination of the moment envelope is all that is necessary to discount any attempt to save steel tonnage by cutting the reinforcing bars. All the positive reinforcement is extended $L_d$ or 19 in. past the centerline of the supports. The bridging at midspan is required where the span is greater than 20 ft. There is no computable load on the bridging and Code does not specifically mention the reinforcement. For this case, the minimum steel ratio $\rho = 0.0035$ is used. Hence for a width of 4 in.,

$$A_s = \rho_{\min} b_w d = 0.0035 \times 4 \times 22 = 0.308 \text{ in}^2$$

Use 1 No. 5 bar at bottom of bridging; assume that temperature reinforcement is adequate reinforcement at the top. The final tee beam is shown below.

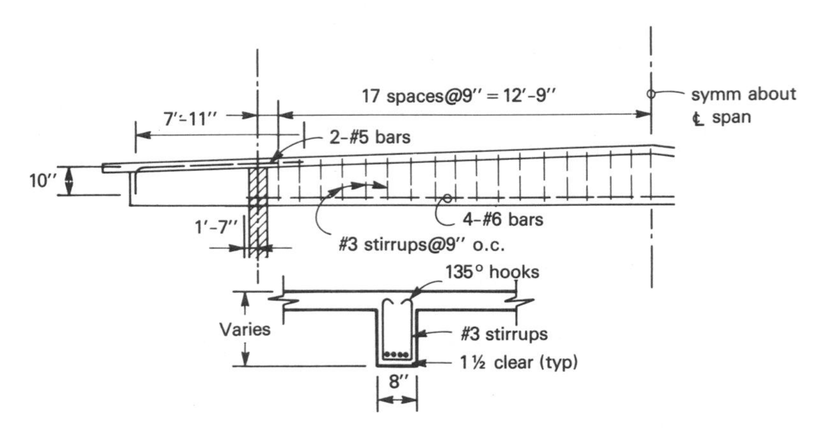


**Figure E9-3(k)** Final Steel Layout.

The next example illustrates the design of a typical floor girder and an interior column in a low (three-story) building. The general structural system is composed of one-way joists for floors and roof, carried by a braced frame structure. The design of one girder at one level of the frame is illustrated; the design of all other girders would follow the same procedure. Similarly, the design of one interior column is illustrated; all other columns would be designed the same way.

**Example 9-4**

*Design of a braced frame structure*

*Given:*
Plan and elevation as shown in the layout sketch
Roof live load: 20 psf
Weight of built-up roofing: 10 psf
Floor live load: 75 psf
Allow a 15-psf floor load for miscellaneous dead load such as ducts, partitions, and so on
Allow 800 lb/ft for perimeter curtain walls and fenestration
$f'_c = 4000$ psi
Grade 60 steel

It is necessary first to establish a trial configuration for the system in order that reasonably accurate dead loads can be estimated. Such a trial configuration is shown in the plan and section sketches for this example. With these estimated sizes known, the dead load can be computed, the moments and shears can be determined, and the required girder sizes and steel ratios established. These computed sizes are then checked back against the estimated sizes; if there is significant difference, the dead loads will be significantly in error, in which case the initial sizes must be corrected and the entire procedure repeated. Design and review of the one-way joist system shown in the layout and configuration sketch would follow the procedures given in Chapter 6 and is not repeated here. Verification that the joist system is adequate is assigned as one of the outside problems given at the end of this chapter.

The column sizes shown in the layout and configuration sketch are estimated using the rule of thumb (Chapter 3)

$$\text{column size} = 12 \text{ in.} + 1 \text{ in. per story above}$$

For this case, the first-floor columns are thus seen to be 12 in. + 2 × 1 in. = 14 in. square. All columns in this example are then made this size.

In the trial configuration, the joists are oriented to span the long direction of 28 ft 0 in. while the girders are oriented in the direction of the shorter span, 20 ft 0 in. The reason for this orientation is to keep the joists and the girders as nearly the same depth as possible, thereby keeping the story heights low. The following comparisons will illustrate.

The depth of a joist system is recognized as being about two-thirds the depth of a conventional tee-beam system. Consequently, the depth of the joist system is estimated at $(2/3)(L_n/12)$ or about 18 in. The depth of the girder is estimated at $L_n/10$ or about 23 in., or roughly 24 in. total depth. These two depths are only 6 in. different, assuring efficient utilization of height. (Throughout these estimates, column sizes are deducted from the span length to obtain clear span length $L_n$). Had the opposite orientation been used, the estimated depth of the joist system would be based on the 20-ft span at about $(2/3)(L_n/12) = 13$

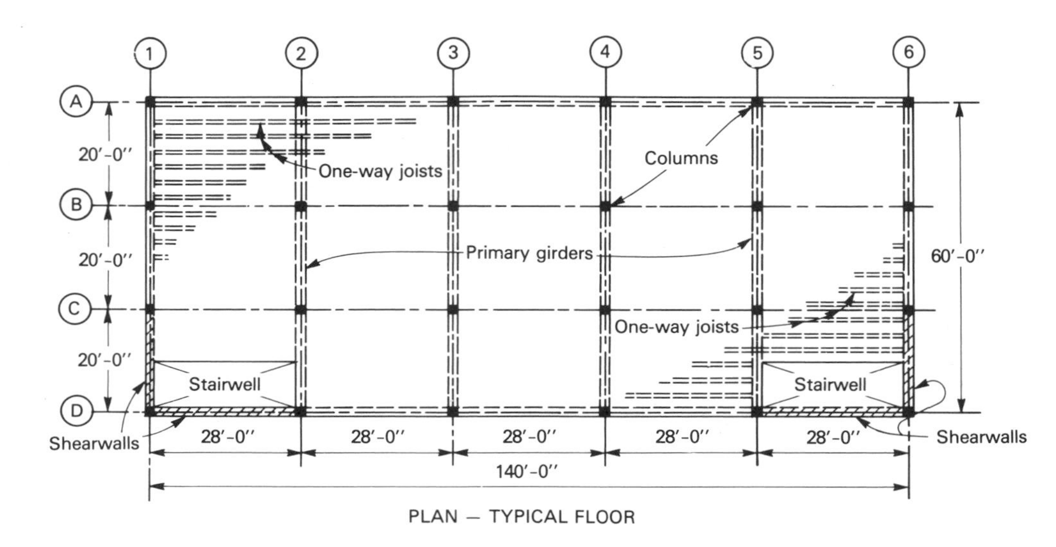


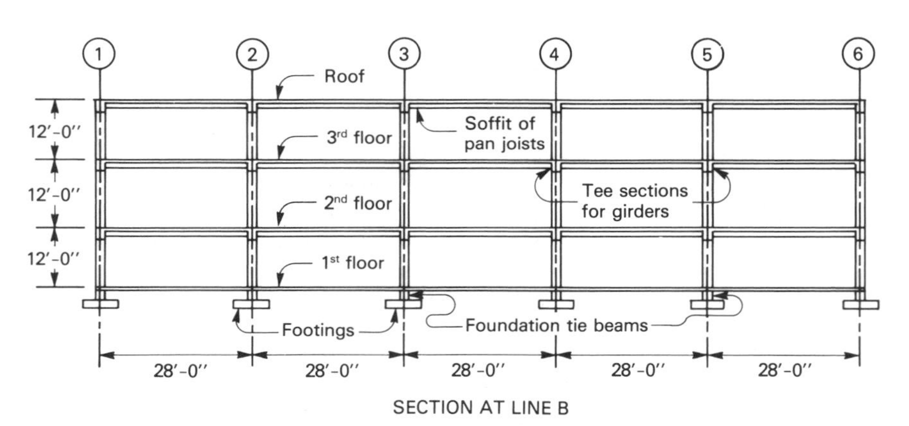


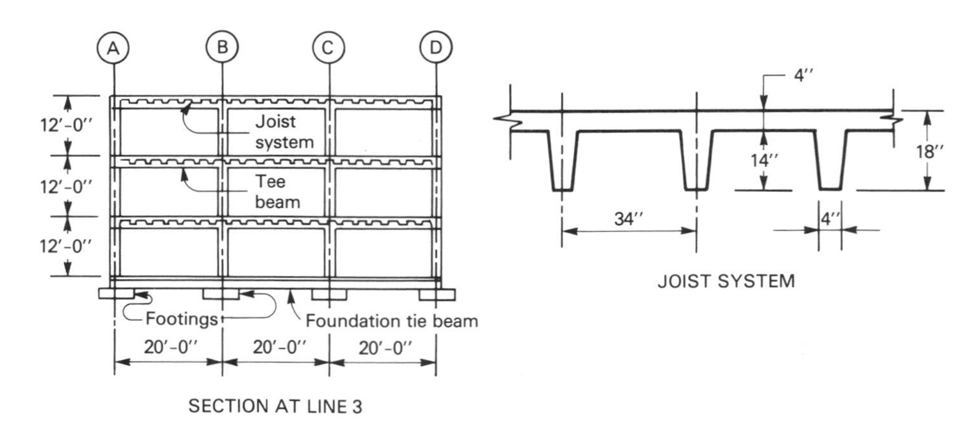


**Figure E9-4(a)** Layout and Configuration.

in., while the depth of the girder would be based on the 28 ft 0 in. span, $L_n/10 = 33$ in. The overall depth of the floor system would then be based on this depth of 33 in., producing a wasted space of some 20 in. under the joists.

The joist system itself is assumed to utilize a 30 in. pan and 4 in. slab, with a stem width of 4 in.

Shear walls are shown at the perimeter of the building to take the lateral loads, thereby providing the "bracing" for the braced frame. Although only three shear walls are actually needed, the use of two shear walls in each direction reduces foundation problems and is usually favored.

Also shown in this building is a foundation tie beam. This tie beam is used to provide a fixed base for the columns. Such a beam is often used in foundation design where the soil does not provide adequate resistance to rotation of the footings.

In the trial configuration, a thin (4 in.) slab is used with a deep (24 in.) girder. It is exactly this circumstance that is described in Chapter 6, warning that there is a possibility that the neutral axis might fall below the bottom of the slab. It will therefore be necessary later to check the final design to assure that the neutral axis falls in the slab.

For a representative frame in the transverse direction, the frame of line 3 is shown in section in the layout and configuration sketch. The girder supporting the second-floor level is removed from that frame and is shown larger in Fig. E9-4(b).

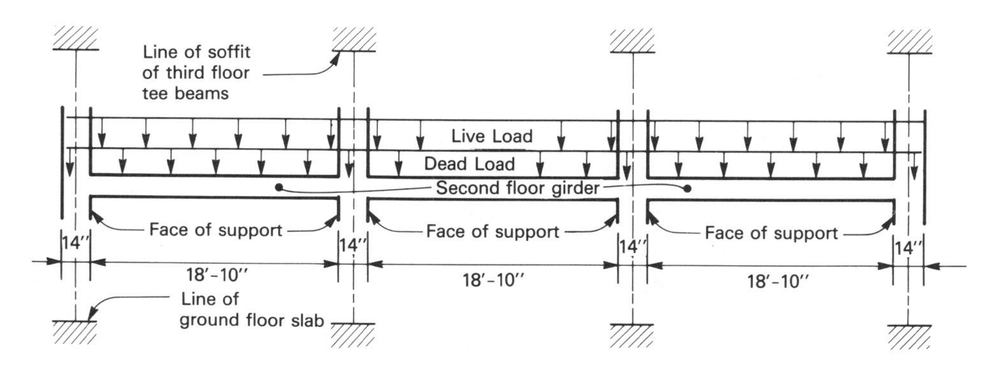


**Figure E9-4(b)** Second Floor Girder.

The load on the girder in the sketch is shown as a uniformly distributed load even though the actual load is a series of discrete concentrated loads, delivered as the end reactions of the joists. Such approximations are commonly used to represent closely spaced repetitive loads. Alternatively, had the floor been designed using widely spaced conventional tee beams, it could become worthwhile to use the actual tee-beam reactions.

The shear and moment envelopes for this second-floor tee section can now be determined, using the estimated sizes of the members for computing the dead load. For this calculation, a stem size for the tee section is estimated to be 20 in. deep and 16 in. wide.

For control of deflections, the minimum overall depth of the girder is $L/18.5$ or 12.2 in. (Table 3-1). Since the required depth for flexure is much greater than this, deflections will be no problem and require no further consideration. The live load on the girder is

$$w_{LL} = 75 \text{ psf} \times 28 \text{ ft} = 2.1 \text{ kips/ft}$$

The dead load of the joist system is found from Table 6-1:

$$w_{DL} = 77 \text{ psf} \times 28 \text{ ft} = 2.2 \text{ kips/ft}$$

The weight of the stem is readily computed, where the weight of regular concrete is taken as 145 pcf:

$$w_{DL} = \left(\frac{16}{12}\right)\left(\frac{20}{12}\right)(145) = 0.32 \text{ kips/ft}$$

The miscellaneous dead load is also readily computed:

$$w_{DL} = 15 \text{ psf} \times 28 \text{ ft} = 0.42 \text{ kips/ft}$$

The final design loads are then, at service levels,

$$w_{LL} = 2.1 \text{ kips/ft}$$

$$w_{DL} = 2.2 + 0.32 + 0.42 = 2.94 \text{ kips/ft}$$

For these loads, the shear and moment values may be found from the ACI coefficients. The ACI coefficients are summarized in the following sketch for this frame. The values shown at the supports are those at the face of support.

Note that the distances to inflection points have been computed and are shown on the sketch. These dimensions will be needed later when cutoff points for flexural steel are being determined. Similarly, the location of the critical shear for determination of shear reinforcement is also shown for future reference. See Fig. E9-4(c).

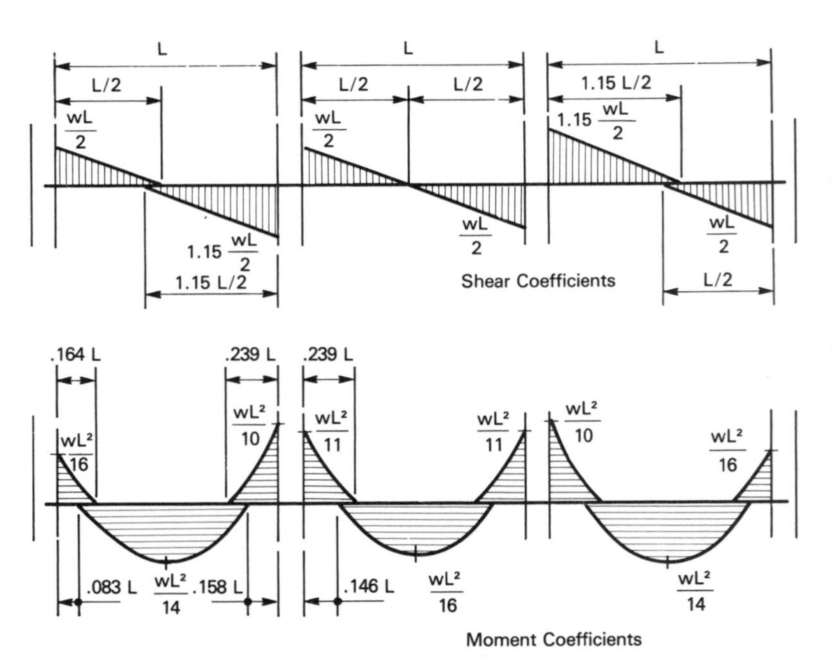


**Figure E9-4(c)** Shear and Moment Coefficients.

The values of live load and estimated dead load computed earlier, $w_{LL} = 2.10$ kips/ft and $w_{DL} = 2.94$ kips/ft, are substituted into the shear and moment equations, yielding the following shear and moment envelopes shown in Fig. E9-4(d).

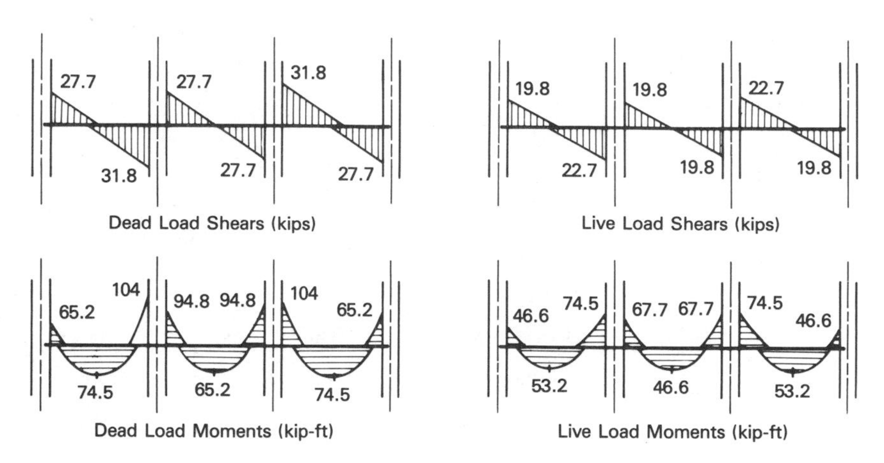


**Figure E9-4(d)** Shear and Moment Envelopes.

The tee section can now be designed using the procedures that were developed in Chapter 6. First, the stem size and reinforcement are selected to suit the maximum negative moment. Then the shear reinforcement (if any) is selected for that stem size. Last, the positive flexural steel is selected and the cutoff points are determined for all flexural reinforcement.

In this girder, the maximum negative moment is seen to occur at the exterior face of the first interior support.

$$M_n = \frac{1.4 \times 104 + 1.7 \times 74.5}{0.9} = 303 \text{ kip-ft}$$

The stem size will be selected for this moment and held constant for all spans. The negative reinforcement for the exterior span will be extended through the support and used as the negative reinforcement for the interior span. The required section modulus is found by interpolating in Table A-6, for $b = 0.66d$, $\rho = 0.008$,

$$\frac{M_n}{0.85 f'_c} = Z_c \quad \frac{303{,}000 \times 12}{0.85 \times 4000} = 0.131 \times bd^2 = 0.131 \times 0.66 \times d^3$$

Solve for $d$: $d = 23.1$ in., hence $b = 15.3$ in. and $A_s = \rho bd = 0.008 \times 15.3 \times 23.1 = 2.83\ \text{in}^2$.

Try section $b = 16$ in., $d = 24$ in., $h = 27$ in., $A_s = 3$ No. 9 bars, $A_s = 3.00\ \text{in}^2$.

The weight of the stem is computed for this trial section:

$$\text{weight of stem} = \left(\frac{16}{12}\right)\left(\frac{23}{12}\right)(145) = 370 > 320 \text{ lb/ft assumed}$$

error is 50 lb/ft. Compared to the 4040 lb/ft total dead load, this error is found to be acceptable. Insofar as flexure is concerned, the section is therefore seen to be adequate. It may now be examined for shear. The critical shear is shown on the shear envelope, given earlier in the calculations.

The critical shear is computed by ratios, where the shear at the face of the first interior support is

$$V_n = \frac{1.4V_{DL} + 1.7V_{LL}}{\phi} = \frac{1.4 \times 31.8 + 1.7 \times 22.7}{0.85} = 97.8$$

By ratios,

$$V_{cr} = 97.8\,\frac{1.15 \times 9.42 - 2}{1.15 \times 9.42} = 79.7 \text{ kips}$$

The capacity of the section in shear is

$$V_c = 2\sqrt{f'_c}\,b_w d = 2\sqrt{4000} \times 16 \times 24 = 48.6 \text{ kips}$$

$$\text{With stirrups at } d/2\text{:} \quad V_n = 3 \times 48.6 = 146 \text{ kips}$$

$$\text{With stirrups at } d/4\text{:} \quad V_n = 5 \times 48.6 = 243 \text{ kips}$$

Since the critical shear of 79.7 kips is greater than the capacity of the concrete alone, 48.6 kips, but less than the capacity with stirrups at $d/2$, 146 kips, stirrups are provided where the shear exceeds 48.6/2 or 24.3 kips. The following sketch shows these locations.

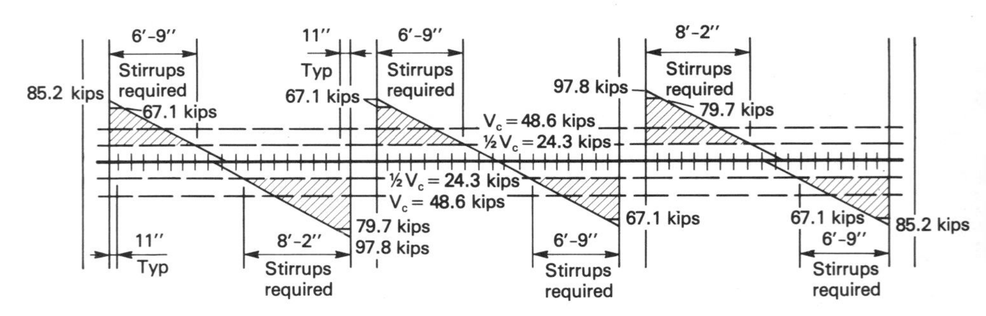


**Figure E9-4(e)** Location of Stirrups.

The size of the stirrups is found using the procedures developed in Chapter 5. The required total area of stirrups is given by the formula

$$V_s = V_n - V_c = \frac{A_v f_y d}{s}$$

Known values are substituted, with $f_y = 40{,}000$ psi:

$$79{,}700 - 48{,}600 = \frac{A_v \times 40{,}000 \times 24}{12}$$

Hence $A_v = 0.395$ in$^2$; $A_v$ per leg $= 0.18$ in$^2$.

Use No. 4 U-stirrups, with 135° hooks at top. Spacing of the stirrups is shown in the foregoing sketch; the first stirrups must fall within a distance $d/2$ or 12 in. from the face of support. It would be the choice of the designer whether to extend the stirrups all the way across the span.

With the design for negative moment and for shear seen to be satisfactory, the reinforcement for positive bending moment can now be determined. Positive moments are shown in the moment envelope developed earlier.

For positive moment at the end spans:

$$M_n = \frac{1.4M_{\mathrm{DL}} + 1.7M_{\mathrm{LL}}}{\phi} = \frac{1.4 \times 74.5 + 1.7 \times 53.2}{0.9}$$

$$= 216 \text{ kip-ft}$$

For positive moment at the interior span:

$$M_n = \frac{1.4M_{\mathrm{DL}} + 1.7M_{\mathrm{LL}}}{\phi} = \frac{1.4 \times 65.2 + 1.7 \times 46.6}{0.9}$$

$$= 189 \text{ kip-ft}$$

Flange width for the tee section in positive bending is prescribed by Code:

$$b': \quad < 2 \times 8 \times \text{slab } t + b = 2 \times 8 \times 4 + 16 = 80 \text{ in.}$$

$$< \text{tee spacing} = 28 \times 12 = 336 \text{ in.}$$

$$< \frac{\text{span}}{4} = \frac{18 \text{ ft 0 in.}}{4} = 57 \text{ in.} \quad (\text{use})$$

At the end span, the required section modulus is found by the usual methods:

$$\frac{M_n}{0.85f'_c} = Z_c \qquad \frac{216{,}000 \times 12}{0.85 \times 4000} = \text{coeff.} \times b'd^2 = \text{coeff.} \times 57 \times 24 \times 24$$

$$\text{minimum coeff.} = 0.023$$

From Table A-6, $\rho = 0.0013$. Hence

$$A_s = \rho b'd = 0.0013 \times 57 \times 24 = 1.78 \text{ in}^2.$$

Similarly at the interior span,

$$\frac{M_n}{0.85f'_c} = Z_c \qquad \frac{189{,}000 \times 12}{0.85 \times 4000} = \text{coeff.} \times b'd^2 \text{ coeff.} \times 57 \times 24 \times 24$$

$$\text{minimum coeff.} = 0.0203 \qquad \rho = 0.00113$$

Hence

$$A_s = \rho b'd = 0.00113 \times 57 \times 24 = 1.55 \text{ in}^2$$

For positive reinforcement, the required area in the end spans is so close to the required area in the interior span that the same reinforcement will be used throughout. Use 4 No. 6 bars in all spans (Table A-3).

A check is now made to see where the neutral axis falls at service levels, where $k_w = 0.13$ (from Table A-6):

$$k_w d = 0.13 \times 24 = 3.12 \text{ in.} < 4 \text{ in. slab thickness} \quad \text{(O.K.)}$$

As suspected earlier, the neutral axis for this tee section falls quite close to the bottom of the slab. The stress diagram is shown in the following sketch.

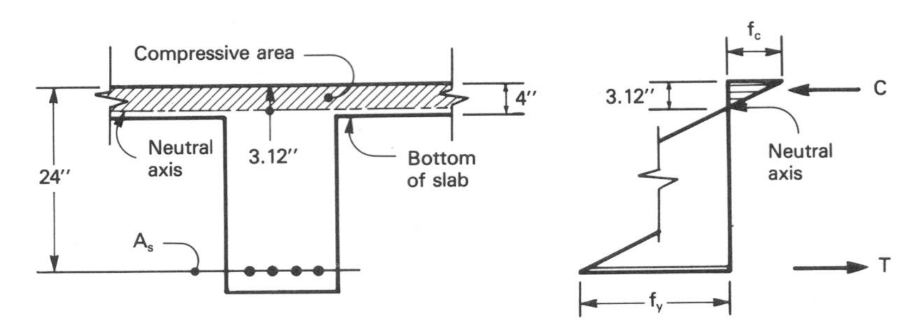


**Figure E9-4(f)** Location of Neutral Axis.

Actually, the neutral axis at ultimate load could fall below the slab a small amount without causing a large amount of error. The fact that it is close to the bottom of the slab should not be construed to mean that the design is marginal in any way.

Cutoff points for positive reinforcement are shown in the following sketch. The inflection points are those computed earlier, shown in the sketch with the moment coefficients. For reference, cutoff lengths $d$, $12d_b$, and span/16 are

$$d = 24 \text{ in.}$$

$$12d_b = 12 \times 0.75 = 9 \text{ in. for No. 6 bars}$$

$$= 12 \times 1.127 = 13.5 \text{ in. for No. 9 bars}$$

$$\frac{\text{span}}{16} = \frac{18 \text{ ft } 10 \text{ in.}}{16} = 14.1 \text{ in.}$$

Use a cutoff 24 in. beyond the inflection points.

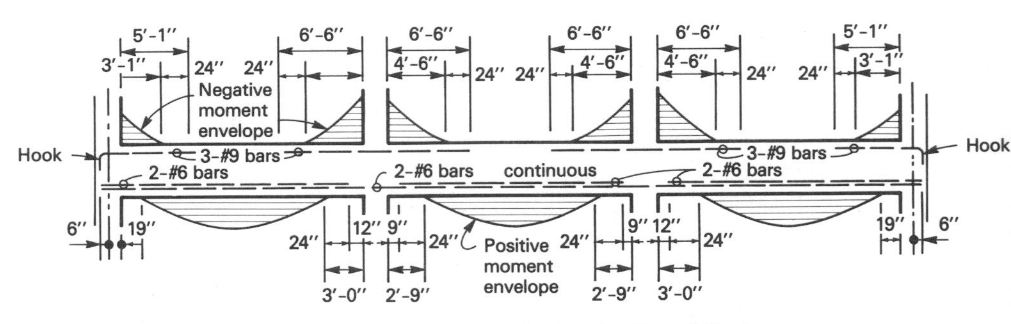


**Figure E9-4(g)** Cutoff Points for Flexural Steel.

The design of the girder is now complete. A sketch of the final design is given at the conclusion of the example, to include the column design. The column design follows.

The columns are designed using the methods developed in Chapter 8. The axial loads are computed assuming that a column carries all loads halfway to the next support on all sides. The flexural load acting on a column is taken as the difference between the girder moments at either side of the column; half this difference is taken by the upper column and half by the lower column. Column B3 shown in the following sketch is taken as a typical interior column.

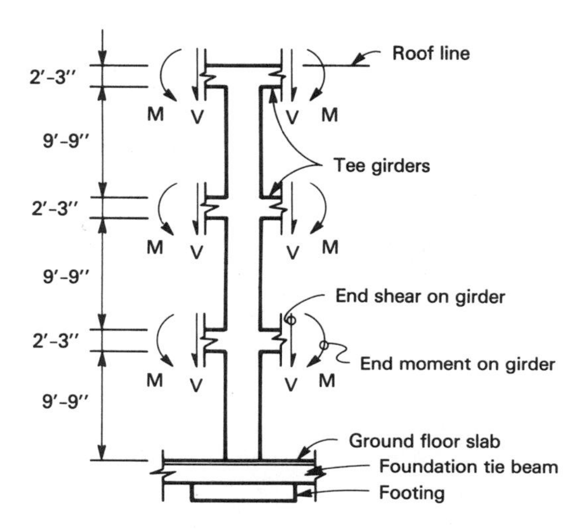


**Figure E9-4(h)** Typical Interior Column.

In computations for the following column loads, the weight of the joist system at all levels is assumed to be the same as that computed earlier for the second floor. Similarly, the girder stem is assumed to be the same size at all levels. In a complete design, these sizes might vary from floor to floor.

| | | |
|---|---|---|
| Roofing load: | $10 \text{ psf} \times 28 \times 20$ | $= 5{,}600$ lb |
| Roof joist system: | $77 \text{ psf} \times 28 \times 20$ | $= 43{,}120$ lb |
| Tee stem: | $320 \text{ plf} \times 20$ | $= 6{,}400$ lb |
| Column weight: | $\left(\frac{14}{12}\right)^2 \times 9.75 \times 145$ | $= 1{,}924$ lb |
| Live load: | $20 \text{ psf} \times 28 \times 20$ | $= 11{,}200$ lb |

At the base of third-floor columns, $P_{DL} = 57$ kips, $P_{LL} = 11$ kips.

| | | |
|---|---|---|
| Third-floor joists: | $77 \text{ psf} \times 28 \times 20$ | $= 43{,}120$ lb |
| Misc. dead load: | $10 \text{ psf} \times 28 \times 20$ | $= 5{,}600$ lb |
| Tee stem: | $320 \text{ plf} \times 20$ | $= 6{,}400$ lb |

| | | |
|---|---|---|
| Column weight: | $\left(\frac{14}{12}\right)^2 \times 9.75 \times 145$ | $= 1{,}924$ lb |
| Live load: | 75 psf × 28 × 20 | = 42,000 lb |

At the base of second-floor columns, $P_{DL} = 114$ kips, $P_{LL} = 53$ kips.

| | | |
|---|---|---|
| Second-floor joists: | 77 psf × 28 × 20 | = 43,120 lb |
| Misc. dead load: | 10 psf × 28 × 20 | = 5,600 lb |
| Tee stem: | 320 plf × 20 | = 6,400 lb |
| Column weight: | $\left(\frac{14}{12}\right)^2 \times 9.75$ | $= 1{,}924$ lb |
| Live load: | 75 psf × 28 × 20 | = 42,000 lb |

At the base of first-floor columns, $P_{DL} = 171$ kips, $P_{LL} = 95$ kips.

The maximum axial load on an interior column is, therefore,

$$P_{DL} = 171 \text{ kips} \qquad P_{LL} = 95 \text{ kips}$$

The minimum axial load on an interior column is then

$$P_{DL} = 171 \text{ kips} \qquad P_{LL} = \tfrac{1}{2}P_{max} = 48 \text{ kips}$$

The moment acting on a column at any level is taken as one-half of the difference between the ACI moment values at that level. The ACI moment values are shown in the moment envelope given earlier:

$$\Delta M = \frac{wL^2}{10} - \frac{wL^2}{11} = \frac{wL^2}{110}$$

$$M_{DL} = \frac{wL^2}{110} = \frac{2.94 \times 18.83^2}{110} = 9.5 \text{ kip-ft}$$

$$M_{LL} = \frac{wL^2}{110} = \frac{2.10 \times 18.83^2}{110} = 6.8 \text{ kip-ft}$$

The nominal ultimate loads are, therefore,

$$\text{maximum } P_n = \frac{1.4 \times 171 + 1.7 \times 95}{0.7} = 573 \text{ kips}$$

$$\text{minimum } P_n = \frac{1.4 \times 171 + 1.7 \times 48}{0.7} = 459 \text{ kip-ft}$$

$$\text{maximum } M_n = \tfrac{1}{2}\frac{(1.4 \times 9.5 + 1.7 \times 6.8)}{0.7} = 18 \text{ kip-ft}$$

The eccentricity ratio at maximum axial load is

$$\frac{e_n}{t} = \frac{M_n/P_n}{t} = \frac{18{,}000 \times 12/573{,}000}{14} = 0.027$$

The eccentricity ratio at minimum axial load is

$$\frac{e_n}{t} = \frac{M_n/P_n}{t} = \frac{18{,}000 \times 12/459{,}000}{14} = 0.034$$

Since both these values are less than 0.10, a minimum value of eccentricity ratio $e_n/t$ of 0.10 will be used.

The stress ratio $R_n$ is computed for entry into Table A-11:

$$R_n = \frac{P_n/bt}{0.85f'_c} = \frac{573{,}000/14 \times 14}{0.85 \times 4000} = 0.86$$

Enter Table A-9 with $e_n/t = 0.1$, $R_n = 0.86$, for steel uniformly distributed around the perimeter. Find the required steel ratio to be about 1%. The steel area is now computed:

$$A_s = \rho bt = 0.01 \times 14 \times 14 = 1.96 \text{ in}^2$$

Use 4 No. 7 bars.

For ties, a bar size of No. 3 is tried. The tie spacing is calculated from Code (7.10.5) provisions:

| | | |
|---|---|---|
| Tie spacing: | < 48 tie diameters or 48 × 0.375 | = 18 in. |
| | < 16 bar diameters or 16 × 0.875 | = 14 in. |
| | < least dimension or | = 14 in. |

Use No. 3 ties at 14 in. spacing the entire height of the column. The final design of the girder and column are shown below.

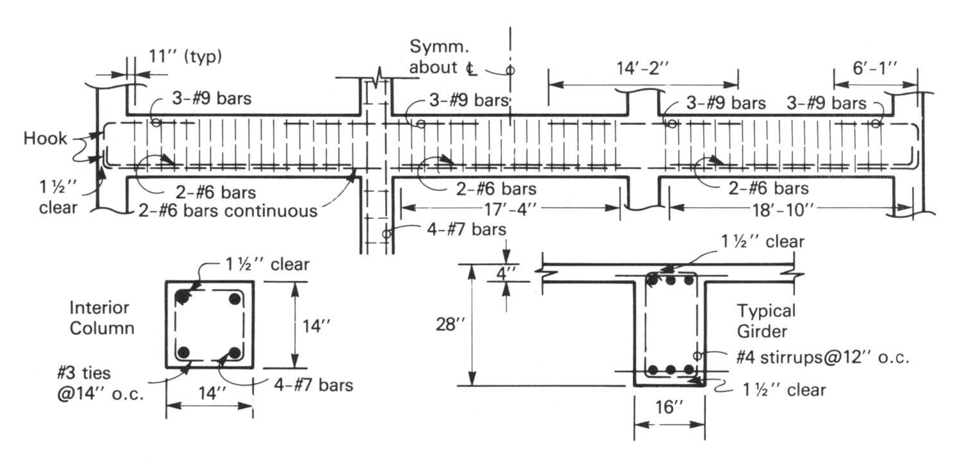


**Figure E9-4(i)** Final Design of Girder and Column.

## REVIEW QUESTIONS

1. Why are shear and moment envelopes a particularly valuable tool in the design of concrete beams but are usually unnecessary in the selection of steel beams?
2. When a continuous floor girder from a rigid frame is "removed" from the frame for analysis and design, how are the continuous columns that frame into the girder treated?
3. Values of moment at a support can vary considerably depending on whether the load is uniform, checkerboarded, or otherwise intermittent. What happens to the values of shear under these conditions?
4. If a frame is braced against lateral movement (or sidesway), how is it that moments can occur in the columns?
5. What are the design load cases for an interior column in a braced frame where the frame is being designed using the ACI coefficients?
6. Why are moments likely to be more severe on an exterior column in a braced frame than on an interior column?
7. Why is the ACI approximate method not likely to be applicable to timber construction?
8. In view of the beam shears that are computed when the ACI coefficients are used, justify the use of simple "tributary areas" to compute the final axial load on a concrete column.
9. What are the conditions of configuration and loading that must exist if the ACI approximate method is to be used?
10. At a continuous interior column of constant size, half the difference in girder moment is distributed to the column below, half to the column above. What happens at a roof girder?
11. In view of the answers to Questions 6 and 10, where is the most critically loaded column in a multistory building likely to be, insofar as moments are concerned?

## PROBLEMS

Sketch the shear and moment envelopes for the following cases, where both the live load and the concentrated load may or may not be in place at a particular time.

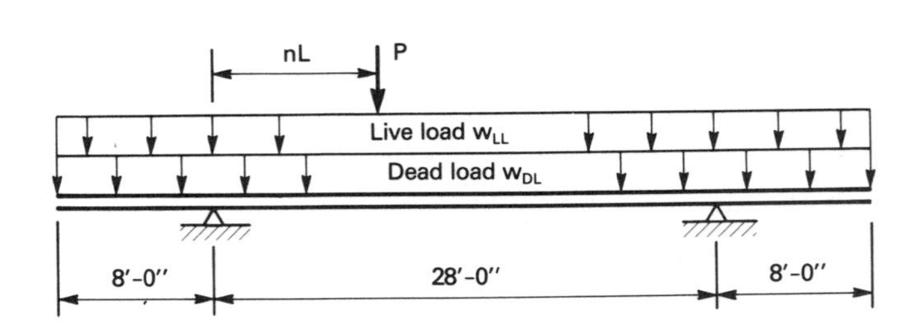

| Problem | $P$ (lb) | $w_{LL}$ (plf) | $w_{DL}$ (plf) | $nL$ |
|---|---|---|---|---|
| **1** | 4500 | 1500 | 2000 | 7′-0″ |
| **2** | 6750 | 2000 | 2000 | 7′-0″ |
| **3** | 6750 | 2000 | 2000 | 14′-0″ |
| **4** | 9000 | 3000 | 2000 | 14′-0″ |

Using the ACI approximate method, sketch the shear and moment envelopes for the end span of the continuous beam shown below.

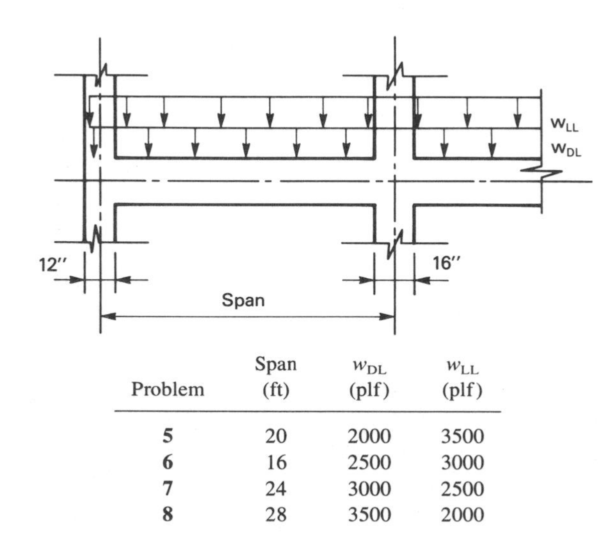


| Problem | Span (ft) | $w_{DL}$ (plf) | $w_{LL}$ (plf) |
|---|---|---|---|
| **5** | 20 | 2000 | 3500 |
| **6** | 16 | 2500 | 3000 |
| **7** | 24 | 3000 | 2500 |
| **8** | 28 | 3500 | 2000 |

**9.** Determine whether the pan-joist system of Example 9-4 is adequate for the given loads.

**10.** For the frame of Example 9-4, design the perimeter ell beam along line 1 at the first floor level. The outside dimensions of the stem should match those of the interior girders.

**11.** For the frame of Example 9-4, design the roof girder along line 3.

**12.** For the frame of Example 9-4, design the ell beam along line A at the first-floor level. Note that the adjacent parallel beam is a joist; it is spaced one ''pan'' width away from the ell beam. For the sake of the fenestration, the outside dimensions of this beam should match those of the perimeter ell beams of Problem 10; as a result, the beam will be much larger than actually needed.

**13.** For the frame of Example 9-4, design the column at grid B1, assuming that all joists and tee-beam stems are the same as those designed in the example for the first floor. The overall dimensions of the column need not match column B-3.

**14.** For the frame of Example 9-4, design the column at grid A1, assuming that all joists and tee-beam stems are the same as those designed in the example for the first floor. The overall dimensions of the column need not match column B-3.

# 10

# Foundations and Slabs on Grade

## GENERAL

Probably the single most common application of concrete in buildings is its use in foundations and slabs on grade. Regardless whether a building is constructed of concrete, masonry, timber, or steel, its foundations will almost certainly be constructed of concrete. Any textbook concerning design in concrete must therefore include at least an introduction to the design of simple foundations.

Foundation loadings are more complex than those used for superstructure design. The upward support for the foundation comes as a result of soil pressures induced by the structure. There is a secondary interaction between the structure and its foundation which can sometimes require a complex, time-consuming solution. The solution for these soil–structure interactions is much simpler for shallow foundations than for deep foundations.

The following discussions are limited to simple shallow foundations. Shallow foundations are defined herein as those founded within about 6 ft of the finished floor. They are by far the most common type of foundation.

## TYPES OF SHALLOW FOUNDATIONS

Three distinct types of shallow foundations shown in Fig. 10-1 are considered in the following discussions,

1. Spread footings under individual columns

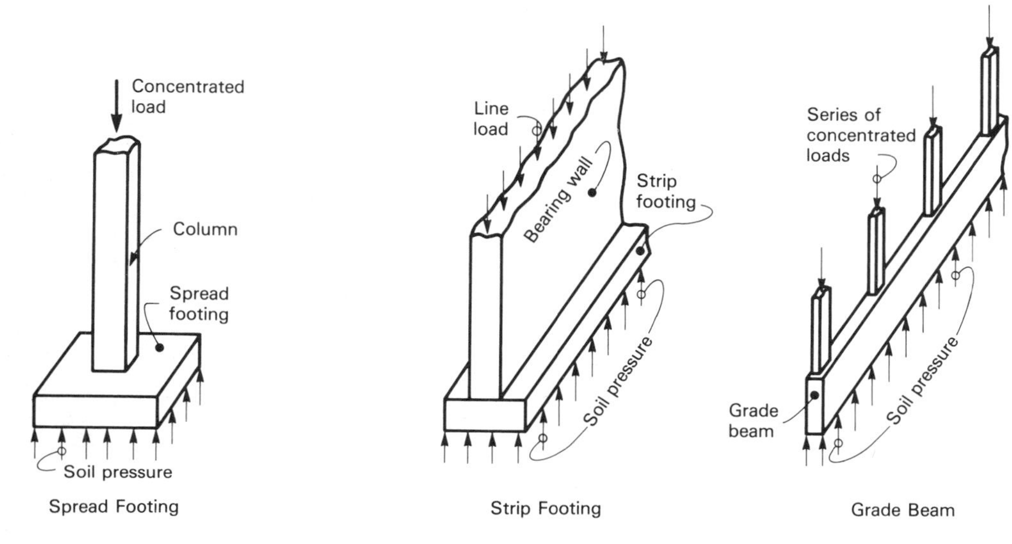


**Figure 10-1** Types of Foundations.

2. Strip footings under bearing walls
3. Grade beams under repetitive columns

All three types of foundations function by distributing a concentrated load over a larger bearing area.

Spread footings, as shown in Fig. 10-1, are subject to flexure as a cantilever in two directions. The column location is kept concentric; any eccentricities will result in a nonuniform distribution of soil pressure under the footing. Where space is restricted at one side, the spread footing may be made rectangular rather than square, but the column is still centered.

Strip footings are subject to flexure only in the outstanding legs; flexural reinforcement is required only in the short direction. There is no stress in the footing in the direction of the wall. The wall itself must be designed to take any variations in loads or settlements along its length.

Grade beams are continuous beams subject to flexure longitudinally along the line of columns they support. They are loaded on the bottom face by the distributed soil pressure. The loading system of the grade beam of Fig. 10-1 may look more familiar if the sketch is viewed upside down.

Footings are rarely reinforced for shear. Since forming costs are minimal, the sections are simply made large enough to take the shear without reinforcement. Although Code permits shear reinforcement in footings, that case is not included in subsequent discussions.

## FOUNDATION LOADING AND FAILURE MODES

Allowable soil pressure under a footing may be limited by one of two considerations: either by the differential settlement between adjacent footings or by the bearing strength of soil. By far the more common limitation is settlement. Even in sandy soils, the progressive crushing or inelastic deformation of friable sands within the first year after construction usually produces the limiting case for footing pressures.

When settlement of the foundation rather than soil strength governs a design, the size of the contact area is based on a reduced value for the allowable soil pressure; this reduced pressure serves to limit settlements under day-to-day service loads. For the other case (i.e., where soil strength governs), the size of the contact area is based on a peak value of soil pressure, but still at working levels. There is no accepted way to correlate test data on soils to ultimate strength criteria, recognizing that the ultimate strength of the soil can fluctuate drastically due to such things as daily weather changes.

Thus, whether the size of the contact area is limited by settlements or by soil strength, it is seen that all soil criteria are specified at working levels or service levels. Ultimate soil pressure is not used. The working dead and live loads must therefore be carried forward in the calculations to be used in establishing the size of the contact area of the foundation.

There is, however, an approximate method that is often seen for converting the column design load at the ultimate level to the footing design load at the working level. As suggested at the beginning of Chapter 8, an approximate order of magnitude for $P_w$ may be computed from the nominal ultimate load $P_n$ by applying a factor of safety of 2.17, that is,

$$P_w = \frac{P_n}{2.17} \tag{8-1}$$

It should be recognized that "conservatism" works backwards in Eq. (8-1). A larger factor of safety would produce a lower estimated value of $P_w$, which in turn would reduce the required size of contact area, which would then produce a higher soil pressure under the actual working loads. The approximation of Eq. (8-1) is therefore not always safe and should be used with care; it should not be used at all where settlements control the design.

## ALLOWABLE SOIL PRESSURES

The allowable soil pressure is the increase in soil pressure that may be applied to a soil at a certain stratum. The soil will safely accept this pressure in addition to its regular overburden pressures. Usually, if the overburden is permanently removed (e.g., basements), the allowable soil pressure may be increased proportionately.

Settlements in soil are time dependent, usually taking place over several months or several years. Consequently, when establishing the size of the contact area that will limit settlements, only those loads may be considered that will exist long enough to produce long-term settlements. Certainly, the dead load of the structure meets this requirement but only about half the live load (furnishings, carpets, files, books, etc.) may be considered long term. The remaining live load comes and goes, causing only elastic deformations in the soil; long-term inelastic settlements have no time to develop.

When establishing the contact area that will limit settlements, it is common practice to use dead load plus about 50% live load for buildings that serve routine architectural functions. The live load in buildings such as libraries may be much higher, as much as 80% of maximum, while in auditoriums the long-term load may contain only 20% of the maximum live load. For simplicity in the subsequent examples, a value of dead load plus 50% live load will be used.

The allowable soil pressure corresponding to this long-term load is specified as a result of the soils investigation (or by Code). This pressure is commonly termed the "reduced allowable soil pressure" and is used to compute the required size of the contact area.

Once the size of the contact area is established, there is no further need for the reduced allowable soil pressure. When the higher peak loads are applied to the building, a higher soil pressure will obviously result. This higher soil pressure, called the peak pressure, is used for all subsequent calculations for stress; the steel and concrete must of course sustain the peak pressure even if only for a short time.

When settlements control the design, the peak soil pressure is not limited to a maximum value. At the time the allowable soil pressure is being determined, all conditions of load, strength and settlement are considered. When the reduced allowable soil pressure is finally established, it may thereafter be assumed that the strength has been found adequate for all peak-load conditions; the soil pressure is limited only by settlements.

For the other case of loading (i.e., when the strength of the soil rather than settlement governs the design), the peak load is the only load ever considered. Combinations of loads that produce peak conditions are discussed with the load factors of Chapter 3. In the subsequent examples, the peak loading will be taken as dead load plus 100% live load.

## FOOTING ROTATIONS

Where soils are so soft that foundations must be designed to limit settlements, the soil can be expected to offer very little resistance to small rotations of the footing. Although the soil–structure interaction of such problems is far beyond the scope of this book, it should be apparent that a soil has only a small fraction of the rigidity of concrete. As a consequence, the footing will simply rotate slightly, similar to a raft floating in water, and thereby relieve any moments that occur at the base of the column. Further, the rotation is so small that no significant changes in soil pressure are induced.

Since the footings rotate so easily compared to the concrete columns, columns founded on isolated spread footings should be assumed to be hinged at their bases. The hinge, however, should be assumed to occur at the contact surface at the underside of

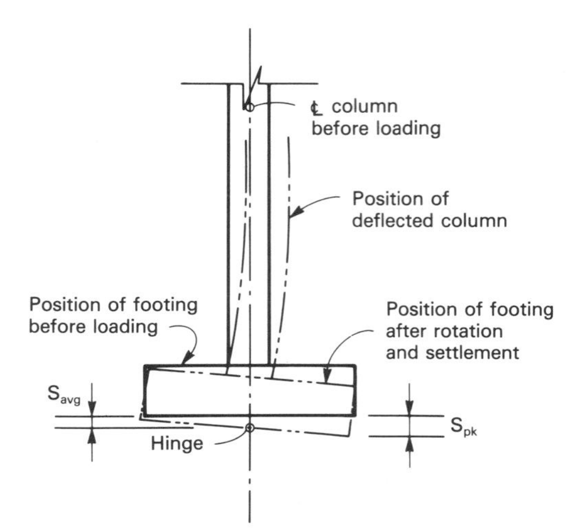


**Figure 10-2** Footing Rotations.

the footing, as shown in Fig. 10-2. The length of the column for calculations is then the distance from the underside of the footing to the underside of the first-floor girders.

Consequently, the effective load delivered to an isolated footing should be taken as the axial load only; moments are simply dissipated. There may, in fact, be some rotations due to the elastic column moments, but the soil pressures can be expected to vary very little due to the small rotations common to an elastic structure. In subsequent discussions, the average soil pressure under a spread footing will be computed as the tributary concentric load divided by the contact area of the footing.

## SPREAD FOOTINGS

Spread footings are rarely, if ever, reinforced for shear. Like slabs, it is far more practical to increase the depth of the footing slightly to increase its shear capacity rather than place numerous stirrups. Since footings are cast directly on soil, formwork is minimal and the extra cost is only that of bulk concrete.

Since shear reinforcement is not used, spread footings can be expected to be quite thick in order to sustain the shear loads. Spans (or overhangs) are comparatively short, so flexure is not usually large enough to require a deeper section than that required for shear. As a general guide, the overall thickness of a spread footing can be expected to be about one-sixth of the maximum dimension, or less, with a minimum thickness of about 12 in. being commonly observed.

When concrete is cast against soil, the minimum cover over the reinforcement is specified by Code (7.7.1) at 3 in. The consequences of foundation failure are so serious, however, that this limit is usually observed as the barest minimum. In subsequent examples, the concrete cover is taken as 4 in. from the centerline of flexural reinforcement, which provides somewhat more than the minimum 3 in. clear cover.

Spread footings are assumed to act as cantilevered beams in two directions. The critical section for bending in a spread footing that supports a concrete column is at the face of the column. For a footing that supports a steel column, the critical section for bending is taken halfway between the face of the steel column and the edge of the base plate. Critical sections are shown in Fig. 10-3.

It should be noted that the bending of the footing is assumed to occur uniformly along the full width of the footing as shown in Fig. 10-3, as if the column were actually a wall extending all the way across the footing. Although this assumption is admittedly a drastic simplification, its success over the years justifies its continued use. For rectangular footings, the procedure is repeated in the other direction to obtain the reinforcement in that direction.

The same simplifying assumption concerning bend lines applies also when designing the spread footing for beam shear. As for other types of beams, the critical section for beam shear is taken at a distance $d$ from the face of the column, shown as lines $AB$ and $CD$ in Fig. 10-3. The ultimate shear stress on the concrete at the critical section is, as usual in beams,

$$v_n = 2\sqrt{f'_c} \tag{5-4a}$$

With no shear reinforcement, the ultimate shear force at the critical section may not be greater than $V_c$, where

$$V_c = v_n b_w d \tag{5-3}$$

and $b_w$ is the total width of the footing at the critical section.

In addition to the foregoing design criteria for beam shear and flexure, Code im-

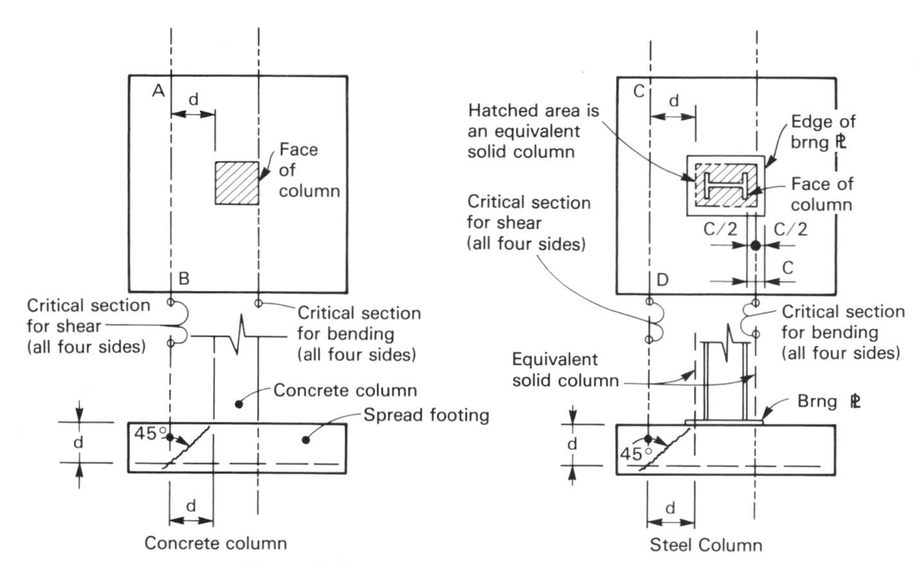


**Figure 10-3** Critical Sections in Spread Footings.

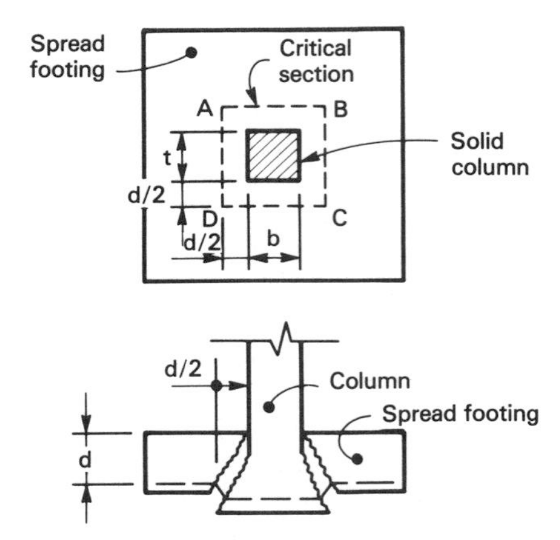


**Figure 10-4** Punching Shear.

poses an additional limitation due to "punching shear." Punching shear produces the pyramidal punch-out shown in Fig. 10-4. Whereas critical beam shear is taken at a distance $d$ from face of support, critical punching shear is taken at a distance $d/2$ from face of support, shown as line *ABCD* in Fig. 10-4. The ultimate shear stress for punching shear in concrete without shear reinforcement is specified by Code (11.11.2.1):

$$v_n = \left(1 + \frac{2}{\beta_c}\right) 2\sqrt{f'_c} \tag{10-1a}$$

but

$$v_n \leq 4\sqrt{f'_c} \tag{10-1b}$$

where $\beta_c$ is the ratio of the long side to the short side $(t/b)$ of the contact area between the column and the footing ($\beta_c = 1$ for square columns and $\beta_c = \infty$ for continuous walls). The ultimate punch-out load to be taken by the critical section is, then,

$$V_c = v_n b_0 d \tag{10-2}$$

where $b_0$ is the perimeter of the critical punch-out shown in Fig. 10-4. For the line *ABCD* in Fig. 10-4,

$$b_0 = 2(b + d) + 2(t + d) \tag{10-3}$$

In addition to flexure, beam shear, and punching shear, the footing must sustain the bearing load presented by the column or wall above. The bearing load on the footing is limited by Code (10.15):

$$P_n = 0.85 f'_c A_g \tag{10-4}$$

where all symbols are as used earlier and the undercapacity factor $\phi$ is 0.7 for bearing.

The allowable bearing load given by Eq. (10-4) may be increased by multiplying by the factor $\sqrt{A_2/A_1}$, but not by more than 2, where $A_2$ is the area of the top of the footing concentric with $A_1$, and for footings, $A_1 = A_g$. In almost all cases where this Code provision is applied to footings, the factor $\sqrt{A_2/A_1}$ will be much greater than 2, with the final result that bearing load will almost always be double that of Eq. (10-4).

For the design of footings, there are thus four conditions to be met: bearing load on top face, punching shear, beam shear, and beam flexure. There are special design tables and graphs for use in designing spread footings; they are useful if one is frequently engaged in such designs. For infrequent use, it is far faster simply to guess the depth of footing and check the four conditions to see if the guess is adequate. That procedure is illustrated in the following examples, first for a square footing and second for a rectangular footing.

**Example 10-1**

*Design of a square spread footing*
*Footing subject only to axial load (no moments)*
*Strength method of design of concrete*

*Given:*
$P_{DL} = 115$ kips
$P_{LL} = 95$ kips
Soil pressure $p_a = 3000$ psf
Column size $16 \times 16$ in.
$f'_c = 3000$ psi
Grade 40 steel

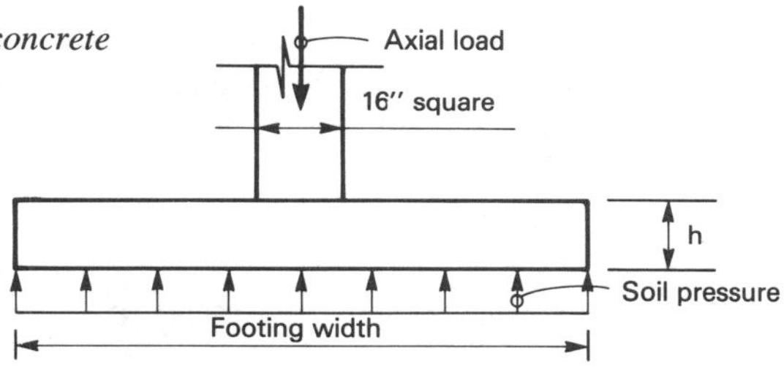

In foundation design, the difference in weight between concrete (145 pcf) and the displaced overburden (115 pcf) is commonly ignored, although some references (Peck, et al, Foundation Engineering, $1^{ST}$ Ed., Wiley, 1953) include it. In subsequent calculations, it is ignored. If it were to be included, the increase in dead load would be estimated and added to the column dead load; at the conclusion of the calculations the accuracy of this initial estimate would be checked and verified.

The required contact area of the footing is the total load, $P_{DL} + P_{LL}$, divided by the allowable soil pressure, 3000 psf:

$$\text{area} = \frac{P_{DL} + P_{LL}}{p_a} = \frac{115 + 95}{3000} = 70 \text{ ft}^2$$

Use square footing, 8 ft 6 in. or 102 in. The actual soil pressure under this footing size is

$$p_{DL} = \frac{P_{DL}}{\text{area}} = \frac{115}{8.5 \times 8.5} = 1.6 \text{ ksf}$$

$$p_{LL} = \frac{P_{LL}}{\text{area}} = \frac{95}{8.5 \times 8.5} = 1.3 \text{ ksf}$$

The first check is for bearing load on top of the footing. The total load $P_n$ delivered to the top of the footing is, where $\phi = 0.7$ for bearing,

$$P_n = \frac{1.4P_{DL} + 1.7P_{LL}}{\phi} = \frac{1.4 \times 115 + 1.7 \times 95}{0.7}$$

$$= 461 \text{ kips}$$

The allowable bearing load on the footing is

$$P_n = 0.85f'_cA_g = 0.85 \times 3000 \times 16 \times 16 = 653 \text{ kips}$$

This allowable load is multiplied by $\sqrt{A_2/A_1}$:

$$\sqrt{\frac{A_2}{A_1}} = \sqrt{\frac{102 \times 102}{16 \times 16}} = 6.38 \text{ but } < 2$$

Hence

$$P_n = 2 \times 653 = 1306 \text{ kips} > 461 \text{ kips} \quad \text{(O.K.)}$$

The bearing capacity of the footing is more than adequate to carry the column load. It remains now to select the thickness of the footing.

A trial thickness $h$ of about one-sixth of the footing width is assumed, or about 18 in. The effective depth $d$ is then taken at 14 in. This trial thickness of 14 in. is first checked for punching shear. The average ultimate punching shear stress occurs at a distance $d/2$ from the face of the column on all four sides of the column, as shown in the following sketch.

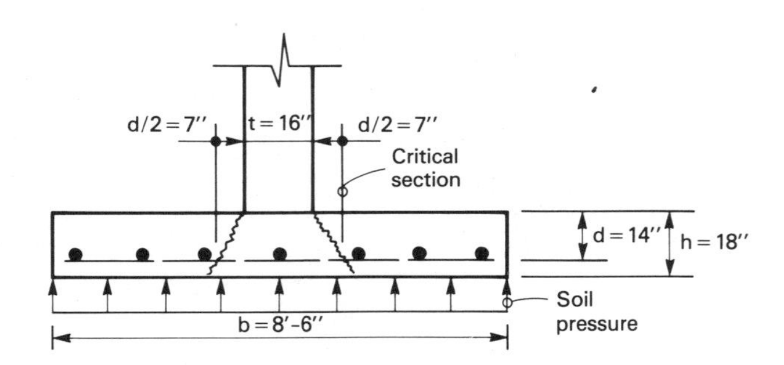


The total punching shear force acting on the section is found by statics:

$$P_{DL} = p_{DL}[b^2 - (t + d)^2] = 1.6(8.5^2 - 2.5^2) = 106 \text{ kips}$$

$$P_{LL} = p_{LL}[b^2 - (t + d)^2] = 1.3(8.5^2 - 2.5^2) = 86 \text{ kips}$$

Hence

$$P_n = \frac{1.4P_{DL} + 1.7P_{LL}}{\phi}$$

$$= \frac{1.4 \times 106 + 1.7 \times 86}{0.85} = 347 \text{ kips}$$

The average punching shear stress at ultimate load is given by Code, where $\beta_c = 1$:

$$v_n = 2\sqrt{f'_c}\left(1 + \frac{2}{\beta_c}\right) = 6\sqrt{f'_c}$$

but $v_n$ must not exceed $4\sqrt{f'_c}$. Hence

$$v_n = 4\sqrt{f'_c} = 220 \text{ psi}$$

The capacity of the footing to carry punching shear is then, where $b_0 = 4(t + d) =$ 120 in.,

$$V_c = v_n b_0 d = 220 \times 120 \times 14 = 370 \geq 347 \text{ kips} \quad \text{(O.K.)}$$

This trial depth of 14 in. is thus seen to be adequate to take the punching shear.

Next the trial depth of 14 in. is checked for its capacity in beam shear. The critical shear force is found at a distance $d$ from the face of the column, as shown in the following sketch.

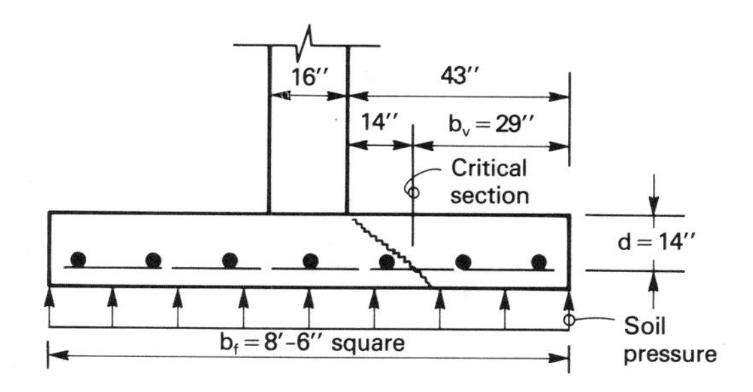


The shear acting on the footing beyond the line of the critical section is

$$V_{\text{DL}} = p_{\text{DL}} \times b_v \times b_f = 1.6 \times \left(\frac{29}{12}\right) \times 8.5 = 33 \text{ kips}$$

$$V_{\text{LL}} = p_{\text{LL}} \times b_v \times b_f = 1.3 \times \left(\frac{29}{12}\right) \times 8.5 = 27 \text{ kips}$$

Hence

$$V_n = \frac{1.4V_{\text{DL}} + 1.7V_{\text{LL}}}{\phi}$$

$$= \frac{1.4 \times 33 + 1.7 \times 27}{0.85} = 108 \text{ kips}$$

The capacity of the concrete in beam shear is

$$V_{\text{cr}} = v_n b_w d = 2\sqrt{f'_c} \times 102 \times 14 = 156 > 108 \text{ kips} \quad \text{(O.K.)}$$

The trial depth of 14 in. is thus seen to be adequate to carry the beam shear on the section.

The trial depth of 14 in. is now checked for beam flexure, where $M = wL_c^2/2$ on the cantilever length $L_c$, as shown in the following sketch.

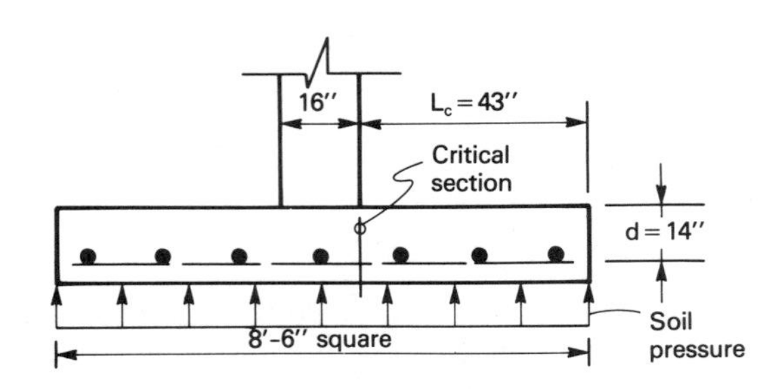


The moment acting on the critical section is

$$M_{DL} = \frac{w_{DL}L_c^2}{2} = \frac{1.6 \times 8.5\left(\frac{43}{12}\right)^2}{2} = 87 \text{ kip-ft}$$

$$M_{LL} = \frac{w_{LL}L_c^2}{2} = \frac{1.3 \times 8.5\left(\frac{43}{12}\right)^2}{2} = 71 \text{ kip-ft}$$

Hence

$$M_n = \frac{1.4M_{DL} + 1.7M_{LL}}{\phi}$$

$$= \frac{1.4 \times 87 + 1.7 \times 71}{0.9} = 269 \text{ kip-ft}$$

The section modulus $Z_c$ is found by the usual means:

$$\frac{M_n}{0.85f'_c} = Z_c \qquad \frac{269{,}000 \times 12}{0.85 \times 3000} = \text{coeff.} \times 102 \times 14 \times 14$$

$$\text{coeff.} = 0.063$$

From Table A-5, select $\rho = 0.0050$ ($\rho_{min}$ for grade 40).

Solve for $A_s$, $A_s = \rho bd = 0.005 \times 102 \times 14 = 7.14 \text{ in}^2$.

Use 12 No. 7 bars each way, evenly spaced ($A_s$ provided $= 7.22 \text{ in}^2$). The trial depth of 14 in. is thus seen to be adequate for flexure, beam shear, punching shear, and bearing; it is selected as the final depth of the footing.

The following sketch shows the final design, to include dowels from the footing to the column above.

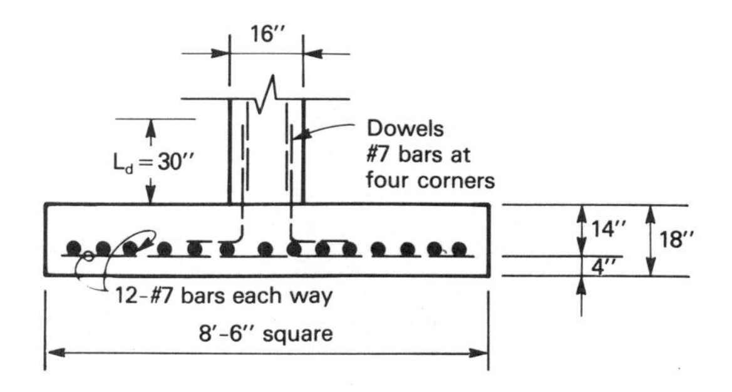


The dowels are chosen rather arbitrarily in this case. Had the applied column load been higher than the allowable bearing capacity of the footing, the excess load would have to be transferred by dowels; the size of the dowels and their embedment would then be designed to carry this excess load. In this footing, however, the footing is adequate to carry the entire load in bearing so theoretically the dowels are not needed. For such cases, the practice is to design the dowels to carry about 10% of the flexural capacity of the column to account for any accidental moments on the footing. This moment is, where $Z_c = bh^2/6$ (approximately),

$$M_n = (0.85 f'_c Z_c)(0.10) = \left(\frac{0.85 \times 3000 \times 16 \times 16^2}{6}\right)(0.10)$$

$$= 174 \text{ kip-in.} = 14.5 \text{ kip-ft}$$

For dowels at the four corners, $\rho' = \rho$ and

$$\frac{M_n}{0.85 f'_c} = Z_c \qquad \frac{174{,}000}{0.85 \times 3000} = \text{coeff.} \times 16 \times 16^2$$

$$\text{coeff.} = 0.0167$$

From Table A-5, use minimum $\rho$; hence

$$A_s = \rho_{\min} b_w d = 0.005 \times 16 \times 14 = 1.12 \text{ in}^2$$

Use No. 7 bars at the four corners.

As a point of interest, the difference between the dead load of the concrete and the weight of the soil in Example 10-1 can now be checked to see how much error is involved in ignoring this difference. At a total depth of 18 in., the additional weight is for a concrete weight of 145 pcf and a soil weight of 115 pcf,

$$\text{weight}/\text{m}^2 = (145 - 115)\left(\frac{18}{12}\right) = 45 \text{ psf} \quad (\text{error})$$

Of the total pressure of 2900 psf, this error is less than 2%.

The next example presents the design of a rectangular footing subjected to a column load. In rectangular footings, Code (15.4.4.2) requires that a major part of the

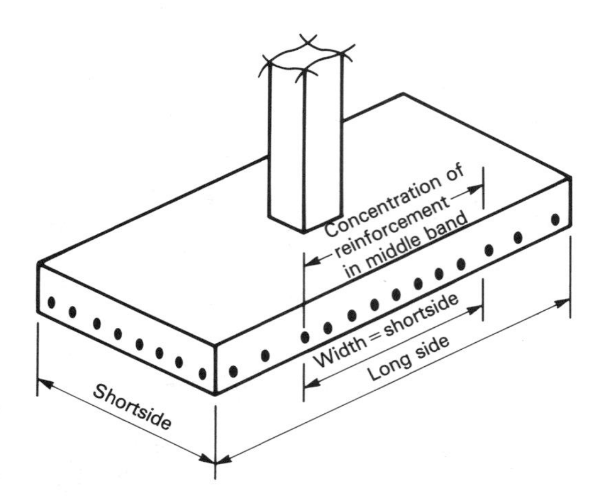


**Figure 10-5** Middle Band in Rectangular Footing.

shorter reinforcement must be located in a middle "band." The width of this middle band is equal to the narrower width of the footing, as shown in Fig. 10-5. The portion of the reinforcement to be concentrated in the middle band width is given by Code (15.4.4.2) by the formula

$$\frac{\text{reinforcement in middle band width}}{\text{total reinforcement in short direction}} = \frac{2}{\beta + 1}$$

where $\beta$ is the ratio of the long side to the short side of the footing. For a square footing, $\beta = 1$ and the equation simply requires equal distribution in both directions. The use of this formula is demonstrated in the following example.

**Example 10-2**

*Design of a rectangular footing*
*Footing subject only to axial load (no moment)*
*Strength method of design of concrete*

*Given:*
Property line as shown
$P_{\text{DL}} = 195$ kips
$P_{\text{LL}} = 110$ kips
Soil pressure $p_a = 4000$ psf
Column size $18 \times 18$ in.
$f'_c = 3000$ psi
Grade 40 steel

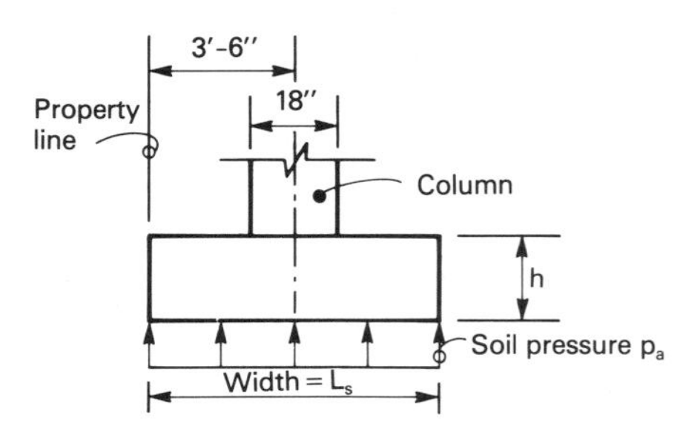


The required contact area of the footing is computed:

$$\text{area} = \frac{P_{\text{DL}} + P_{\text{LL}}}{p_a} = \frac{195 + 110}{4} = 76.3 \text{ ft}^2$$

Use a footing 7 ft × 11 ft.

The actual soil pressure is then

$$p_{\text{DL}} = \frac{P_{\text{DL}}}{\text{area}} = \frac{195}{7 \times 11} = 2.53 \text{ ksf}$$

$$p_{\text{LL}} = \frac{P_{\text{LL}}}{\text{area}} = \frac{110}{7 \times 11} = 1.43 \text{ ksf}$$

The bearing load on the footing is checked first. The load delivered to the top of the footing is

$$P_{\text{n}} = \frac{1.4P_{\text{DL}} + 1.7P_{\text{LL}}}{\phi}$$

$$= \frac{1.4 \times 195 + 1.7 \times 110}{0.7} = 657 \text{ kips}$$

The allowable load on the footing is

$$P_n = 0.85f'_cA_g = 0.85 \times 3000 \times 18 \times 18 = 826 \text{ kips}$$

This allowable load is multiplied by the factor $\sqrt{A_2/A_1} = \sqrt{7 \times 11/1.5 \times 1.5} = 5.85$, but may not be more than doubled. Hence, the allowable load is

$$P_n = 2 \times 826 = 1652 > 660 \text{ kips} \quad \text{(O.K.)}$$

The footing is therefore adequate to receive the column load from above.

The footing is checked next for punching shear. Punching shear occurs along the pyramidal punch-out shown in the following sketch.

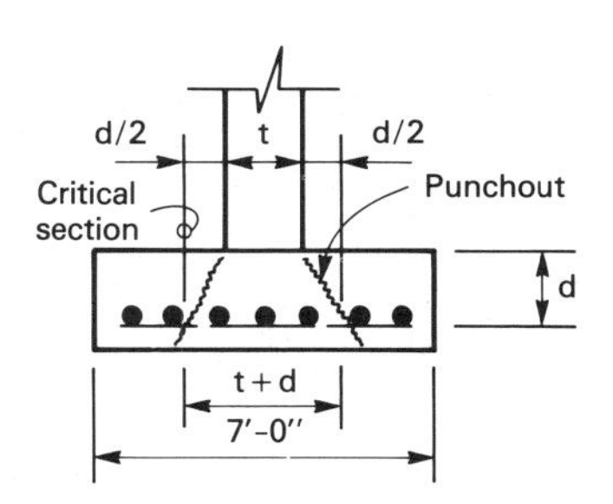


A trial thickness of about one-sixth of the longer side is assumed; hence $h = 11/6 = 24$ in. The effective depth $d$ is taken at 20 in., $t + d = 38$ in. $= 3.17$ ft. The punching load is found by simple statics:

$$P_{\text{DL}} = p_{\text{DL}}[L_L \times L_s - (t + d)^2] = 2.53(7 \times 11 - 3.17^2) = 169 \text{ kips}$$

$$P_{\text{LL}} = p_{\text{LL}}[L_L \times L_s - (t + d)^2] = 1.43(7 \times 11 - 3.17^2) = 96 \text{ kips}$$

Hence

$$P_n = \frac{1.4P_{\text{DL}} + 1.7P_{\text{LL}}}{\phi}$$

$$= \frac{1.4 \times 169 + 1.7 \times 96}{0.85} = 470 \text{ kips}$$

The punching shear stress in the concrete at ultimate load is

$$v_n = 2\sqrt{f'_c}\left(1 + \frac{2}{\beta_c}\right) \qquad \text{where } \beta_c = 1$$

$$= 2\sqrt{f'_c}\,(3) = 6\sqrt{f'_c} \text{ but} < 4f'_c$$

Hence

$$v_n = 4\sqrt{f'_c} = 4\sqrt{3000} = 219 \text{ psi}$$

The capacity of the footing in punching shear is then

$$V_c = v_n b_w d = 219(38 \times 4)20 = 666 > 470 \text{ kips} \quad \text{(O.K.)}$$

The trial depth of 20 in. is seen to be adequate for the punching shear.

The trial depth of 20 in. is next checked for beam shear at a distance $d$ from the face of column, first in the long direction, then in the short direction. The locations of the critical sections for beam shear are shown in the following sketch:

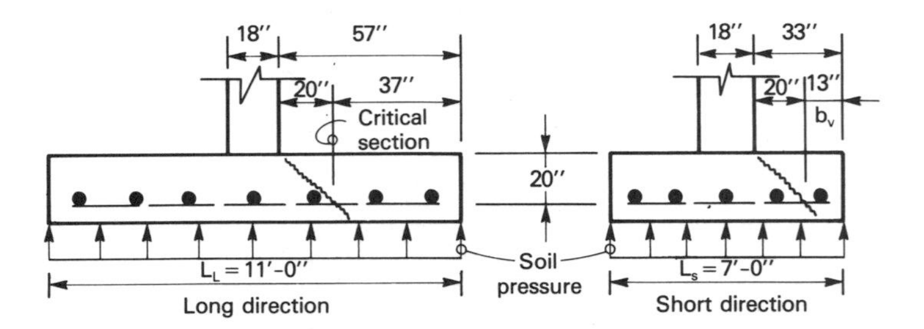


The shear force acting on the footing beyond the critical section in the long direction is found by statics:

$$V_{\text{DL}} = p_{\text{DL}} b_v L_s = 2.53\left(\frac{37}{12}\right)7.0 = 54.6 \text{ kips}$$

$$V_{\text{LL}} = p_{\text{LL}} b_v L_s = 1.43\left(\frac{37}{12}\right)7.0 = 30.9 \text{ kips}$$

Hence

$$V_n = \frac{1.4V_{\text{DL}} + 1.7V_{\text{LL}}}{\phi}$$

$$= \frac{1.4 \times 54.6 + 1.7 \times 30.9}{0.85} = 152 \text{ kips}$$

The capacity of the concrete in beam shear is

$$V_{cr} = v_n b_w d = 2\sqrt{3000} \times 84 \times 20 = 184 > 152 \text{ kips} \quad \text{(O.K.)}$$

The section is thus seen to be adequate in the long direction. In the short direction, the shear force is

$$V_{DL} = p_{DL} b_v L_L = 2.53\left(\frac{13}{12}\right)11 = 30 \text{ kips}$$

$$V_{LL} = p_{LL} b_v L_L = 1.43\left(\frac{13}{12}\right)11 = 17 \text{ kips}$$

Hence

$$V_n = \frac{1.4V_{DL} + 1.7V_{LL}}{\phi}$$

$$= \frac{1.4 \times 30 + 1.7 \times 17}{0.85} = 83 \text{ kips}$$

The capacity of the concrete in beam shear is

$$V_{cr} = v_n b_w d = 2\sqrt{3000} \times 132 \times 20 = 289 > 83 \quad \text{(O.K.)}$$

The trial depth of 20 in. is thus seen to be adequate for beam shear in both directions.

The trial depth of 20 in. is now used to design for beam flexure on the cantilever, first in the long direction, then in the short direction. The dimensions in the two directions are shown in the following sketch.

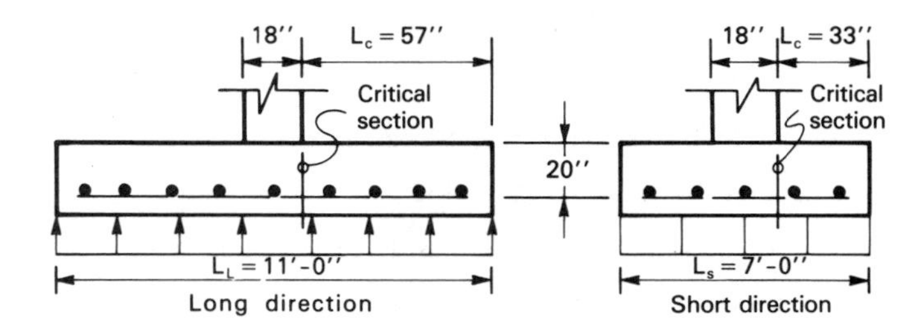


In the long direction, the moment at the critical section is

$$M_{DL} = \frac{w_{DL} L_c^2}{2} = \frac{2.53 \times 7(57/12)^2}{2} = 200 \text{ kip-ft}$$

$$M_{LL} = \frac{w_{LL} L_c^2}{2} = \frac{1.43 \times 7(57/12)^2}{2} = 113 \text{ kip-ft}$$

Hence

$$M_n = \frac{1.4M_{DL} + 1.7M_{LL}}{\phi}$$

$$= \frac{1.4 \times 200 + 1.7 \times 113}{0.9} = 525 \text{ kip-ft}$$

The plastic section modulus is computed:

$$\frac{M_n}{0.85f'_c} = Z_c \qquad \frac{525{,}000 \times 12}{0.85 \times 3000} = \text{coeff.} \times 84 \times 20^2$$

$$\text{coeff.} = 0.0735$$

From Table A-5, select $\rho = 0.0050$ (minimum $\rho$).

Compute $A_s = \rho bd = 0.005 \times 84 \times 20 = 8.4 \text{ in}^2$.

Use 11 No. 8 bars (long direction).

In the short direction, the moment at the critical section is

$$M_{\text{DL}} = \frac{w_{\text{LL}}L_c^2}{2} = \frac{2.53 \times 11(33/12)^2}{2} = 105 \text{ kip-ft}$$

$$M_{\text{LL}} = \frac{w_{\text{LL}}L_c^2}{2} = \frac{1.43 \times 11(33/12)^2}{2} = 59 \text{ kip-ft}$$

Hence

$$M_n = \frac{1.4M_{\text{DL}} + 1.7M_{\text{LL}}}{\phi}$$

$$= \frac{1.4 \times 105 + 1.7 \times 59}{0.9} = 275 \text{ kip-ft}$$

The plastic section modulus is computed:

$$\frac{M_n}{0.85f'_c} = Z_c \qquad \frac{275{,}000 \times 12}{0.85 \times 3000} = \text{coeff.} \times 132 \times 20^2$$

$$\text{coeff.} = 0.0245$$

From Table A-5, select $\rho = 0.005$ (minimum $\rho$).

Compute $A_s = \rho bd = 0.005 \times 132 \times 20 = 13.2 \text{ in}^2$.

The portion of the total $A_s$ that must be placed in the middle band, 7 ft 0 in. wide,

$$\frac{A_s \text{ in middle band}}{\text{total } A_s} = \frac{2}{(11/7) + 1} = 0.78$$

$$A_s \text{ in middle band} = 0.78 \times 13.2 = 10.3 \text{ in}^2$$

Use 19 No. 8 bars, 13 in the middle 7 ft, with 3 additional bars in each end. The following sketch shows the final steel arrangement.

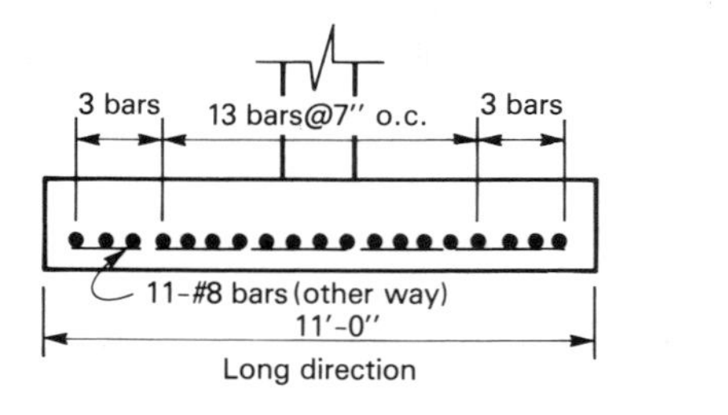


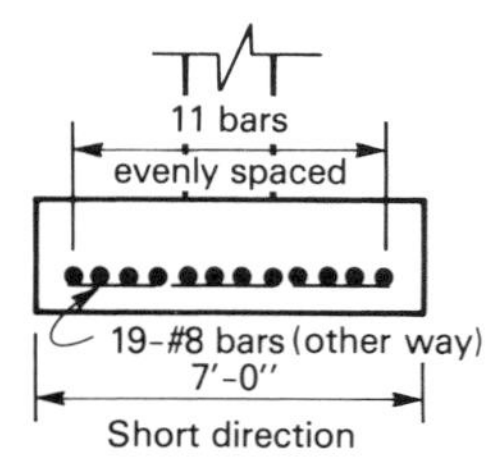

## STRIP FOOTINGS

A strip footing is assumed to act as a cantilevered slab projecting from under the wall it supports. For strip footings supporting a concrete wall, the critical section for bending is at the face of the wall. If the wall is masonry, an equivalent concrete wall is used which is half the thickness of the masonry wall. The critical section for bending is then at the face of the equivalent concrete wall, as shown in Fig. 10-6.

Although Code (15.5.2) permits the critical section for beam shear to be measured from the actual face of the masonry wall, it is common practice to measure the critical shear from the face of the equivalent wall. In such an approach, the entire design of the foundation is thus performed for an equivalent wall, affording a consistent treatment of the design. The ultimate strength of the equivalent wall in this approach is taken at twice the ultimate strength of the masonry, $f'_m$.

As with spread footings, shear reinforcement is rarely if ever used in strip footings. The strip footing is simply made thicker to take any excess shear. The thickness of strip footings may be expected to be somewhat less than the thickness of spread footings, but a thickness less than 12 in. is rarely seen.

There is no punching shear on a strip footing. A strip footing is designed only for beam shear and beam moment on the outstanding legs. There are no other special considerations in the design of a strip footing. It is simply a double cantilevered beam. A brief example will illustrate the procedure.

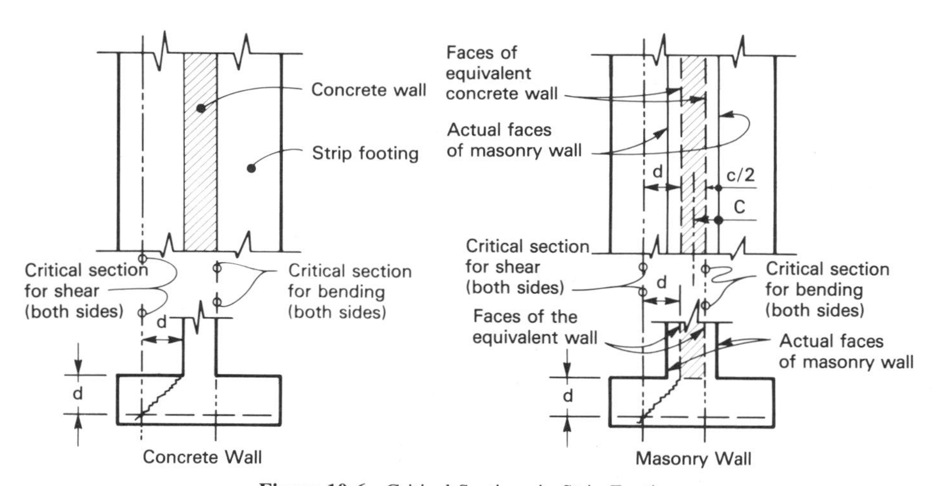


**Figure 10-6** Critical Sections in Strip Footings.

**Example 10-3**

*Design of strip footing for a masonry wall*
*Footing subject only to axial load (no moment)*
*Strength method for design of concrete*

*Given:*
$P_{DL} = 8$ kips/ft
$P_{LL} = 5.5$ kips/ft
Soil pressure $p_a = 2000$ psf
Wall thickness 12 in.
$f'_m = 1350$ psi
$f'_c = 3000$ psi
Grade 60 steel

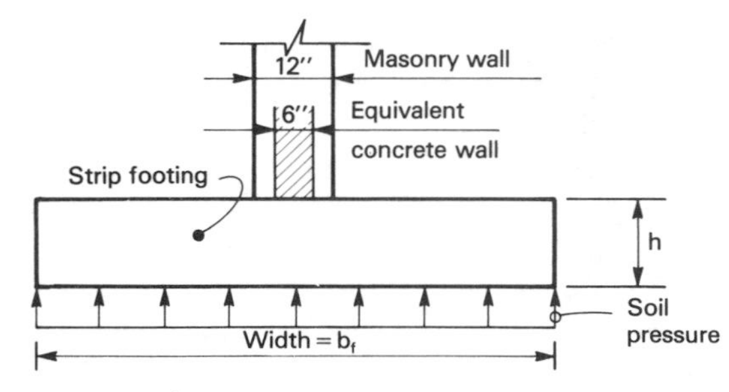


The footing will be designed for an equivalent concrete wall 6 in. thick, with $f'_c = 2f'_m = 2700$ psi. The relative dimensions are shown in the sketch above. Compute the required contact area:

$$\text{area} = \frac{P_{DL} + P_{LL}}{p_a} = \frac{8 + 5.5}{2000} = 6.75 \text{ ft}$$

Use a footing width of 7 ft.

Compute the actual soil pressures:

$$p_{DL} = \frac{P_{DL}}{b_f} = \frac{8}{7} = 1.14 \text{ kips/ft}$$

$$p_{LL} = \frac{P_{LL}}{b_f} = \frac{5.5}{7} = 0.79 \text{ kips/ft}$$

Although not usually necessary for walls, the bearing load on the top of the footing will now be checked. The load delivered to the top of the footing is

$$P_n = \frac{1.4P_{DL} + 1.7P_{LL}}{\phi}$$

$$= \frac{1.4 \times 8 + 1.7 \times 5.5}{0.7} = 29.4 \text{ kips/ft}$$

The allowable bearing load on the footing is

$$P_n = 0.85f'_cA_g = 0.85 \times 3000 \times 12 \times 6 = 184 \text{ kips/ft}$$

The allowable load is multiplied by the factor $\sqrt{A_2/A_1} = \sqrt{7 \times 1/0.5 \times 1} = 4$; use 2.

$$P_n = 2 \times 184 = 368 \gg 29.4 \text{ kips/ft} \quad \text{(O.K.)}$$

The width of 7 ft is thus seen to be more than adequate for bearing capacity.

A trial thickness is now assumed for the footing. As for spread footings, a thickness of one-sixth to one-eighth of the total width in flexure is usually a good start. For a total width of 7 ft 0 in., the estimated overall thickness $h$ is then

$$h = \frac{\text{width}}{6} = \frac{7 \times 12}{6} = 14 \text{ in.}$$

The effective depth $d$ is taken at 10 in. This trial depth of 10 in. for $d$ is used to check the beam shear at a distance $d$ from the face of the equivalent wall. The dimensions are shown in the following sketch.

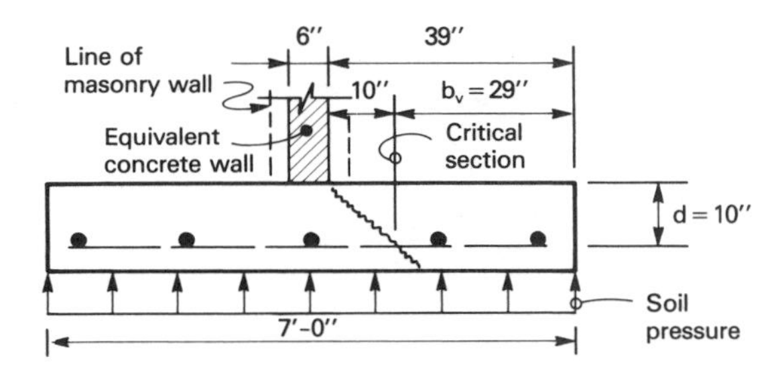


The shear on the section is, per foot of length,

$$V_{\text{DL}} = P_{\text{DL}} b_v L = 1.14\left(\frac{29}{12}\right) \times 1 = 2.76 \text{ kips/ft}$$

$$V_{\text{LL}} = P_{\text{LL}} b_v L = 0.79\left(\frac{29}{12}\right) \times 1 = 1.91 \text{ kips/ft}$$

Hence

$$V_n = \frac{1.4V_{\text{DL}} + 1.7V_{\text{LL}}}{\phi}$$

$$= \frac{1.4 \times 2.76 + 1.7 \times 1.91}{0.85} = 8.37 \text{ kips/ft}$$

The capacity of the section is, per foot of length,

$$V_{\text{cr}} = v_n b_w d = 2\sqrt{3000} \times 12 \times 10 = 13 > 8.37 \text{ kips/ft} \quad \text{(O.K.)}$$

The trial depth of 10 in. is thus seen to be adequate to take the beam shear.

The trial depth of 10 in. for $d$ is now used for the design of the member in flexure. The critical section for flexure is at the face of the equivalent wall, as shown in the following sketch.

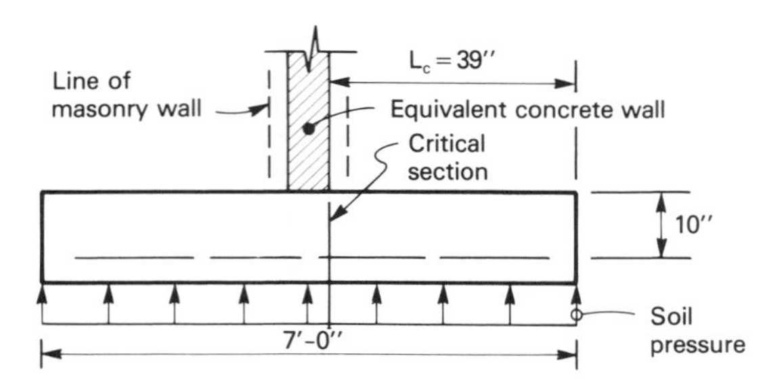


The moment acting on the critical section is

$$M_{\text{DL}} = \frac{P_{\text{DL}}L_c^2}{2} = \frac{1.14 \times \left(\frac{39}{12}\right)^2}{2} = 6.02 \text{ kip-ft/ft}$$

$$M_{\text{LL}} = \frac{P_{\text{LL}}L_c^2}{2} = \frac{0.79 \times \left(\frac{39}{12}\right)^2}{2} = 4.17 \text{ kip-ft/ft}$$

Hence

$$M_n = \frac{(1.4M_{\text{DL}} + 1.7M_{\text{LL}})}{\phi}$$

$$= \frac{(1.4 \times 6.02 + 1.7 \times 4.17)}{0.9} = 17.2 \text{ kip-ft/ft}$$

The section modulus $Z_c$ is found by the usual means:

$$\frac{M_n}{0.85f'_c} = Z_c \qquad \frac{17{,}200 \times 12}{0.85 \times 3000} = \text{coeff.} \times 12 \times 10^2$$

$$\text{coeff.} = 0.0675$$

From Table A-5, select $\rho = 0.0035$ (minimum).

Solve for $A_s = \rho bd = 0.0035 \times 12 \times 10 = 0.420 \text{ in}^2/\text{ft}$.

Use No. 6 bars at 12 in. o.c.

In the longitudinal direction, there is no computable load on the footing. The only requirement for steel is that for temperature and shrinkage, which in this case is $0.0018bh$. The required area of steel in the cross section is

$$A_s = 0.0018bt = 0.0018 \times 84 \times 14 = 2.12 \text{ in}^2$$

Use 7 No. 5 bars, evenly spaced.

A sketch of the final design is shown below. The actual masonry wall is shown rather than the equivalent wall used throughout the foregoing calculations.

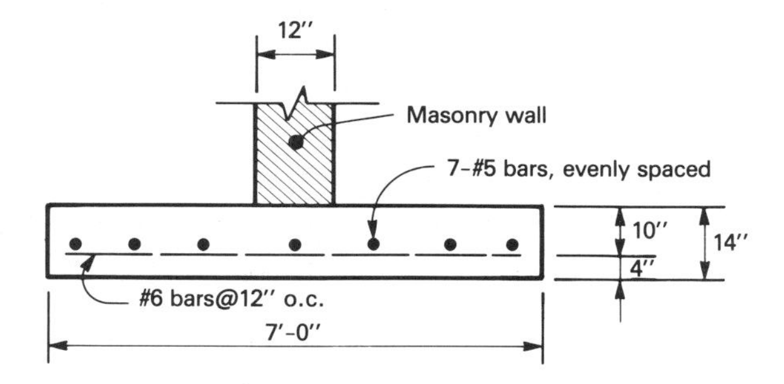


The dowels connecting the wall to its footing are as arbitrarily chosen for strip footings as they are for spread footings. If the wall is reinforced, it is common practice simply to match the wall reinforcement for size and location of dowels. If the wall is unreinforced, dowels may also be omitted.

## GRADE BEAMS

Grade beams are one of the more economical foundations in common use for light buildings. They are used quite often for the foundations of lightweight prefabricated steel buildings as well as for one- or two-story concrete frames. They are not suited for heavy loads, however, and are rarely feasible for buildings having more than two stories.

A grade beam may be designed the same way as any other continuous rectangular beam. The ACI coefficients introduced in Chapter 9 are applicable. As noted earlier in this chapter, a grade beam is like any other continuous beam except that loads are upside down.

It may be argued (successfully) that a grade beam is an elastic beam on an elastic foundation. Consequently, when the beam at midspan deflects upward due to soil pressure, the soil pressure decreases somewhat. Approximately, a sinusoidal soil pressure distribution is eventually produced, as shown in Fig. 10-7. The proposed use of a uniform pressure distribution and the ACI coefficients can therefore be seen to be in error by an undetermined amount.

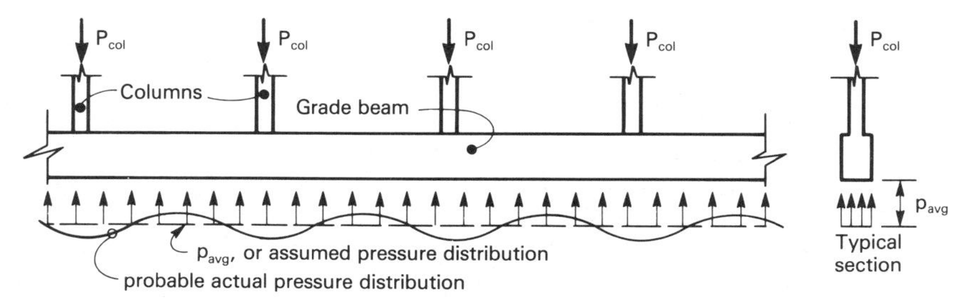


**Figure 10-7** Soil Pressures on a Grade Beam.

Although the foregoing argument is true, the magnitude of error is not prohibitively high. In comparing the sinusoidal pressure distribution of Fig. 10-7 with the uniform distribution, it is seen that in all cases the load decreases at midspan and increases at the column points. Thus the moments both at midspan and at the column points decrease while the shears at the column points remain almost constant.

It is evident that the ACI coefficients will yield a conservative value of moment and a reasonable value for shear. Considering the inaccuracies in both the soils analysis and the concrete analysis, the additional error is not considered prohibitive, particularly since it is always on the ''safe'' side.

In the example following this discussion, the ACI coefficients are used with averaged distributed soil pressures. The soil–structure interaction that would produce the sinusoidal pressure distribution is left to more advanced study.

As with other types of footings, shear reinforcement is rarely used in grade beams. Since falsework and shoring are minimal, it is far more practical to increase the size of the beam slightly rather than to fabricate and place a large number of stirrups. As a consequence, the size of a grade beam is usually much deeper than that of other concrete beams having a similar span.

In addition, usually no effort is made to cut off the longitudinal reinforcement in a grade beam. The maximum required area of flexural steel, both top and bottom, is simply made continuous throughout the length of the grade beam. For such a design, only the highest numerical values of positive and negative moments are therefore needed when flexural reinforcement is selected.

Very often, the required width of the grade beam is so narrow that the column is wider than this width; the column would have to overhang the grade beam if it were to be centered on the grade beam. A simple solution is to make the grade beam as wide as the column; the column would then be properly seated on the grade beam. In such cases, the resulting soil pressure may be considerably less than the allowable pressure, but the overall suitability of the design justifies the configuration. The design used in the following example includes such circumstances.

**Example 10-4**

*Design of a grade beam foundation*
*Repetitive column loads at uniform spacing*
*Strength method of design of concrete*

*Given:*

Load on the end column is roughly half the load on an interior column

Column size 12 in. × 12 in. for all columns

$P_{DL}$ = 30 kips and $P_{LL}$ = 24 kips (each column)

Allowable soil pressure $p_a$ = 3000 psf

$f'_c$ = 3000 psi

Grade 60 steel

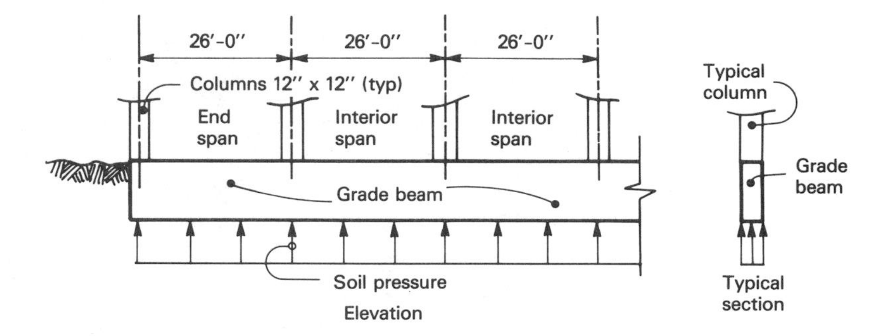


In this example, the difference in weight between the concrete grade beam and the soil it displaces are ignored. The required contact area under the grade beam is computed by simple statics, assuming that the grade beam has a bearing area 26 ft long.

$$\text{area} = \frac{P_{DL} + P_{LL}}{p_a} = \frac{30 + 24}{3000}$$

$$= 18 \text{ ft}^2 \text{ per bay}$$

$$\text{width} = \frac{\text{area}}{\text{length}} = \frac{18}{26} = 0.69 \text{ ft} = 8\tfrac{1}{2} \text{ in.}$$

If the footing width were to be made $8\frac{1}{2}$ in. as indicated in the foregoing calculations, the column with its width at 12 in. would be eccentric on the grade beam. To keep the column load concentric on the grade beam, the grade beam is simply made 12 in. wide, recognizing that the soil pressure will be somewhat less than the allowable:

$$p_{DL} = \frac{P_{DL}}{\text{area}} = \frac{30}{1 \times 26} = 1.15 \text{ kips/ft}^2$$

$$p_{LL} = \frac{P_{LL}}{\text{area}} = \frac{24}{1 \times 26} = 0.92 \text{ kips/ft}^2$$

$$\text{Total } p = 1.15 + 0.92 = 2070 \text{ psf} < 3000 \text{ psf allowed}$$

The grade beam will now be checked for beam shear, using the ACI coefficients for maximum shear. At an end span:

$$V_{DL} = 1.15 p_{DL} L_n = \frac{1.15(1.15 \times 25)}{2} = 16.5 \text{ kips}$$

$$V_{LL} = 1.15 p_{LL} L_n = \frac{1.15(0.92 \times 25)}{2} = 13.2 \text{ kips}$$

Hence

$$V_n = \frac{1.4 V_{DL} + 1.7 V_{LL}}{\phi}$$

$$= \frac{1.4 \times 16.5 + 1.7 \times 13.2}{0.85} = 53.6 \text{ kips}$$

A trial depth of about $L/12$ or 26 in. is adopted for the grade beam, with a corresponding value of 22 in. for the effective depth $d$. This tall slender section of $12 \times 26$ in. is now reviewed for its adequacy in shear. The critical section for shear is shown on the following partial shear diagram.

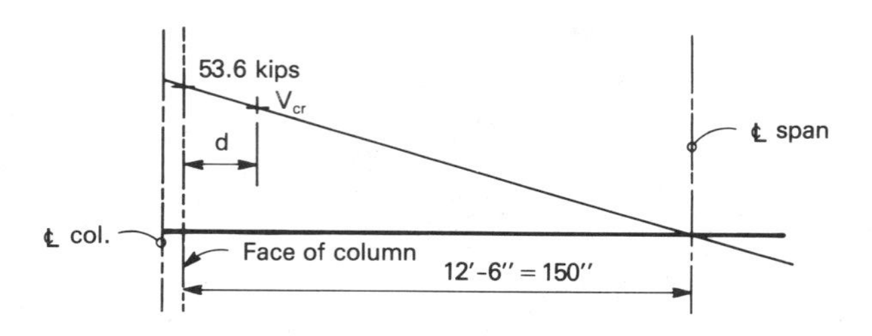


Critical shear on the section is found by ratios:

$$V_{\rm cr} = \frac{150 - 22}{150} 53.6 = 45.7 \text{ kips}$$

The capacity of the section in shear is

$$V_c = v_n b_w d = 2\sqrt{3000} \times 12 \times 22 = 29 \text{ kips} < 45.7 \text{ kips} \quad \text{(no good)}$$

This section is not adequate for critical shear. A larger section, 38 in. deep, will be tried, $d = 34$ in. The critical shear on the new section, 34 in. deep, is again computed by ratios:

$$V_{\rm cr} = \frac{150 - 34}{150} 53.6 = 41.5 \text{ kips}$$

The capacity of this deeper section in shear is

$$V_c = v_n b_w d = 2\sqrt{3000} \times 12 \times 34 = 44.7 > 41.5 \quad \text{(O.K.)}$$

The trial depth of 34 in. is seen to be adequate for beam shear.

This trial depth of 34 in. is now reviewed for flexure. The maximum negative moment acting on the section is taken from the ACI coefficients:

$$M_{\rm DL} = \frac{w_{\rm DL} \times L_n^2}{10} = \frac{(1.15 \times 1) \times 25^2}{10} = 71.9 \text{ kip-ft}$$

$$M_{\rm LL} = \frac{w_{\rm LL} \times L_n^2}{10} = \frac{(0.92 \times 1) \times 25^2}{10} = 57.5 \text{ kip-ft}$$

Hence

$$M_n = \frac{1.4M_{\rm DL} + 1.7M_{\rm LL}}{\phi}$$

$$= \frac{1.4 \times 71.9 + 1.7 \times 57.5}{0.9} = 220 \text{ kip-ft}$$

The required section modulus is computed as usual:

$$\frac{M_n}{0.85f'_c} = Z_c \qquad \frac{220{,}000 \times 12}{0.85 \times 3000} = \text{coeff.} \times 12 \times 34^2$$

$$\text{coeff.} = 0.075$$

From Table A-5, select $\rho = 0.0035$.

Compute $A_s = \rho bd = 0.0035 \times 12 \times 34 = 1.428 \text{ in}^2$.

Since the maximum positive moment is less than the maximum negative moment, the minimum value of $\rho$ will be required both for positive and for negative steel. Use 2 No. 8 bars top and bottom continuous. Since the distance along the side of the beam between positive and negative reinforcement is more than 18 in., use an extra No. 8 bar at middepth.

It should be noted that this tall slender beam may be unstable, tending to "roll" under load. Bridging for grade beams is not specifically prescribed by Code, but the criteria for joists (also tall, slender beams) gives a clue to the bridging requirements. For spans 20 ft to 30 ft, extra bracing is required at midspan (i.e., bridging is required at column lines and at midspan). Consequently, bridging is provided for this grade beam at 13 ft on center, at column lines and midspan. The stiffeners are selected as beams 12 in. wide × 20 in. deep ($d$ = 16 in.), with minimum ratio of steel, $\rho = 0.0035$.

$$A_s = \rho_{\min} bd = 0.0035 \times 12 \times 16 = 0.672 \text{ in}^2$$

Use 2 No. 6 bars top and bottom. The final design of the grade beam is shown in the following sketch.

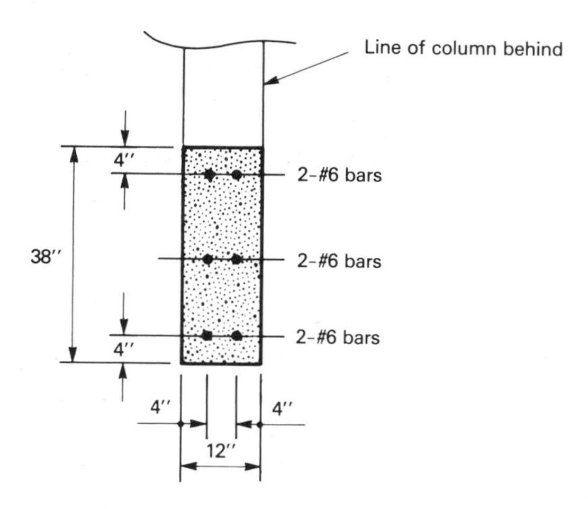


## OTHER TYPES OF FOOTINGS

Other types of shallow foundations in common use are usually only variations of the preceding types. A common type is the combined footing shown in Fig. 10-8. The combined footing can have almost any reasonable shape as long as the resultant of column loads is collinear with the resultant of a uniform soil pressure.

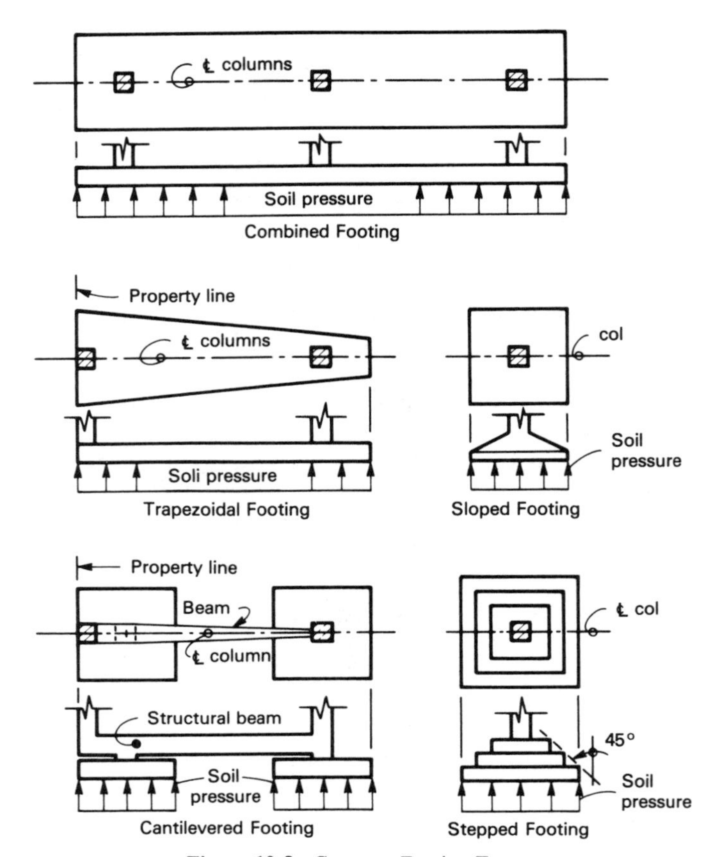


**Figure 10-8** Common Footing Types.

Another common type of arrangement is the cantilever footing, also shown in Fig. 10-8. It is frequently used where it is necessary to place a column so close to the property line that there is no way to keep the column concentric. It can also be a useful temporary foundation when underpinning a building for foundation repairs.

Foundations of unreinforced concrete or unreinforced masonry have fallen from use in recent years. They are, however, very satisfactory foundations when properly utilized. As suggested by the sketch in Fig. 10-8, there is no tensile stress where the side slopes are kept within the natural shear planes.

Unreinforced foundations can be especially useful when the foundation is submerged in groundwater containing chlorides (salt). The chlorides can attack reinforcement but have little effect otherwise on the concrete. The use of unreinforced footings serves to eliminate the risk of corrosion of reinforcement.

All the footings shown in Fig. 10-8 may be designed by the methods presented in the foregoing sections. A great deal of number juggling may be necessary in some cases to match the foundation to the loads. It should be recognized that the combined footing shown in Fig. 10-8 acts as an elastic beam on an elastic foundation, the same way that a grade beam does.

## SLABS ON GRADE

The design of concrete slabs on grade is prescribed in ACI 302, Recommended Practice for Concrete Floor and Slab Construction (American Concrete Institute, 1980). The design method for the slab thickness is again empirical, based on wheel load, tire pressure, and concrete strength. A summary of some typical results is presented in Table 10-1.

In addition to presenting a method of design of slabs on grade, ACI 302 also presents criteria for isolation joints, construction joints, and control joints. Typical isolation joints are shown in Fig. 10-9. Isolation joints are used to isolate structural walls and columns from the floor slab such that "hard points" of lateral restraint are not created at the floor line.

Construction joints are placed in the floor slab to permit the casting to be interrupted and to be continued at a later time. When the interruption lasts long enough for the concrete to set, one of the construction joints of Fig. 10-10 may be used. The joints shown in Fig. 10-10 are taken from ACI 302; there are others in common use.

Control joints create stress raisers that serve to control the location of shrinkage and thermal cracking. Cracking still occurs, but it is much more likely to occur along a control joint than elsewhere. Typical control joints are shown in Fig. 10-11. Some common practices in using control joints to control temperature and shrinkage cracking are given in Chapter 3.

Vapor barriers are also discussed in ACI 302. A vapor barrier is a plastic membrane placed under the floor slab, usually specified as polyethylene film. Rather elaborate

**TABLE 10-1 MINIMUM SLAB THICKNESSES[a]**

| | |
|---|---|
| Light foot traffic: residential use or tile covered | 4 in. minimum |
| Foot traffic | |
| Office, churches, schools, hospitals | 4 in. minimum (5 in. preferred) |
| Sidewalks in residential areas | 5 in. minimum |
| Automobile wheels: automobile driveways, garages | 5 in. minimum |
| Forklift and light truck loads: light industrial, commercial | 5 in. minimum |
| Forklift and medium truck loads: Abrasive wear, industrial/commercial | 6 in. minimum |

[a]If trucks are to pass over isolation joints that have no provision for load transfer, such as doorways, the slab should be thickened approximately 50% and tapered to the required thickness at a slope not more than 1 in 10.

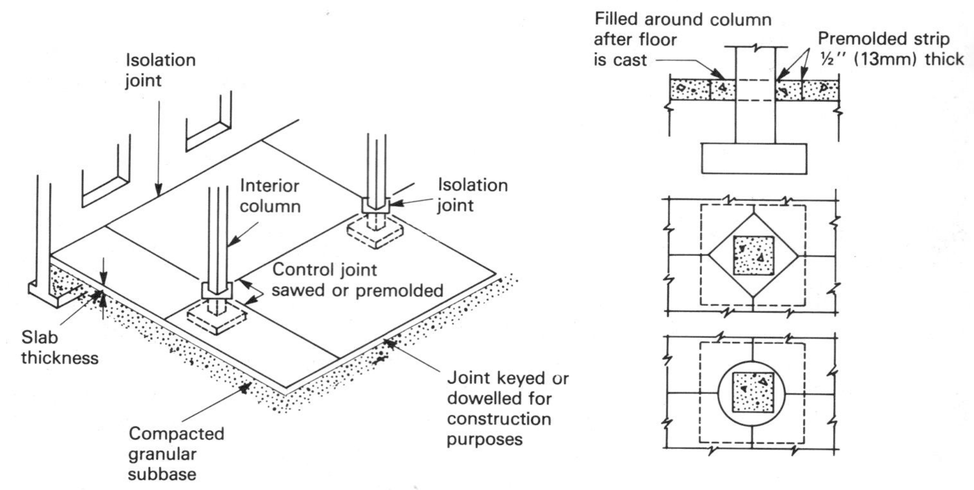


**Figure 10-9** Isolation Joints.

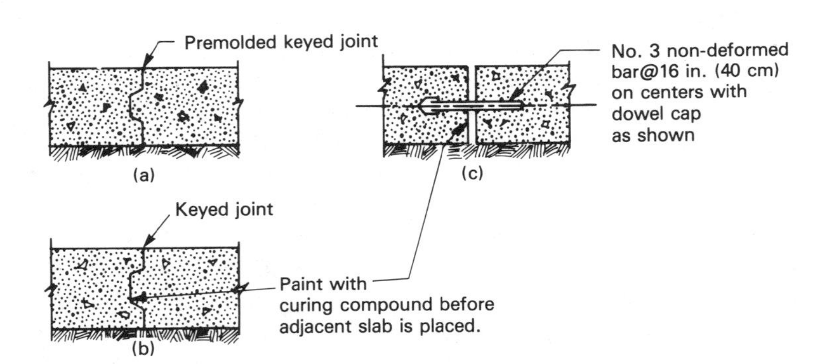


**Figure 10-10** Construction Joints.

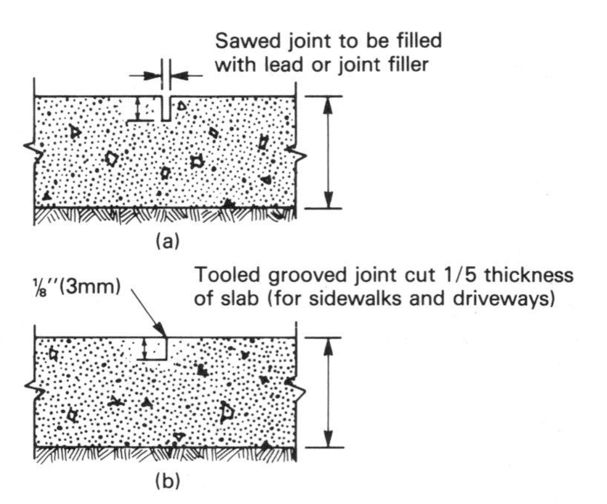


**Figure 10-11** Control Joints.

precautions are required to protect the film from being punctured while the concrete is being cast.

As long as the vapor barrier is intact, it will presumably prevent "vapors" (whatever those are) from percolating upward through the slab. Assuming that it works, it will also prevent "vapors" (whatever those are) from percolating downward. ACI 302 suggests an alternative: A coarse granular fill about 4 in. thick will provide an effective water barrier under a slab.

Slabs on grade are one of the more common features of construction in the industry and are used in all types of buildings. Details such as placement of joints, embedment of pipes, isolation of ducts, reinforcement of edges, prevention of undercutting, minimization of snow and ice hazards, and a wealth of other details are presented in ACI 302. A thorough knowledge of the accepted practices in ACI 302 is recommended to anyone designing in concrete.

## REVIEW QUESTIONS

1. Why is it so essential to keep a column concentric on its foundation?
2. Define "allowable soil pressure."
3. Why is the size of the contact area determined at service levels rather than at ultimate load?
4. Compare the design of a foundation where strength limits the allowable soil pressure to the design where settlements limit the allowable pressure.
5. For buildings having ordinary architectural functions and occupancies, how much of the live load can be expected to be in place long enough to produce foundation settlements?
6. Why don't large moments form at the base of a column on an isolated footing?
7. When a footing rotates due to deformations of the structure above, what is the effect on soil pressures under the footing?
8. What is the minimum cover over the reinforcement for concrete foundations cast directly against the soil? For concrete foundations cast against forms?
9. In a spread footing carrying a column load, where is the critical section for beam shear? For moment? For punching shear?
10. On a spread footing supporting a column, what is the limiting load that can be delivered to the footing by direct bearing of the column?
11. What are the load conditions to be checked when designing a spread footing for a column load?
12. What are the load conditions to be checked when designing a strip footing for a wall load?
13. Compute the ultimate average shear stress in Eqs. (10-1) for a strip footing carrying a wall load, where the ultimate strength of concrete is 4000 psi.
14. How is a strip footing designed to support a masonry wall?
15. Where is the critical section for punching shear on a strip footing supporting a masonry wall?

16. Under what types and configurations of buildings are grade beams usually a feasible foundation?
17. Why does the soil pressure under a grade beam become sinusoidal?
18. Compared to the exacting design of structural steel or timber members, the design of concrete foundations is treated almost as a rough approximation. Why?
19. In soils where corrosion of reinforcement is recognized as a serious hazard, how can a concrete foundation be designed for maximum durability at minimum risk?
20. How thick should a concrete slab be made when it is used as a parking stand for an ordinary passenger car?
21. Why are isolation joints sometimes necessary where columns penetrate a floor slab?
22. How may capillarity be interrupted under a floor slab?

## PROBLEMS

Design an isolated spread footing for the given columns under the given limitations.

| Problem | Column size (in.) | Axial $P_{DL}$ (kips) | Load $P_{LL}$ (kips) | Soil pressure (psf) | $f'_c$ (psi) | Steel grade | Remarks |
|---|---|---|---|---|---|---|---|
| **1** | 12 × 12 | 81 | 54 | 4000 | 3000 | 40 | — |
| **2** | 10 × 10 | 16 | 17 | 2000 | 3000 | 40 | Property line 1.5 ft from centerline of column |
| **3** | 16 × 16 | 122 | 99 | 3000 | 4000 | 50 | — |
| **4** | 10 × 10 | 37 | 28 | 2000 | 4000 | 50 | Obstruction 2 ft from centerline column |
| **5** | 12 × 12 | 40 | 27 | 2900 | 3000 | 60 | Obstruction 1.5 ft from centerline column |
| **6** | 14 × 14 | 79 | 37 | 3000 | 4000 | 60 | — |
| **7** | 12 × 12 | 50 | 34 | 3000 | 4000 | 40 | — |
| **8** | 12 × 12 | 69 | 67 | 2000 | 3000 | 60 | Property line 3.5 ft from centerline of column |

Design a strip footing for the following wall loads.

| Problem | Wall size (in.) | Axial $P_{DL}$ (k/ft.) | Load $P_{LL}$ (k/ft.) | Soil pressure (psf) | $f'_c$ (psi) | $f'_m$ (psi) | Steel grade | Remarks |
|---|---|---|---|---|---|---|---|---|
| **9** | 8 | 14 | 12 | 4000 | 3000 | 1500 | 40 | Masonry wall |
| **10** | 12 | 7 | 9 | 2000 | 3000 | — | 50 | Concrete wall |
| **11** | 12 | 12 | 7 | 3000 | 4000 | — | 50 | Concrete wall |
| **12** | 16 | 15 | 12 | 3000 | 4000 | 1500 | 60 | Masonry wall |

Design a grade beam foundation for a repetitive column load as shown. The columns are square. The end column load may be assumed to be roughly half that of an interior column load.

| Problem | Column size (in.) | Bay length (ft) | Axial $P_{DL}$ (kips) | Load $P_{LL}$ (kips) | Soil pressure (psf) | $f'_c$ (psi) | Steel grade |
|---|---|---|---|---|---|---|---|
| **13** | 12 | 20 | 32 | 36 | 3000 | 3000 | 40 |
| **14** | 12 | 16 | 24 | 32 | 5000 | 4000 | 50 |
| **15** | 10 | 25 | 14 | 20 | 4000 | 4000 | 40 |
| **16** | 16 | 30 | 46 | 46 | 5000 | 3000 | 60 |

# Selected References


ACI Committee 408, "Bond Stress—The State of the Art," *ACI Journal, Proceedings*, 63, November 1966.

American Concrete Institute, *Reinforced Concrete Design Handbook*, Publication SP-3. Detroit: ACI. 1969.

American Concrete Institute, *Recommended Practice for Concrete Floor and Slab Construction*, ACI-302. Detroit: ACI, 1980.

American Concrete Institute, *Manual of Standard Practice for Detailing Reinforced Concrete Structures*, ACI 315, Detroit: ACI, 1980.

American Concrete Institute, *Building Code Requirements for Reinforced Concrete*, ACI 318-83. Detroit: ACI, 1983.

American Concrete Institute, *Recommendations for Design of Beam-Column Joints in Monolithic Reinforced Concrete Structures*, ACI-352. Detroit: ACI, 1985.

American Institute of Steel Construction, *Steel Construction Manual*. New York: AISC, 1980.

American Society for Testing and Materials, *Standard Specification for Deformed and Plain Billet-Steel Bars for Concrete Reinforcement*, ASTM A615-76a. Philadelphia: ASTM, 1976.

Concrete Reinforcing Steel Institute, *CRSI Handbook*, Chicago: CRSI, 1984.

Corps of Engineers, *Handbook for Concrete and Cement*, Vicksburg, Miss.: U.S. Army Waterways Experiment Station, 1986.

FERGUSON, P. M., *Reinforced Concrete Fundamentals*, New York: Wiley, 1958.

ORANGUN, C. O., J. O. JIRSA, and J. E. BREEN, "A Reevaluation of Test Data on Development Length and Splices," *ACI Journal Proceedings*, 74, March 1977.

Portland Cement Association, *Design and Control of Concrete Mixes*, Chicago: PCA, 1979.

ROGERS, PAUL, "Simplified Method of Stirrup Spacing," *Concrete International*, January 1979.

TAYLOR, H. P. J., "The Fundamental Behavior of Reinforced Concrete Beams in Bending and Shear," *Shear in Reinforced Concrete*, Vol. 1 (SP-42). Detroit: American Concrete Institute, 1974.

UNESCO, *Reinforced Concrete: An International Manual*. London: Butterworth, 1971.
U.S. Bureau of Reclamation, *Concrete Manual*, 8th ed., Denver, Colo.: USBR, 1975.
WANG, CHU-KIA, and CHARLES G. SALMON, *Reinforced Concrete Design*, 3rd ed. New York: Harper & Row, 1979.
WHITNEY, C. S., and EDWARD COHEN, "Guide for Ultimate Strength Design of Reinforced Concrete," *ACI Journal, Proceedings*, 53, November 1956.
Wire Reinforcement Institute, *Building Design Handbook*, Washington, D.C.: WRI, 1960.

# APPENDIX: Design Tables

The tables given in this Appendix are adequate for the design problems introduced in this book. For more detailed analysis or for larger projects, the use of more comprehensive design aids is recommended. The tables herein have been reduced as much as possible where such reduction would not incur an undue amount of interpolation.

The beam tables (Tables A-5 through A-7) and the column tables (Tables A-8 through A-10) are dimensionless and homogeneous. The coefficients are therefore the same for both SI units and Imperial units. The remaining tables are in Imperial units; their counterparts in SI units may be found in numerous other handbooks and pamphlets.

The development of the tables is described in the corresponding text presented in earlier chapters of the book. The equations presented therein are complete and may be used to expand or enlarge these tables. General agreement with standard ACI tables has been noted in the relevant discussions in the text.

The column tables are somewhat at variance with the ACI interaction curves where values of e/t are higher than those corresponding to a ''balanced'' design. The maximum variation has been found to be less than 8%, which is considered to be acceptable for calculations in concrete. Elsewhere, the tables coincide exactly with the ACI interaction curves.

The tables of development lengths, Tables A-11 through A-13, are sharply limited. For larger sizes of bars, for wire fabric, for unusual conditions of cover or imbedment, or under any circumstances where these tables are not exactly applicable, a more comprehensive set of tables should be consulted. For some circumstances, the development of bond prescribed by ACI 318-63 may offer some insight; that method is briefly discussed toward the end of Chapter 7.

**Table A-1(a)** Allowable Stresses in Concrete.

| TYPE OF STRESS NORMAL WEIGHT CONCRETE: 145 pcf | ACI318-83 | STRENGTH in psi 3000 | 4000 | 5000 |
|---|---|---|---|---|
| MODULUS OF ELASTICITY, $E_c$ | $57000\sqrt{f'_c}$ | $3.1\times10^6$ | $3.6\times10^6$ | $4.0\times10^6$ |
| Modular ratio $n = E_s/E_c$ | $509/\sqrt{f'_c}$ | 9.29 | 8.04 | 7.20 |
| FLEXURE STRESS, $f_c$ | | | | |
| Extreme compression fiber | $.45f'_c$ | 1350 | 1800 | 2250 |
| SHEAR STRESS, v | | | | |
| Beams, walls, one-way slabs | $1.1\sqrt{f'_c}$ | 60 | 70 | 78 |
| Joist floors (stems) | $1.2\sqrt{f'_c}$ | 67 | 76 | 85 |
| *Two-way slabs and footings | $(1+2/\beta_c)\sqrt{f'_c}$ but $< 2\sqrt{f'_c}$ | 110 | 126 | 141 |
| BEARING STRESS, $f_c$ | | | | |
| On full area | $.3f'_c$ | 900 | 1200 | 1500 |
| **Less than full area (all sides) | $.3f'_c\sqrt{A_2/A_1}$ | | | |

* $\beta_c$ is the ratio of long side to short side of concentrated load or reaction area.

** $A_1$ is the loaded area.

$A_2$ is the maximum area of the portion of the supporting surface that is geometrically similar to and concentric with the loaded area.

**Table A-1(b)** Allowable Stress in Reinforcement.

| | |
|---|---|
| MODULUS OF ELASTICITY OF ALL STEEL: | $E_s$ = 29,000,000 psi |
| TENSILE AND COMPRESSIVE STRESSES: | |
| Grade 40 steel ($f_y$ = 40,000 psi).......... | $f_s$ = 20,000 psi |
| Grade 50 steel ($f_y$ = 50,000 psi).......... | $f_s$ = 20,000 psi |
| Grade 60 steel ($f_y$ = 60,000 psi).......... | $f_s$ = 24,000 psi |

**Table A-2(a)** Area of Steel in a Section 1 ft. Wide.

| SPACING (in.) | BAR SIZES #3 | #4 | #5 | #6 | #7 | #8 | #9 | #10 | #11 |
|---|---|---|---|---|---|---|---|---|---|
| 2 | .66 | 1.18 | 1.84 | 2.65 | 3.61 | 4.71 | | | |
| 2½ | .53 | .94 | 1.47 | 2.12 | 2.89 | 3.77 | 4.80 | | |
| 3 | .44 | .79 | 1.23 | 1.77 | 2.41 | 3.14 | 4.00 | 5.07 | 6.25 |
| 3½ | .38 | .67 | 1.05 | 1.51 | 2.06 | 2.69 | 3.43 | 4.34 | 5.35 |
| 4 | .33 | .59 | .92 | 1.33 | 1.80 | 2.36 | 3.00 | 3.80 | 4.68 |
| 4½ | .29 | .52 | .82 | 1.18 | 1.60 | 2.09 | 2.66 | 3.38 | 4.16 |
| 5 | .27 | .47 | .74 | 1.06 | 1.44 | 1.88 | 2.40 | 3.04 | 3.75 |
| 5½ | .24 | .43 | .67 | .96 | 1.31 | 1.71 | 2.18 | 2.76 | 3.41 |
| 6 | .22 | .39 | .61 | .88 | 1.20 | 1.57 | 2.00 | 2.53 | 3.12 |
| 6½ | .20 | .36 | .57 | .82 | 1.11 | 1.45 | 1.84 | 2.34 | 2.88 |
| 7 | .19 | .34 | .53 | .76 | 1.03 | 1.35 | 1.71 | 2.17 | 2.68 |
| 7½ | .18 | .31 | .49 | .71 | .96 | 1.26 | 1.60 | 2.03 | 2.50 |
| 8 | .17 | .29 | .46 | .66 | .90 | 1.18 | 1.50 | 1.90 | 2.34 |
| 9 | .15 | .26 | .41 | .59 | .80 | 1.05 | 1.33 | 1.69 | 2.08 |
| 10 | .13 | .24 | .37 | .53 | .72 | .94 | 1.20 | 1.52 | 1.87 |
| 11 | .12 | .21 | .33 | .48 | .66 | .86 | 1.09 | 1.38 | 1.70 |
| 12 | .11 | .20 | .31 | .44 | .60 | .79 | 1.00 | 1.27 | 1.56 |
| 13 | .10 | .18 | .28 | .41 | .56 | .72 | .92 | 1.17 | 1.44 |
| 14 | .09 | .17 | .26 | .38 | .52 | .67 | .86 | 1.09 | 1.34 |
| 15 | .09 | .16 | .25 | .35 | .48 | .63 | .80 | 1.01 | 1.25 |
| 16 | .08 | .15 | .23 | .33 | .45 | .59 | .75 | .95 | 1.17 |
| 17 | .08 | .14 | .22 | .31 | .42 | .55 | .71 | .89 | 1.10 |
| 18 | .07 | .13 | .20 | .29 | .40 | .52 | .67 | .84 | 1.04 |

**Table A-2(b)** Properties of Reinforcing Bars.

| BAR SIZE | #3 | #4 | #5 | #6 | #7 | #8 | #9 | #10 | #11 |
|---|---|---|---|---|---|---|---|---|---|
| DIAMETER, inches | .375 | .500 | .625 | .750 | .875 | 1.00 | 1.13 | 1.27 | 1.41 |
| AREA, sq.inches | 0.11 | 0.20 | 0.31 | 0.44 | 0.60 | 0.79 | 1.00 | 1.27 | 1.56 |
| PERIMETER, inches | 1.18 | 1.57 | 1.96 | 2.36 | 2.75 | 3.14 | 3.55 | 3.99 | 4.43 |
| WEIGHT, Lb./Ft. | 0.38 | 0.67 | 1.04 | 1.50 | 2.04 | 2.67 | 3.40 | 4.30 | 5.31 |

**Table A-3** Steel Area in Square Inches for Combinations of Bar Sizes.

STEEL AREA IN SQUARE INCHES FOR COMBINATIONS OF BAR SIZES

| NO. | AREA | NUMBER OF BARS 1 | 2 | 3 | 4 | NUMBER OF BARS 1 | 2 | 3 | 4 | NUMBER OF BARS 1 | 2 | 3 | 4 |
|---|---|---|---|---|---|---|---|---|---|---|---|---|---|
| #3 | | | | | | | | | | | | | |
| 1 | .11 | | | | | | | | | | | | |
| 2 | .22 | | | | | | | | | | | | |
| 3 | .33 | | | | | | | | | | | | |
| 4 | .44 | | | | | | | | | | | | |
| 5 | .55 | | | | | | | | | | | | |
| 6 | .66 | | | | | | | | | | | | |
| #4 | | | | | | | | | | | | | |
| 1 | .20 | | | | | | | | | | | | |
| 2 | .39 | | | | | | | | | | | | |
| 3 | .59 | | | | | | | | | | | | |
| 4 | .79 | | | | | | | | | | | | |
| 5 | .98 | | | | | | | | | | | | |
| 6 | 1.18 | | | | | | | | | | | | |
| #5 | | #5 PLUS #4 | | | | | | | | | | | |
| 1 | .31 | .50 | .70 | .90 | 1.09 | | | | | | | | |
| 2 | .61 | .81 | 1.01 | 1.20 | 1.40 | | | | | | | | |
| 3 | .92 | 1.12 | 1.31 | 1.51 | 1.71 | | | | | | | | |
| 4 | 1.23 | 1.42 | 1.62 | 1.82 | 2.01 | | | | | | | | |
| 5 | 1.53 | 1.73 | 1.93 | 2.12 | 2.32 | | | | | | | | |
| 6 | 1.84 | 2.04 | 2.23 | 2.43 | 2.63 | | | | | | | | |
| #6 | | #6 PLUS #5 | | | | #6 PLUS #4 | | | | | | | |
| 1 | .44 | .75 | 1.06 | 1.36 | 1.67 | .64 | .83 | 1.03 | 1.23 | | | | |
| 2 | .88 | 1.19 | 1.50 | 1.80 | 2.11 | 1.08 | 1.28 | 1.47 | 1.67 | | | | |
| 3 | 1.33 | 1.63 | 1.94 | 2.25 | 2.55 | 1.52 | 1.72 | 1.91 | 2.11 | | | | |
| 4 | 1.77 | 2.07 | 2.38 | 2.69 | 2.99 | 1.96 | 2.16 | 2.36 | 2.55 | | | | |
| 5 | 2.21 | 2.52 | 2.82 | 3.13 | 3.44 | 2.41 | 2.60 | 2.80 | 2.99 | | | | |
| 6 | 2.65 | 2.96 | 3.26 | 3.57 | 3.88 | 2.85 | 3.04 | 3.24 | 3.44 | | | | |
| #7 | | #7 PLUS #6 | | | | #7 PLUS #5 | | | | #7 PLUS #4 | | | |
| 1 | .60 | 1.04 | 1.48 | 1.93 | 2.37 | .91 | 1.21 | 1.52 | 1.83 | .80 | .99 | 1.19 | 1.39 |
| 2 | 1.20 | 1.64 | 2.09 | 2.53 | 2.97 | 1.51 | 1.82 | 2.12 | 2.43 | 1.40 | 1.60 | 1.79 | 1.99 |
| 3 | 1.80 | 2.25 | 2.69 | 3.13 | 3.57 | 2.11 | 2.42 | 2.72 | 3.03 | 2.00 | 2.20 | 2.39 | 2.59 |
| 4 | 2.41 | 2.85 | 3.29 | 3.73 | 4.17 | 2.71 | 3.02 | 3.33 | 3.63 | 2.60 | 2.80 | 2.99 | 3.19 |
| 5 | 3.01 | 3.45 | 3.89 | 4.33 | 4.77 | 3.31 | 3.62 | 3.93 | 4.23 | 3.20 | 3.40 | 3.60 | 3.79 |
| 6 | 3.61 | 4.05 | 4.49 | 4.93 | 5.38 | 3.91 | 4.22 | 4.53 | 4.84 | 3.80 | 4.00 | 4.20 | 4.39 |
| #8 | | #8 PLUS #7 | | | | #8 PLUS #6 | | | | #8 PLUS #5 | | | |
| 1 | .79 | 1.39 | 1.99 | 2.59 | 3.19 | 1.23 | 1.67 | 2.11 | 2.55 | 1.09 | 1.40 | 1.71 | 2.01 |
| 2 | 1.57 | 2.17 | 2.77 | 3.37 | 3.98 | 2.01 | 2.45 | 2.90 | 3.34 | 1.88 | 2.18 | 2.49 | 2.80 |
| 3 | 2.36 | 2.96 | 3.56 | 4.16 | 4.76 | 2.80 | 3.24 | 3.68 | 4.12 | 2.66 | 2.97 | 3.28 | 3.58 |
| 4 | 3.14 | 3.74 | 4.34 | 4.95 | 5.55 | 3.58 | 4.03 | 4.47 | 4.91 | 3.45 | 3.76 | 4.06 | 4.37 |
| 5 | 3.93 | 4.53 | 5.13 | 5.73 | 6.33 | 4.37 | 4.81 | 5.25 | 5.69 | 4.23 | 4.54 | 4.85 | 5.15 |
| 6 | 4.71 | 5.31 | 5.92 | 6.52 | 7.12 | 5.15 | 5.60 | 6.04 | 6.48 | 5.02 | 5.33 | 5.63 | 5.94 |
| #9 | | #9 PLUS #8 | | | | #9 PLUS #7 | | | | #9 PLUS #6 | | | |
| 1 | 1.00 | 1.79 | 2.57 | 3.36 | 4.14 | 1.60 | 2.20 | 2.80 | 3.41 | 1.44 | 1.88 | 2.33 | 2.77 |
| 2 | 2.00 | 2.79 | 3.57 | 4.36 | 5.14 | 2.60 | 3.20 | 3.80 | 4.41 | 2.44 | 2.88 | 3.33 | 3.77 |
| 3 | 3.00 | 3.79 | 4.57 | 5.36 | 6.14 | 3.60 | 4.20 | 4.80 | 5.41 | 3.44 | 3.88 | 4.33 | 4.77 |
| 4 | 4.00 | 4.79 | 5.57 | 6.36 | 7.14 | 4.60 | 5.20 | 5.80 | 6.41 | 4.44 | 4.88 | 5.33 | 5.77 |
| 5 | 5.00 | 5.79 | 6.57 | 7.36 | 8.14 | 5.60 | 6.20 | 6.80 | 7.41 | 5.44 | 5.88 | 6.33 | 6.77 |
| 6 | 6.00 | 6.79 | 7.57 | 8.36 | 9.14 | 6.60 | 7.20 | 7.80 | 8.41 | 6.44 | 6.88 | 7.33 | 7.77 |
| #10 | | #10 PLUS #9 | | | | #10 PLUS #8 | | | | #10 PLUS #7 | | | |
| 1 | 1.27 | 2.27 | 3.27 | 4.27 | 5.27 | 2.05 | 2.84 | 3.62 | 4.41 | 1.87 | 2.47 | 3.07 | 3.67 |
| 2 | 2.53 | 3.53 | 4.53 | 5.53 | 6.53 | 3.32 | 4.10 | 4.89 | 5.68 | 3.13 | 3.74 | 4.34 | 4.94 |
| 3 | 3.80 | 4.80 | 5.80 | 6.80 | 7.80 | 4.59 | 5.37 | 6.16 | 6.94 | 4.40 | 5.00 | 5.60 | 6.21 |
| 4 | 5.07 | 6.07 | 7.07 | 8.07 | 9.07 | 5.85 | 6.64 | 7.42 | 8.21 | 5.67 | 6.27 | 6.87 | 7.47 |
| 5 | 6.33 | 7.33 | 8.33 | 9.33 | 10.3 | 7.12 | 7.90 | 8.69 | 9.48 | 6.94 | 7.54 | 8.14 | 8.74 |
| 6 | 7.60 | 8.60 | 9.60 | 10.6 | 11.6 | 8.39 | 9.17 | 9.96 | 10.7 | 8.20 | 8.80 | 9.40 | 10.0 |
| #11 | | #11 PLUS #10 | | | | #11 PLUS #9 | | | | #11 PLUS #8 | | | |
| 1 | 1.56 | 2.83 | 4.09 | 5.36 | 6.63 | 2.56 | 3.56 | 4.56 | 5.56 | 2.35 | 3.13 | 3.92 | 4.70 |
| 2 | 3.12 | 4.39 | 5.66 | 6.92 | 8.19 | 4.12 | 5.12 | 6.12 | 7.12 | 3.91 | 4.69 | 5.48 | 6.26 |
| 3 | 4.68 | 5.95 | 7.22 | 8.48 | 9.75 | 5.68 | 6.68 | 7.68 | 8.68 | 5.47 | 6.26 | 7.04 | 7.83 |
| 4 | 6.25 | 7.51 | 8.78 | 10.1 | 11.3 | 7.25 | 8.25 | 9.25 | 10.3 | 7.03 | 7.82 | 8.60 | 9.39 |
| 5 | 7.81 | 9.07 | 10.3 | 11.6 | 12.9 | 8.81 | 9.81 | 10.8 | 11.8 | 8.59 | 9.38 | 10.2 | 11.0 |
| 6 | 9.37 | 10.6 | 11.9 | 13.2 | 14.4 | 10.4 | 11.4 | 12.4 | 13.4 | 10.2 | 10.9 | 11.7 | 12.5 |

**Table A-4** Design Guidelines.

MINIMUM WIDTHS OF BEAMS OR STEMS
LOCATED AT INTERIOR EXPOSURES*
(Maximum aggregate size ¾ in.)

| Bar Diam. | With #3 stirrups** | | | | | | | With no stirrups | |
|---|---|---|---|---|---|---|---|---|---|
| | Number of bars in a single layer | | | | | | | No. bars | |
| | 2 | 3 | 4 | 5 | 6 | 7 | 8 | 2 | 3 |
| #4 | 6.1 | 7.6 | 9.1 | 10.6 | 12.1 | 13.6 | 15.1 | 5.00 | 6.50 |
| #5 | 6.3 | 7.9 | 9.6 | 11.2 | 12.8 | 14.4 | 16.1 | 5.25 | 6.88 |
| #6 | 6.5 | 8.3 | 10.0 | 11.8 | 13.5 | 15.3 | 17.0 | 5.50 | 7.25 |
| #7 | 6.7 | 8.6 | 10.5 | 12.4 | 14.2 | 16.1 | 18.0 | 5.75 | 7.63 |
| #8 | 6.9 | 8.9 | 10.9 | 12.9 | 14.9 | 16.9 | 18.9 | 6.00 | 8.00 |
| #9 | 7.3 | 9.5 | 11.8 | 14.0 | 16.3 | 18.6 | 20.8 | 6.25 | 8.38 |
| #10 | 7.7 | 10.2 | 12.8 | 15.3 | 17.8 | 20.4 | 22.9 | 6.50 | 8.75 |
| #11 | 8.0 | 10.8 | 13.7 | 16.5 | 19.3 | 22.1 | 24.9 | 6.75 | 9.13 |

*For exterior exposures, bars #6 and larger, add ¾ inch.
**For #4 stirrups, add ¾ inch; for #5 stirrups add 1½ inch.

Clearances at Interior Exposures

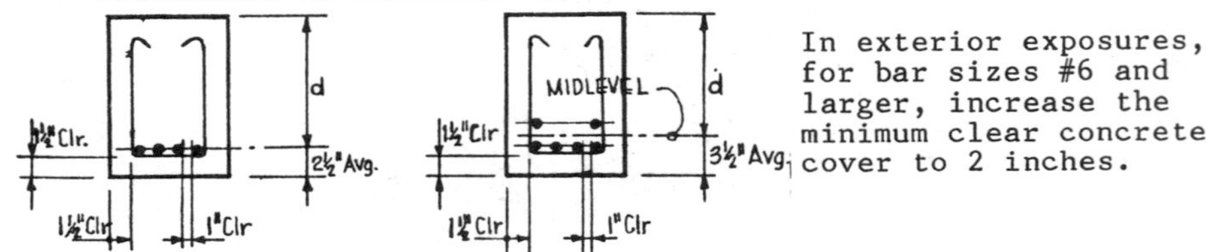


In exterior exposures, for bar sizes #6 and larger, increase the minimum clear concrete cover to 2 inches.

General guidelines:

1. For rectangular beams, width of the beam should be between .5d and .6d if there are no constraints.
2. For tee beams, the depth d should be about 5 times the slab thickness and the width b should be about .5d.
3. Increments of overall dimensions for beams should be not less than ½ inch.
4. Increments of overall slab thickness should be not less than ½ inch with a minimum thickness of 3½ inches.
5. Minimum clear cover is 1½ inch from outside of stirrups for interior exposures.
6. No less than 2 reinforcing bars should be used as longitudinal reinforcement in beams.
7. Reinforcing bars should always be symmetrical about the vertical axis of the section.
8. Bar sizes greater than #11 should not be used in beams and no more than 2 bar sizes should be used.
9. Bars should be placed in one layer if possible; if in multiple layers, heavier sizes should be closest to the outer face.

**Table A-5** Coefficients for Section Constants.*

$f'_c$ = 3000psi (20.7N/sqmm)
g = .125, beta = .85, n = 9.36

| $A_s/bd$ | AT ULTIMATE LOADS: GRADE 40 STEEL $f_y$ = 40,000psi $Z_c$ | $k\beta$ | $y_n$ | svc $f_c$ | GRADE 50 STEEL $f_y$ = 50,000psi $Z_c$ | $k\beta$ | $y_n$ | svc $f_c$ | GRADE 60 STEEL $f_y$ = 60,000psi $Z_c$ | $k\beta$ | $y_n$ | svc $f_c$ | AT SERVICE LOADS: ELASTIC CONSTANTS FOR ALL STEELS $S_c$ | $k_w$ | $y_w$ | $I_o$ | $f_s/f_c$ |
|---|---|---|---|---|---|---|---|---|---|---|---|---|---|---|---|---|---|
| | | | | | COMPRESSIVE STEEL AREA 0.0% OF TENSILE STEEL AREA | | | | | | | | | | | | |
| .0005 | .008 | .008 | .559 | 2: 262 | .010 | .010 | .558 | 2: 328 | .012 | .012 | .557 | 3: 393 | .045 | .092 | .532 | .004 | 92.18 |
| .0010 | .016 | .016 | .555 | 3: 382 | .019 | .020 | .553 | 3: 476 | .023 | .024 | .551 | 4: 570 | .061 | .128 | .520 | .008 | 63.89 |
| .0015 | .023 | .024 | .551 | 3: 477 | .029 | .029 | .548 | 4: 595 | .035 | .035 | .545 | 5: 712 | .073 | .154 | .511 | .011 | 51.37 |
| .0020 | .031 | .031 | .547 | 4: 560 | .039 | .039 | .543 | 5: 698 | .046 | .047 | .539 | 6: 834 | .083 | .176 | .504 | .015 | 43.92 |
| .0025 | .039 | .039 | .543 | 4: 635 | .048 | .049 | .538 | 5: 790 | .057 | .059 | .533 | 7: 943 | .091 | .194 | .498 | .018 | 38.84 |
| .0030 | .046 | .047 | .539 | 5: 704 | .057 | .059 | .533 | 6: 875 | .068 | .071 | .527 | 7:1044 | .098 | .211 | .492 | .021 | 35.09 |
| .0035 | .053 | .055 | .535 | 5: 769 | .066 | .069 | .528 | 7: 955 | .079 | .082 | .521 | 8:1137 | .104 | .225 | .487 | .023 | 32.19 |
| .0040 | .061 | .063 | .531 | 6: 830 | .075 | .079 | .523 | 7:1029 | .090 | .094 | .515 | 8:1225 | .110 | .239 | .483 | .026 | 29.84 |
| .0050 | .075 | .079 | .523 | 7: 944 | .093 | .098 | .513 | 8:1167 | .111 | .118 | .504 | 10:1386 | .120 | .263 | .475 | .031 | 26.27 |
| .0060 | .090 | .094 | .515 | 7:1048 | .111 | .118 | .504 | 9:1294 | .131 | .141 | .492 | 11:1533 | .128 | .284 | .468 | .036 | 23.64 |
| .0070 | .104 | .110 | .508 | 8:1146 | .128 | .137 | .494 | 10:1411 | .151 | .165 | .480 | 12:1668 | .136 | .302 | .462 | .041 | 21.60 |
| .0080 | .118 | .126 | .500 | 9:1237 | .145 | .157 | .484 | 10:1521 | .171 | .189 | .468 | 12:1794 | .143 | .319 | .456 | .046 | 19.96 |
| .0090 | .131 | .141 | .492 | 9:1324 | .161 | .177 | .474 | 11:1623 | .190 | .212 | .456 | 13:1910 | .149 | .335 | .451 | .050 | 18.60 |
| .0100 | .145 | .157 | .484 | 10:1406 | .177 | .196 | .464 | 12:1721 | .208 | .236 | .445 | 14:2020 | .154 | .349 | .446 | .054 | 17.45 |
| .0110 | .158 | .173 | .476 | 10:1485 | .193 | .216 | .454 | 13:1813 | .226 | .259 | .433 | 15:2122 | .159 | .362 | .442 | .058 | 16.47 |
| .0120 | .171 | .189 | .468 | 11:1560 | .208 | .236 | .445 | 13:1900 | .243 | .283 | .421 | 15:2219 | .164 | .375 | .438 | .061 | 15.62 |
| .0140 | .196 | .220 | .453 | 12:1702 | .237 | .275 | .425 | 14:2061 | .276 | .330 | .398 | 17:2395 | .172 | .397 | .430 | .069 | 14.19 |
| .0160 | .220 | .251 | .437 | 13:1832 | .265 | .314 | .405 | 15:2208 | .306 | .377 | .374 | 18:2551 | .180 | .418 | .423 | .075 | 13.05 |
| .0180 | .243 | .283 | .421 | 13:1953 | .291 | .354 | .386 | 16:2341 | | | | | .186 | .436 | .417 | .081 | 12.11 |
| .0200 | .265 | .314 | .405 | 14:2065 | .316 | .393 | .366 | 17:2461 | | | | | .192 | .453 | .412 | .087 | 11.32 |
| .0220 | .286 | .346 | .390 | 15:2169 | | | | | | | | | .198 | .468 | .406 | .092 | 10.64 |
| .0240 | .306 | .377 | .374 | 16:2265 | | | | | | | | | .202 | .482 | .402 | .098 | 10.05 |
| .0260 | .325 | .409 | .358 | 16:2355 | | | | | | | | | .207 | .496 | .397 | .102 | 9.53 |
| .0278 | .341 | .437 | .344 | 17:2429 | | | | | | | | | .211 | .507 | .394 | .107 | 9.12 |
| | | | | | COMPRESSIVE STEEL AREA 50% OF TENSILE STEEL AREA | | | | | | | | | | | | |
| .0005 | .010 | .030 | .548 | 2: 345 | .012 | .031 | .547 | 3: 407 | .014 | .032 | .547 | 3: 469 | .044 | .094 | .531 | .004 | 90.64 |
| .0010 | .018 | .042 | .542 | 3: 443 | .022 | .044 | .541 | 4: 531 | .025 | .046 | .540 | 4: 619 | .061 | .128 | .520 | .008 | 63.99 |
| .0015 | .026 | .051 | .537 | 4: 519 | .031 | .054 | .535 | 4: 628 | .036 | .057 | .534 | 5: 736 | .074 | .152 | .512 | .011 | 52.23 |
| .0020 | .033 | .059 | .533 | 4: 584 | .040 | .063 | .531 | 5: 709 | .047 | .067 | .529 | 6: 835 | .085 | .171 | .505 | .015 | 45.25 |
| .0025 | .040 | .066 | .530 | 4: 641 | .049 | .071 | .527 | 5: 782 | .058 | .076 | .525 | 6: 923 | .094 | .188 | .500 | .018 | 40.50 |
| .0030 | .048 | .072 | .527 | 5: 693 | .058 | .078 | .524 | 6: 847 | .069 | .084 | .520 | 7:1002 | .103 | .202 | .495 | .021 | 37.01 |
| .0035 | .055 | .077 | .524 | 5: 741 | .067 | .084 | .520 | 6: 908 | .079 | .092 | .517 | 7:1074 | .111 | .214 | .491 | .024 | 34.31 |
| .0040 | .062 | .083 | .521 | 5: 785 | .076 | .090 | .517 | 7: 964 | .090 | .099 | .513 | 8:1141 | .118 | .226 | .487 | .026 | 32.14 |
| .0050 | .076 | .092 | .517 | 6: 866 | .093 | .102 | .512 | 7:1065 | .111 | .113 | .506 | 9:1263 | .131 | .245 | .481 | .032 | 28.84 |
| .0060 | .090 | .100 | .512 | 6: 939 | .111 | .112 | .506 | 8:1156 | .132 | .126 | .500 | 9:1372 | .144 | .262 | .475 | .037 | 26.42 |
| .0070 | .104 | .108 | .508 | 7:1005 | .128 | .122 | .501 | 9:1238 | .152 | .138 | .494 | 10:1470 | .155 | .276 | .471 | .042 | 24.56 |
| .0080 | .118 | .115 | .505 | 7:1065 | .145 | .131 | .497 | 9:1314 | .172 | .149 | .488 | 11:1561 | .166 | .289 | .466 | .047 | 23.07 |
| .0090 | .132 | .122 | .502 | 8:1122 | .162 | .140 | .493 | 10:1384 | .193 | .160 | .482 | 11:1644 | .176 | .300 | .462 | .051 | 21.84 |
| .0100 | .145 | .128 | .498 | 8:1174 | .179 | .148 | .488 | 10:1450 | .213 | .171 | .477 | 12:1721 | .185 | .310 | .459 | .056 | 20.81 |
| .0110 | .159 | .134 | .496 | 8:1223 | .196 | .156 | .484 | 10:1511 | .233 | .181 | .472 | 12:1793 | .195 | .320 | .456 | .060 | 19.92 |
| .0120 | .173 | .140 | .493 | 9:1270 | .213 | .164 | .481 | 11:1568 | .253 | .192 | .467 | 13:1861 | .204 | .328 | .453 | .065 | 19.16 |
| .0140 | .200 | .150 | .487 | 9:1355 | .247 | .178 | .473 | 12:1674 | .293 | .211 | .457 | 14:1985 | .221 | .343 | .448 | .073 | 17.90 |
| .0160 | .227 | .160 | .483 | 10:1432 | .281 | .192 | .466 | 12:1769 | .332 | .230 | .447 | 14:2095 | .238 | .357 | .444 | .081 | 16.89 |
| .0180 | .254 | .169 | .478 | 10:1503 | .314 | .205 | .460 | 13:1855 | | | | | .254 | .368 | .440 | .088 | 16.07 |
| .0200 | .281 | .177 | .474 | 11:1567 | .347 | .218 | .454 | 13:1933 | | | | | .269 | .378 | .436 | .096 | 15.39 |
| .0220 | .308 | .185 | .470 | 11:1626 | | | | | | | | | .284 | .387 | .433 | .103 | 14.80 |
| .0240 | .335 | .193 | .466 | 12:1681 | | | | | | | | | .299 | .396 | .431 | .110 | 14.30 |
| .0260 | .362 | .204 | .460 | 12:1731 | | | | | | | | | .314 | .403 | .428 | .117 | 13.86 |
| .0278 | .386 | .218 | .453 | 12:1770 | | | | | | | | | .326 | .409 | .426 | .123 | 13.52 |
| | | | | | COMPRESSIVE STEEL AREA 100% OF TENSILE STEEL AREA | | | | | | | | | | | | |
| .0005 | .011 | .038 | .543 | 3: 376 | .013 | .039 | .543 | 3: 438 | .015 | .040 | .542 | 3: 500 | .043 | .095 | .531 | .004 | 89.28 |
| .0010 | .019 | .052 | .537 | 3: 461 | .022 | .053 | .536 | 4: 548 | .026 | .055 | .535 | 4: 635 | .061 | .127 | .520 | .008 | 64.08 |
| .0015 | .026 | .061 | .532 | 4: 526 | .032 | .064 | .531 | 4: 632 | .037 | .066 | .529 | 5: 738 | .075 | .150 | .512 | .011 | 53.00 |
| .0020 | .034 | .069 | .528 | 4: 580 | .041 | .072 | .527 | 5: 702 | .048 | .075 | .525 | 6: 823 | .087 | .168 | .507 | .015 | 46.45 |
| .0025 | .041 | .075 | .525 | 4: 627 | .050 | .079 | .523 | 5: 762 | .058 | .083 | .521 | 6: 896 | .098 | .182 | .502 | .018 | 42.01 |
| .0030 | .048 | .080 | .522 | 5: 669 | .059 | .085 | .520 | 6: 816 | .069 | .090 | .518 | 7: 962 | .108 | .195 | .498 | .021 | 38.76 |
| .0035 | .055 | .085 | .520 | 5: 707 | .067 | .090 | .517 | 6: 864 | .080 | .096 | .515 | 7:1020 | .117 | .205 | .494 | .024 | 36.25 |
| .0040 | .062 | .089 | .518 | 5: 742 | .076 | .095 | .515 | 6: 908 | .090 | .101 | .512 | 7:1074 | .126 | .215 | .491 | .027 | 34.24 |
| .0050 | .076 | .096 | .514 | 6: 803 | .093 | .103 | .511 | 7: 986 | .111 | .111 | .507 | 8:1169 | .142 | .231 | .486 | .032 | 31.20 |
| .0060 | .090 | .102 | .511 | 6: 857 | .111 | .110 | .507 | 7:1054 | .132 | .119 | .503 | 9:1251 | .158 | .244 | .481 | .038 | 28.99 |
| .0070 | .104 | .107 | .509 | 6: 904 | .128 | .116 | .504 | 8:1114 | .152 | .127 | .499 | 9:1324 | .172 | .255 | .477 | .043 | 27.29 |
| .0080 | .118 | .112 | .507 | 7: 946 | .145 | .122 | .502 | 8:1168 | .173 | .133 | .496 | 10:1388 | .186 | .265 | .474 | .048 | 25.94 |
| .0090 | .131 | .116 | .505 | 7: 984 | .162 | .127 | .499 | 8:1216 | .193 | .139 | .493 | 10:1447 | .200 | .274 | .471 | .053 | 24.84 |
| .0100 | .145 | .119 | .503 | 7:1019 | .180 | .131 | .497 | 9:1260 | .214 | .145 | .490 | 10:1500 | .214 | .281 | .469 | .058 | 23.91 |
| .0110 | .159 | .123 | .501 | 7:1051 | .197 | .135 | .495 | 9:1300 | .234 | .150 | .488 | 11:1548 | .227 | .288 | .466 | .063 | 23.13 |
| .0120 | .173 | .126 | .500 | 7:1080 | .214 | .139 | .493 | 9:1337 | .255 | .155 | .485 | 11:1593 | .240 | .294 | .464 | .067 | 22.45 |
| .0140 | .200 | .131 | .497 | 8:1132 | .248 | .146 | .490 | 10:1403 | .296 | .163 | .481 | 12:1672 | .265 | .305 | .461 | .077 | 21.34 |
| .0160 | .228 | .135 | .495 | 8:1178 | .282 | .151 | .487 | 10:1460 | .337 | .171 | .477 | 12:1741 | .290 | .314 | .458 | .086 | 20.47 |
| .0180 | .255 | .139 | .493 | 8:1217 | .317 | .156 | .484 | 10:1511 | | | | | .314 | .321 | .455 | .094 | 19.76 |
| .0200 | .283 | .142 | .491 | 9:1253 | .351 | .161 | .482 | 11:1555 | | | | | .338 | .328 | .453 | .103 | 19.17 |
| .0220 | .310 | .145 | .490 | 9:1284 | | | | | | | | | .362 | .334 | .451 | .112 | 18.67 |
| .0240 | .338 | .148 | .489 | 9:1313 | | | | | | | | | .385 | .339 | .449 | .120 | 18.25 |
| .0260 | .365 | .150 | .488 | 9:1338 | | | | | | | | | .409 | .344 | .448 | .129 | 17.88 |
| .0278 | .390 | .152 | .487 | 9:1359 | | | | | | | | | .429 | .347 | .447 | .136 | 17.59 |

*TABLED VALUES ARE COEFFICIENTS FOR THE SECTION CONSTANTS: $Z_c$ = coeff.x $bd^2$; $y_n$ = coeff.x d;
$S_c$ = coeff.x $bd^2$; $y_w$ = coeff.x d; $I_o$ = coeff.x $bd^3$; $k\beta$ and $k_w$ are pure coefficients.
Units for 'svc $f_c$' are given first in N/sqmm, then in lb/sqin.
VALUE OF $I_o$ IS ONLY FOR SHORT-TERM LIVE LOADS. FOR LONG-TERM LOADS, USE $\frac{1}{2}E_c$ WITH $I_o$.
COEFFICIENTS ABOVE THE INTERIOR LINE APPLY ONLY TO TEE SHAPES.

**Table A-6** Coefficients for Section Constants. *

$f'_c$ = 4000psi (27.6N/sqmm)
g = .125, beta = .85, n = 8.11

| $A_s/bd$ | AT ULTIMATE LOADS | | | | | | | | | | | | AT SERVICE LOADS | | | | |
|---|---|---|---|---|---|---|---|---|---|---|---|---|---|---|---|---|---|
| | GRADE 40 STEEL $f_y$ = 40,000psi | | | | GRADE 50 STEEL $f_y$ = 50,000psi | | | | GRADE 60 STEEL $f_y$ = 60,000psi | | | | ELASTIC CONSTANTS FOR ALL STEELS | | | | |
| | $Z_c$ | $k\beta$ | $y_n$ | svc $f_c$ | $Z_c$ | $k\beta$ | $y_n$ | svc $f_c$ | $Z_c$ | $k\beta$ | $y_n$ | svc $f_c$ | $S_c$ | $k_w$ | $y_w$ | $I_o$ | $f_s/f_c$ |
| | COMPRESSIVE STEEL AREA 0.0% OF TENSILE STEEL AREA | | | | | | | | | | | | | | | | |
| .0005 | .006 | .006 | .560 | 2: 281 | .007 | .007 | .559 | 2: 351 | .009 | .009 | .558 | 3: 421 | .042 | .086 | .534 | .004 | 86.07 |
| .0010 | .012 | .012 | .557 | 3: 408 | .015 | .015 | .555 | 4: 509 | .018 | .018 | .554 | 4: 610 | .057 | .119 | .523 | .007 | 59.74 |
| .0015 | .018 | .018 | .554 | 4: 510 | .022 | .022 | .551 | 4: 636 | .026 | .027 | .549 | 5: 761 | .069 | .144 | .514 | .010 | 48.09 |
| .0020 | .023 | .024 | .551 | 4: 598 | .029 | .029 | .548 | 5: 746 | .035 | .035 | .545 | 6: 892 | .078 | .165 | .508 | .013 | 41.15 |
| .0025 | .029 | .029 | .548 | 5: 678 | .036 | .037 | .544 | 6: 844 | .043 | .044 | .540 | 7:1010 | .086 | .182 | .502 | .016 | 36.42 |
| .0030 | .035 | .035 | .545 | 5: 752 | .043 | .044 | .540 | 6: 936 | .052 | .053 | .536 | 8:1118 | .092 | .198 | .497 | .018 | 32.93 |
| .0035 | .040 | .041 | .542 | 6: 821 | .050 | .052 | .537 | 7:1021 | .060 | .062 | .532 | 8:1218 | .098 | .212 | .492 | .021 | 30.22 |
| .0040 | .046 | .047 | .539 | 6: 886 | .057 | .059 | .533 | 8:1101 | .068 | .071 | .527 | 9:1313 | .104 | .224 | .488 | .023 | 28.04 |
| .0050 | .057 | .059 | .533 | 7:1008 | .071 | .074 | .526 | 9:1250 | .084 | .088 | .518 | 10:1489 | .113 | .247 | .480 | .028 | 24.71 |
| .0060 | .068 | .071 | .527 | 8:1120 | .084 | .088 | .518 | 10:1387 | .100 | .106 | .509 | 11:1650 | .122 | .267 | .473 | .032 | 22.25 |
| .0080 | .090 | .094 | .515 | 9:1325 | .111 | .118 | .504 | 11:1636 | .131 | .141 | .492 | 13:1938 | .135 | .301 | .462 | .041 | 18.82 |
| .0100 | .111 | .118 | .504 | 10:1510 | .136 | .147 | .489 | 13:1858 | .161 | .177 | .474 | 15:2194 | .147 | .330 | .453 | .048 | 16.48 |
| .0120 | .131 | .141 | .492 | 12:1680 | .161 | .177 | .474 | 14:2060 | .190 | .212 | .456 | 17:2424 | .156 | .354 | .444 | .055 | 14.77 |
| .0140 | .151 | .165 | .480 | 13:1838 | .185 | .206 | .459 | 15:2246 | .217 | .247 | .439 | 18:2633 | .165 | .376 | .437 | .062 | 13.44 |
| .0160 | .171 | .189 | .468 | 14:1986 | .208 | .236 | .445 | 17:2417 | .243 | .283 | .421 | 19:2823 | .172 | .396 | .431 | .068 | 12.37 |
| .0180 | .190 | .212 | .456 | 15:2124 | .230 | .265 | .430 | 18:2577 | .268 | .318 | .403 | 21:2998 | .178 | .414 | .425 | .074 | 11.49 |
| .0200 | .208 | .236 | .445 | 16:2255 | .251 | .295 | .415 | 19:2725 | .291 | .354 | .386 | 22:3157 | .184 | .430 | .419 | .079 | 10.75 |
| .0220 | .226 | .259 | .433 | 16:2379 | .272 | .324 | .400 | 20:2863 | | | | | .189 | .445 | .414 | .084 | 10.11 |
| .0240 | .243 | .283 | .421 | 17:2496 | .291 | .354 | .386 | 21:2992 | | | | | .194 | .459 | .410 | .089 | 9.56 |
| .0260 | .259 | .306 | .409 | 18:2607 | .310 | .383 | .371 | 21:3111 | | | | | .199 | .472 | .405 | .094 | 9.07 |
| .0280 | .276 | .330 | .398 | 19:2712 | | | | | | | | | .203 | .484 | .401 | .098 | 8.64 |
| .0320 | .306 | .377 | .374 | 20:2906 | | | | | | | | | .210 | .506 | .394 | .106 | 7.91 |
| .0360 | .334 | .424 | .350 | 21:3079 | | | | | | | | | .217 | .526 | .387 | .114 | 7.31 |
| .0371 | .341 | .437 | .344 | 22:3121 | | | | | | | | | .218 | .531 | .386 | .116 | 7.16 |
| | COMPRESSIVE STEEL AREA 50% OF TENSILE STEEL AREA | | | | | | | | | | | | | | | | |
| .0005 | .008 | .026 | .550 | 3: 391 | .009 | .027 | .549 | 3: 457 | .011 | .027 | .549 | 4: 524 | .041 | .088 | .533 | .004 | 84.39 |
| .0010 | .014 | .036 | .544 | 3: 496 | .017 | .038 | .544 | 4: 590 | .020 | .039 | .543 | 5: 685 | .057 | .120 | .523 | .007 | 59.56 |
| .0015 | .020 | .044 | .540 | 4: 577 | .024 | .047 | .539 | 5: 694 | .028 | .049 | .538 | 6: 810 | .069 | .143 | .515 | .010 | 48.60 |
| .0020 | .026 | .051 | .537 | 4: 647 | .031 | .054 | .535 | 5: 782 | .036 | .057 | .534 | 6: 917 | .079 | .161 | .509 | .013 | 42.09 |
| .0025 | .031 | .057 | .534 | 5: 708 | .038 | .061 | .532 | 6: 860 | .045 | .065 | .530 | 7:1011 | .088 | .177 | .503 | .016 | 37.66 |
| .0030 | .037 | .062 | .531 | 5: 764 | .045 | .067 | .529 | 6: 930 | .053 | .072 | .527 | 8:1097 | .096 | .191 | .499 | .018 | 34.40 |
| .0035 | .042 | .067 | .529 | 6: 816 | .052 | .072 | .526 | 7: 995 | .061 | .078 | .523 | 8:1175 | .103 | .203 | .495 | .021 | 31.88 |
| .0040 | .048 | .072 | .527 | 6: 863 | .058 | .078 | .524 | 7:1056 | .069 | .084 | .520 | 9:1248 | .110 | .214 | .491 | .023 | 29.85 |
| .0050 | .058 | .080 | .523 | 7: 951 | .072 | .087 | .519 | 8:1166 | .085 | .096 | .515 | 10:1380 | .123 | .232 | .485 | .028 | 26.76 |
| .0060 | .069 | .087 | .519 | 7:1029 | .085 | .096 | .514 | 9:1265 | .100 | .106 | .509 | 10:1499 | .134 | .249 | .480 | .033 | 24.50 |
| .0080 | .090 | .100 | .512 | 8:1167 | .111 | .112 | .506 | 10:1438 | .132 | .126 | .500 | 12:1706 | .154 | .275 | .471 | .042 | 21.35 |
| .0100 | .111 | .112 | .507 | 9:1286 | .137 | .127 | .499 | 11:1587 | .162 | .144 | .491 | 13:1884 | .172 | .297 | .464 | .050 | 19.23 |
| .0120 | .132 | .122 | .502 | 10:1392 | .162 | .140 | .493 | 12:1718 | .193 | .160 | .482 | 14:2040 | .189 | .314 | .458 | .058 | 17.68 |
| .0140 | .152 | .131 | .497 | 10:1486 | .188 | .152 | .486 | 13:1835 | .223 | .176 | .474 | 15:2179 | .205 | .330 | .453 | .065 | 16.49 |
| .0160 | .173 | .140 | .493 | 11:1572 | .213 | .164 | .481 | 13:1942 | .253 | .192 | .467 | 16:2304 | .219 | .343 | .448 | .072 | 15.54 |
| .0180 | .193 | .148 | .489 | 11:1650 | .239 | .175 | .475 | 14:2038 | .283 | .206 | .459 | 17:2417 | .234 | .354 | .444 | .079 | 14.77 |
| .0200 | .214 | .155 | .485 | 12:1723 | .264 | .185 | .470 | 15:2127 | .313 | .221 | .452 | 17:2521 | .248 | .365 | .441 | .086 | 14.12 |
| .0220 | .234 | .162 | .481 | 12:1789 | .289 | .195 | .465 | 15:2209 | | | | | .261 | .374 | .438 | .092 | 13.57 |
| .0240 | .254 | .169 | .478 | 13:1852 | .314 | .205 | .460 | 16:2286 | | | | | .274 | .382 | .435 | .098 | 13.09 |
| .0260 | .275 | .175 | .475 | 13:1910 | .339 | .215 | .455 | 16:2356 | | | | | .287 | .390 | .432 | .105 | 12.67 |
| .0280 | .295 | .181 | .472 | 14:1964 | | | | | | | | | .300 | .397 | .430 | .111 | 12.30 |
| .0320 | .335 | .193 | .466 | 14:2063 | | | | | | | | | .325 | .410 | .426 | .123 | 11.68 |
| .0360 | .375 | .212 | .456 | 15:2148 | | | | | | | | | .349 | .420 | .422 | .134 | 11.18 |
| .0371 | .386 | .218 | .453 | 15:2167 | | | | | | | | | .355 | .423 | .422 | .137 | 11.06 |
| | COMPRESSIVE STEEL AREA 100% OF TENSILE STEEL AREA | | | | | | | | | | | | | | | | |
| .0005 | .009 | .034 | .546 | 3: 433 | .010 | .034 | .545 | 3: 500 | .011 | .035 | .545 | 4: 567 | .041 | .089 | .533 | .004 | 82.92 |
| .0010 | .015 | .046 | .540 | 4: 524 | .018 | .047 | .539 | 4: 618 | .020 | .048 | .538 | 5: 712 | .057 | .120 | .522 | .007 | 59.41 |
| .0015 | .021 | .054 | .535 | 4: 594 | .025 | .056 | .534 | 5: 708 | .029 | .058 | .533 | 6: 822 | .070 | .142 | .515 | .010 | 49.07 |
| .0020 | .026 | .061 | .532 | 4: 652 | .032 | .064 | .531 | 5: 783 | .037 | .066 | .529 | 6: 915 | .081 | .159 | .510 | .013 | 42.94 |
| .0025 | .032 | .067 | .529 | 5: 703 | .039 | .070 | .528 | 6: 849 | .045 | .073 | .526 | 7: 994 | .091 | .173 | .505 | .016 | 38.79 |
| .0030 | .037 | .072 | .527 | 5: 748 | .045 | .075 | .525 | 6: 907 | .053 | .079 | .523 | 7:1066 | .100 | .185 | .501 | .018 | 35.74 |
| .0035 | .043 | .076 | .524 | 5: 789 | .052 | .080 | .522 | 7: 960 | .061 | .085 | .520 | 8:1130 | .108 | .195 | .497 | .021 | 33.39 |
| .0040 | .048 | .080 | .522 | 6: 827 | .059 | .085 | .520 | 7:1008 | .069 | .090 | .518 | 8:1189 | .116 | .205 | .494 | .024 | 31.50 |
| .0050 | .059 | .087 | .519 | 6: 894 | .072 | .093 | .516 | 8:1094 | .085 | .098 | .513 | 9:1293 | .131 | .221 | .489 | .029 | 28.65 |
| .0060 | .069 | .093 | .516 | 7: 953 | .085 | .099 | .513 | 8:1169 | .100 | .106 | .509 | 10:1384 | .145 | .234 | .485 | .033 | 26.56 |
| .0080 | .090 | .102 | .511 | 7:1052 | .111 | .110 | .507 | 9:1295 | .132 | .119 | .503 | 11:1537 | .171 | .255 | .478 | .042 | 23.69 |
| .0100 | .111 | .110 | .508 | 8:1134 | .137 | .119 | .503 | 10:1399 | .162 | .130 | .498 | 11:1663 | .195 | .271 | .472 | .051 | 21.77 |
| .0120 | .131 | .116 | .505 | 8:1203 | .162 | .127 | .499 | 10:1487 | .193 | .139 | .493 | 12:1769 | .218 | .285 | .468 | .060 | 20.38 |
| .0140 | .152 | .121 | .502 | 9:1263 | .188 | .133 | .496 | 11:1562 | .224 | .147 | .489 | 13:1860 | .241 | .295 | .464 | .068 | 19.33 |
| .0160 | .173 | .126 | .500 | 9:1315 | .214 | .139 | .493 | 11:1628 | .255 | .155 | .485 | 13:1939 | .263 | .305 | .461 | .076 | 18.50 |
| .0180 | .193 | .129 | .498 | 9:1361 | .240 | .144 | .490 | 12:1686 | .286 | .161 | .482 | 14:2010 | .284 | .313 | .458 | .084 | 17.82 |
| .0200 | .214 | .133 | .496 | 10:1402 | .265 | .149 | .488 | 12:1738 | .316 | .167 | .479 | 14:2072 | .305 | .320 | .456 | .091 | 17.25 |
| .0220 | .235 | .136 | .495 | 10:1439 | .291 | .153 | .486 | 12:1785 | | | | | .326 | .326 | .454 | .099 | 16.78 |
| .0240 | .255 | .139 | .493 | 10:1473 | .317 | .156 | .484 | 13:1827 | | | | | .346 | .331 | .452 | .106 | 16.37 |
| .0260 | .276 | .141 | .492 | 10:1503 | .342 | .160 | .483 | 13:1866 | | | | | .367 | .336 | .450 | .114 | 16.02 |
| .0280 | .296 | .144 | .491 | 11:1531 | | | | | | | | | .387 | .340 | .449 | .121 | 15.71 |
| .0320 | .338 | .148 | .489 | 11:1580 | | | | | | | | | .427 | .348 | .446 | .136 | 15.19 |
| .0360 | .379 | .151 | .487 | 11:1622 | | | | | | | | | .466 | .354 | .444 | .150 | 14.77 |
| .0371 | .390 | .152 | .487 | 11:1632 | | | | | | | | | .477 | .356 | .444 | .154 | 14.67 |

*TABLED VALUES ARE COEFFICIENTS FOR THE SECTION CONSTANTS: $Z_c$ = coeff.x $bd^2$; $y_n$ = coeff.x d; $S_c$ = coeff.x $bd^2$; $y_w$ = coeff.x d; $I_o$ = coeff.x $bd^3$; $k\beta$ and $k_w$ are pure coefficients.
Units for 'svc $f_c$' are given first in N/sqmm, then in lb/sqin.
VALUE OF $I_o$ IS ONLY FOR SHORT-TERM LIVE LOADS. FOR LONG-TERM LOADS, USE $\frac{1}{2}E_c$ WITH $I_o$.
COEFFICIENTS ABOVE THE INTERIOR LINE APPLY ONLY TO TEE SHAPES.

**Table A-7** Coefficients for Section Constants.*

$f'_c$ = 5000psi (34.4N/sqmm)
g = .125, beta = .81, n = 7.25

| $A_s/bd$ | AT ULTIMATE LOADS: GRADE 40 STEEL $f_y$ = 40,000psi | | | | GRADE 50 STEEL $f_y$ = 50,000psi | | | | GRADE 60 STEEL $f_y$ = 60,000psi | | | | AT SERVICE LOADS: ELASTIC CONSTANTS FOR ALL STEELS | | | | |
|---|---|---|---|---|---|---|---|---|---|---|---|---|---|---|---|---|---|
| | $Z_c$ | $k\beta$ | $y_n$ | svc $f_c$ | $Z_c$ | $k\beta$ | $y_n$ | svc $f_c$ | $Z_c$ | $k\beta$ | $y_n$ | svc $f_c$ | $S_c$ | $k_w$ | $y_w$ | $I_o$ | $f_s/f_c$ |
| | COMPRESSIVE STEEL AREA 0.0% OF TENSILE STEEL AREA | | | | | | | | | | | | | | | | |
| .0005 | .005 | .005 | .560 | 2: 296 | .006 | .006 | .560 | 3: 370 | .007 | .007 | .559 | 3: 443 | .040 | .082 | .535 | .003 | 81.60 |
| .0010 | .009 | .009 | .558 | 3: 430 | .012 | .012 | .557 | 4: 536 | .014 | .014 | .555 | 4: 643 | .055 | .113 | .525 | .006 | 56.69 |
| .0015 | .014 | .014 | .555 | 4: 536 | .018 | .018 | .554 | 5: 669 | .021 | .021 | .552 | 6: 802 | .065 | .137 | .517 | .009 | 45.67 |
| .0020 | .019 | .019 | .553 | 4: 629 | .023 | .024 | .551 | 5: 785 | .028 | .028 | .548 | 6: 939 | .074 | .156 | .510 | .012 | 39.10 |
| .0025 | .023 | .024 | .551 | 5: 713 | .029 | .029 | .548 | 6: 889 | .035 | .035 | .545 | 7:1063 | .082 | .173 | .505 | .014 | 34.63 |
| .0030 | .028 | .028 | .548 | 5: 790 | .035 | .035 | .545 | 7: 984 | .042 | .042 | .541 | 8:1177 | .088 | .188 | .500 | .017 | 31.33 |
| .0035 | .032 | .033 | .546 | 6: 863 | .040 | .041 | .542 | 7:1074 | .048 | .049 | .538 | 9:1283 | .094 | .201 | .495 | .019 | 28.76 |
| .0040 | .037 | .038 | .544 | 6: 931 | .046 | .047 | .539 | 8:1159 | .055 | .057 | .534 | 10:1384 | .099 | .214 | .491 | .021 | 26.70 |
| .0050 | .046 | .047 | .539 | 7:1059 | .057 | .059 | .533 | 9:1316 | .068 | .071 | .527 | 11:1570 | .108 | .235 | .484 | .026 | 23.54 |
| .0060 | .055 | .057 | .534 | 8:1178 | .068 | .071 | .527 | 10:1462 | .081 | .085 | .520 | 12:1741 | .117 | .255 | .478 | .030 | 21.22 |
| .0080 | .073 | .075 | .525 | 10:1394 | .090 | .094 | .515 | 12:1726 | .107 | .113 | .506 | 14:2051 | .130 | .288 | .467 | .037 | 17.97 |
| .0100 | .090 | .094 | .515 | 11:1591 | .111 | .118 | .504 | 14:1964 | .131 | .141 | .492 | 16:2327 | .141 | .315 | .457 | .044 | 15.76 |
| .0120 | .107 | .113 | .506 | 12:1772 | .131 | .141 | .492 | 15:2182 | .155 | .170 | .478 | 18:2579 | .150 | .339 | .449 | .051 | 14.13 |
| .0140 | .123 | .132 | .497 | 13:1942 | .151 | .165 | .480 | 16:2384 | .178 | .198 | .464 | 19:2810 | .159 | .360 | .442 | .057 | 12.87 |
| .0160 | .139 | .151 | .487 | 14:2101 | .171 | .189 | .468 | 18:2573 | .201 | .226 | .449 | 21:3024 | .166 | .379 | .436 | .063 | 11.86 |
| .0180 | .155 | .170 | .478 | 16:2253 | .190 | .212 | .456 | 19:2750 | .222 | .255 | .435 | 22:3222 | .172 | .397 | .430 | .068 | 11.02 |
| .0200 | .171 | .189 | .468 | 17:2396 | .208 | .236 | .445 | 20:2917 | .243 | .283 | .421 | 23:3407 | .178 | .413 | .425 | .073 | 10.32 |
| .0240 | .201 | .226 | .449 | 18:2664 | .243 | .283 | .421 | 22:3224 | .282 | .339 | .393 | 26:3741 | .188 | .441 | .415 | .083 | 9.19 |
| .0280 | .229 | .264 | .431 | 20:2908 | .276 | .330 | .398 | 24:3497 | | | | | .197 | .466 | .407 | .092 | 8.32 |
| .0320 | .256 | .302 | .412 | 22:3133 | .306 | .377 | .374 | 26:3742 | | | | | .204 | .488 | .400 | .100 | 7.62 |
| .0360 | .282 | .339 | .393 | 23:3339 | | | | | | | | | .211 | .507 | .393 | .107 | 7.04 |
| .0400 | .306 | .377 | .374 | 24:3529 | | | | | | | | | .217 | .525 | .388 | .114 | 6.56 |
| .0440 | .329 | .415 | .355 | 26:3702 | | | | | | | | | .222 | .541 | .382 | .120 | 6.15 |
| .0444 | .331 | .418 | .353 | 26:3718 | | | | | | | | | .222 | .543 | .382 | .121 | 6.11 |
| | COMPRESSIVE STEEL AREA 50% OF TENSILE STEEL AREA | | | | | | | | | | | | | | | | |
| .0005 | .007 | .023 | .551 | 3: 429 | .008 | .023 | .551 | 3: 499 | .009 | .024 | .551 | 4: 569 | .039 | .083 | .535 | .003 | 79.85 |
| .0010 | .012 | .032 | .547 | 4: 540 | .014 | .033 | .546 | 4: 640 | .016 | .034 | .545 | 5: 740 | .054 | .114 | .524 | .006 | 56.35 |
| .0015 | .016 | .039 | .543 | 4: 626 | .020 | .041 | .542 | 5: 749 | .023 | .043 | .541 | 6: 873 | .066 | .136 | .517 | .009 | 45.97 |
| .0020 | .021 | .045 | .540 | 5: 699 | .025 | .047 | .539 | 6: 842 | .030 | .050 | .538 | 7: 986 | .075 | .154 | .511 | .012 | 39.80 |
| .0025 | .026 | .050 | .537 | 5: 765 | .031 | .053 | .536 | 6: 925 | .036 | .056 | .534 | 7:1085 | .084 | .169 | .506 | .014 | 35.60 |
| .0030 | .030 | .055 | .535 | 6: 824 | .037 | .058 | .533 | 7:1000 | .043 | .062 | .531 | 8:1176 | .091 | .182 | .502 | .017 | 32.51 |
| .0035 | .035 | .059 | .533 | 6: 878 | .042 | .063 | .531 | 7:1069 | .049 | .068 | .529 | 9:1259 | .098 | .194 | .498 | .019 | 30.12 |
| .0040 | .039 | .063 | .531 | 6: 929 | .047 | .068 | .529 | 8:1133 | .056 | .073 | .526 | 9:1337 | .104 | .205 | .494 | .021 | 28.19 |
| .0050 | .048 | .070 | .527 | 7:1022 | .058 | .076 | .524 | 9:1250 | .069 | .083 | .521 | 10:1478 | .116 | .223 | .488 | .026 | 25.27 |
| .0060 | .056 | .077 | .524 | 8:1106 | .069 | .084 | .521 | 9:1356 | .081 | .092 | .517 | 11:1605 | .127 | .239 | .483 | .030 | 23.11 |
| .0080 | .073 | .088 | .518 | 9:1253 | .090 | .098 | .514 | 11:1541 | .107 | .108 | .508 | 13:1828 | .146 | .265 | .474 | .038 | 20.12 |
| .0100 | .090 | .098 | .513 | 10:1381 | .111 | .110 | .507 | 12:1702 | .132 | .123 | .501 | 14:2020 | .163 | .286 | .467 | .046 | 18.10 |
| .0120 | .107 | .107 | .509 | 10:1495 | .132 | .121 | .502 | 13:1844 | .156 | .138 | .494 | 15:2189 | .178 | .304 | .461 | .053 | 16.63 |
| .0140 | .123 | .115 | .505 | 11:1597 | .152 | .132 | .496 | 14:1971 | .181 | .151 | .487 | 16:2341 | .193 | .319 | .456 | .060 | 15.49 |
| .0160 | .140 | .123 | .501 | 12:1690 | .173 | .142 | .491 | 14:2087 | .205 | .164 | .480 | 17:2478 | .207 | .332 | .452 | .066 | 14.59 |
| .0180 | .156 | .130 | .498 | 12:1775 | .193 | .152 | .487 | 15:2193 | .229 | .176 | .474 | 18:2604 | .220 | .344 | .448 | .072 | 13.84 |
| .0200 | .173 | .136 | .494 | 13:1854 | .213 | .161 | .482 | 16:2291 | .253 | .189 | .468 | 19:2719 | .233 | .354 | .444 | .079 | 13.22 |
| .0240 | .205 | .149 | .488 | 14:1996 | .254 | .178 | .474 | 17:2466 | .301 | .212 | .457 | 20:2923 | .257 | .372 | .438 | .090 | 12.23 |
| .0280 | .238 | .160 | .483 | 15:2120 | .294 | .194 | .466 | 18:2618 | | | | | .281 | .387 | .433 | .102 | 11.48 |
| .0320 | .271 | .170 | .477 | 15:2230 | .334 | .209 | .458 | 19:2753 | | | | | .303 | .400 | .429 | .112 | 10.89 |
| .0360 | .303 | .180 | .473 | 16:2329 | | | | | | | | | .325 | .411 | .426 | .123 | 10.40 |
| .0400 | .336 | .189 | .468 | 17:2418 | | | | | | | | | .347 | .420 | .422 | .133 | 10.00 |
| .0440 | .367 | .207 | .459 | 17:2493 | | | | | | | | | .368 | .429 | .420 | .143 | 9.66 |
| .0444 | .370 | .209 | .458 | 17:2500 | | | | | | | | | .370 | .430 | .419 | .144 | 9.63 |
| | COMPRESSIVE STEEL AREA 100% OF TENSILE STEEL AREA | | | | | | | | | | | | | | | | |
| .0005 | .007 | .030 | .548 | 3: 481 | .008 | .030 | .547 | 4: 552 | .010 | .031 | .547 | 4: 622 | .038 | .085 | .534 | .003 | 78.32 |
| .0010 | .012 | .040 | .542 | 4: 577 | .015 | .042 | .542 | 5: 677 | .017 | .043 | .541 | 5: 776 | .054 | .115 | .524 | .006 | 56.04 |
| .0015 | .017 | .048 | .538 | 4: 651 | .020 | .050 | .538 | 5: 773 | .024 | .051 | .537 | 6: 894 | .066 | .136 | .517 | .009 | 46.24 |
| .0020 | .022 | .054 | .535 | 5: 713 | .026 | .056 | .534 | 6: 853 | .030 | .059 | .533 | 7: 992 | .076 | .152 | .512 | .012 | 40.42 |
| .0025 | .026 | .060 | .533 | 5: 767 | .032 | .062 | .531 | 6: 922 | .037 | .065 | .530 | 7:1077 | .086 | .166 | .507 | .014 | 36.48 |
| .0030 | .031 | .064 | .530 | 6: 816 | .037 | .067 | .529 | 7: 985 | .043 | .070 | .527 | 8:1154 | .094 | .178 | .503 | .017 | 33.59 |
| .0035 | .035 | .068 | .528 | 6: 860 | .042 | .072 | .527 | 7:1041 | .050 | .075 | .525 | 8:1223 | .102 | .188 | .500 | .019 | 31.35 |
| .0040 | .039 | .072 | .527 | 6: 900 | .048 | .076 | .525 | 8:1093 | .056 | .080 | .523 | 9:1286 | .109 | .197 | .497 | .021 | 29.56 |
| .0050 | .048 | .078 | .523 | 7: 972 | .058 | .083 | .521 | 8:1185 | .069 | .087 | .519 | 10:1398 | .123 | .213 | .492 | .026 | 26.84 |
| .0060 | .056 | .084 | .521 | 7:1036 | .069 | .089 | .518 | 9:1267 | .082 | .094 | .515 | 10:1497 | .136 | .226 | .487 | .030 | 24.85 |
| .0080 | .073 | .092 | .516 | 8:1143 | .090 | .099 | .513 | 10:1404 | .107 | .106 | .509 | 11:1664 | .160 | .247 | .480 | .039 | 22.11 |
| .0100 | .090 | .099 | .513 | 9:1233 | .111 | .107 | .509 | 10:1518 | .132 | .116 | .504 | 12:1802 | .182 | .263 | .475 | .047 | 20.27 |
| .0120 | .107 | .105 | .510 | 9:1309 | .131 | .114 | .505 | 11:1614 | .156 | .125 | .500 | 13:1919 | .203 | .277 | .470 | .054 | 18.94 |
| .0140 | .123 | .110 | .507 | 9:1375 | .152 | .120 | .502 | 12:1698 | .181 | .132 | .496 | 14:2020 | .224 | .288 | .467 | .062 | 17.93 |
| .0160 | .140 | .114 | .505 | 10:1433 | .173 | .126 | .500 | 12:1771 | .206 | .139 | .493 | 15:2109 | .244 | .297 | .463 | .069 | 17.12 |
| .0180 | .156 | .118 | .503 | 10:1484 | .193 | .131 | .497 | 13:1837 | .230 | .145 | .490 | 15:2187 | .263 | .306 | .461 | .076 | 16.47 |
| .0200 | .173 | .122 | .502 | 11:1530 | .214 | .135 | .495 | 13:1895 | .255 | .150 | .487 | 16:2258 | .282 | .313 | .458 | .083 | 15.93 |
| .0240 | .206 | .127 | .499 | 11:1610 | .255 | .142 | .491 | 14:1996 | .304 | .160 | .482 | 16:2379 | .319 | .325 | .454 | .097 | 15.07 |
| .0280 | .239 | .132 | .496 | 12:1676 | .296 | .149 | .488 | 14:2079 | | | | | .356 | .334 | .451 | .110 | 14.43 |
| .0320 | .272 | .136 | .494 | 12:1732 | .337 | .154 | .485 | 15:2151 | | | | | .392 | .342 | .448 | .123 | 13.92 |
| .0360 | .305 | .140 | .493 | 12:1781 | | | | | | | | | .427 | .349 | .446 | .136 | 13.52 |
| .0400 | .338 | .143 | .491 | 13:1823 | | | | | | | | | .463 | .355 | .444 | .149 | 13.18 |
| .0440 | .371 | .145 | .490 | 13:1860 | | | | | | | | | .498 | .360 | .443 | .161 | 12.90 |
| .0444 | .374 | .145 | .490 | 13:1863 | | | | | | | | | .501 | .360 | .442 | .163 | 12.88 |

*TABLED VALUES ARE COEFFICIENTS FOR THE SECTION CONSTANTS: $Z_c$ = coeff.x $bd^2$; $y_n$ = coeff.x d; $S_c$ = coeff.x $bd^2$; $y_w$ = coeff.x d; $I_o$ = coeff.x $bd^3$; $k\beta$ and $k_w$ are pure coefficients.
Units for 'svc $f_c$' are given first in N/sqmm, then in lb/sqin.
VALUE OF $I_o$ IS ONLY FOR SHORT-TERM LIVE LOADS. FOR LONG-TERM LOADS, USE $\frac{1}{2}E_c$ WITH $I_o$.
COEFFICIENTS ABOVE THE INTERIOR LINE APPLY ONLY TO TEE SHAPES.

**Table A-8** Stress Ratios for Columns.

$f'_c$ = 3000psi (20.7 N/sqmm)
n = 9.36; g = .125

$R_n = \dfrac{P_n/bt}{.85f'_c}$

$R_w = \dfrac{P_w/bt}{f_c}$

$e_n = M_n/P_n$

$e_w = M_w/P_w$

STEEL AT FLEXURE FACES ONLY (b, t, gt, ½ $A_s$ each face)

STEEL UNIFORMLY DISTRIBUTED (b, t, gt, ¼ $A_s$ each face)

**$R_n$ AT ULTIMATE LOAD**

| At Ultimate Loads | $e_n/t$ | Flexure .01 | .02 | .03 | .04 | .05 | .06 | .07 | .08 | Uniform .01 | .02 | .03 | .04 | .05 | .06 | .07 | .08 |
|---|---|---|---|---|---|---|---|---|---|---|---|---|---|---|---|---|---|
| GRADE 40 STEEL | .1 | .91 | 1.03 | 1.15 | 1.26 | 1.38 | 1.49 | 1.61 | 1.73 | .91 | 1.02 | 1.14 | 1.25 | 1.36 | 1.48 | 1.59 | 1.70 |
| | .2 | .72 | .83 | .93 | 1.03 | 1.13 | 1.23 | 1.33 | 1.42 | .70 | .80 | .90 | .99 | 1.08 | 1.18 | 1.27 | 1.36 |
| | .3 | .58 | .68 | .77 | .86 | .95 | 1.03 | 1.12 | 1.20 | .54 | .63 | .72 | .80 | .88 | .96 | 1.04 | 1.11 |
| | .4 | .45 | .57 | .65 | .73 | .81 | .88 | .96 | 1.03 | .41 | .51 | .59 | .67 | .74 | .80 | .87 | .94 |
| | .5 | .34 | .48 | .56 | .64 | .71 | .77 | .84 | .91 | .30 | .41 | .50 | .57 | .63 | .69 | .75 | .81 |
| | .6 | .26 | .40 | .50 | .56 | .63 | .69 | .75 | .81 | .23 | .33 | .41 | .48 | .55 | .60 | .66 | .71 |
| | .7 | .20 | .33 | .43 | .50 | .56 | .62 | .67 | .73 | .18 | .28 | .35 | .42 | .47 | .53 | .58 | .63 |
| | .8 | .16 | .27 | .37 | .45 | .51 | .56 | .61 | .66 | .15 | .24 | .30 | .36 | .42 | .47 | .52 | .56 |
| | .9 | .13 | .23 | .32 | .40 | .46 | .51 | .56 | .60 | .13 | .20 | .27 | .32 | .37 | .42 | .46 | .51 |
| | 1.0 | .12 | .20 | .28 | .35 | .42 | .47 | .52 | .56 | .11 | .18 | .24 | .29 | .33 | .38 | .42 | .46 |
| | 2.0 | .05 | .09 | .13 | .16 | .19 | .23 | .26 | .29 | .05 | .08 | .11 | .14 | .16 | .18 | .21 | .23 |
| | 3.0 | .03 | .06 | .08 | .11 | .13 | .15 | .17 | .19 | .03 | .05 | .08 | .09 | .11 | .13 | .14 | .15 |
| | 4.0 | $k_n<g$ | .04 | .06 | .08 | .10 | .11 | .13 | .14 | .02 | .04 | .06 | .07 | .08 | .10 | .11 | .12 |
| GRADE 50 STEEL | .1 | .94 | 1.09 | 1.24 | 1.39 | 1.53 | 1.68 | 1.83 | 1.97 | .93 | 1.08 | 1.22 | 1.36 | 1.50 | 1.64 | 1.78 | 1.92 |
| | .2 | .75 | .89 | 1.01 | 1.14 | 1.26 | 1.39 | 1.51 | 1.63 | .72 | .84 | .96 | 1.08 | 1.19 | 1.31 | 1.42 | 1.53 |
| | .3 | .60 | .72 | .84 | .95 | 1.06 | 1.16 | 1.27 | 1.38 | .56 | .67 | .77 | .87 | .96 | 1.06 | 1.16 | 1.25 |
| | .4 | .49 | .61 | .71 | .81 | .90 | 1.00 | 1.09 | 1.18 | .43 | .54 | .63 | .72 | .80 | .89 | .97 | 1.05 |
| | .5 | .38 | .52 | .61 | .70 | .79 | .87 | .96 | 1.04 | .33 | .45 | .54 | .61 | .69 | .76 | .83 | .90 |
| | .6 | .30 | .45 | .54 | .62 | .70 | .77 | .85 | .92 | .26 | .37 | .46 | .53 | .60 | .66 | .73 | .79 |
| | .7 | .23 | .38 | .48 | .56 | .63 | .70 | .76 | .83 | .21 | .31 | .39 | .47 | .53 | .59 | .64 | .70 |
| | .8 | .19 | .32 | .43 | .50 | .57 | .63 | .69 | .76 | .17 | .27 | .34 | .41 | .47 | .53 | .58 | .63 |
| | .9 | .16 | .28 | .38 | .46 | .52 | .58 | .64 | .69 | .15 | .23 | .30 | .36 | .42 | .48 | .53 | .57 |
| | 1.0 | .14 | .24 | .33 | .42 | .48 | .53 | .59 | .64 | .13 | .21 | .27 | .33 | .38 | .43 | .48 | .53 |
| | 2.0 | .06 | .11 | .15 | .19 | .23 | .28 | .32 | .36 | .06 | .10 | .13 | .16 | .19 | .21 | .24 | .27 |
| | 3.0 | .04 | .07 | .10 | .13 | .15 | .18 | .20 | .23 | .04 | .06 | .09 | .11 | .13 | .15 | .16 | .18 |
| | 4.0 | .03 | .05 | .07 | .09 | .11 | .13 | .15 | .17 | .03 | .05 | .07 | .08 | .10 | .11 | .13 | .14 |
| GRADE 60 STEEL | .1 | .98 | 1.15 | 1.33 | 1.51 | 1.69 | 1.86 | 2.04 | 2.22 | .96 | 1.12 | 1.29 | 1.46 | 1.62 | 1.79 | 1.96 | 2.13 |
| | .2 | .78 | .94 | 1.09 | 1.24 | 1.39 | 1.54 | 1.69 | 1.84 | .74 | .88 | 1.02 | 1.15 | 1.29 | 1.42 | 1.55 | 1.69 |
| | .3 | .62 | .77 | .91 | 1.04 | 1.17 | 1.30 | 1.42 | 1.55 | .57 | .70 | .81 | .93 | 1.04 | 1.15 | 1.26 | 1.37 |
| | .4 | .51 | .64 | .77 | .88 | 1.00 | 1.11 | 1.22 | 1.34 | .45 | .57 | .67 | .77 | .86 | .96 | 1.05 | 1.15 |
| | .5 | .42 | .55 | .66 | .77 | .87 | .97 | 1.07 | 1.17 | .35 | .47 | .57 | .65 | .74 | .82 | .90 | .98 |
| | .6 | .33 | .48 | .58 | .68 | .77 | .86 | .95 | 1.04 | .28 | .40 | .49 | .57 | .64 | .71 | .79 | .86 |
| | .7 | .26 | .43 | .52 | .61 | .69 | .77 | .86 | .94 | .23 | .34 | .43 | .50 | .57 | .63 | .70 | .76 |
| | .8 | .21 | .37 | .47 | .55 | .63 | .70 | .78 | .85 | .19 | .29 | .38 | .45 | .51 | .57 | .63 | .68 |
| | .9 | .18 | .32 | .43 | .50 | .57 | .64 | .71 | .78 | .16 | .26 | .33 | .40 | .46 | .51 | .57 | .62 |
| | 1.0 | .15 | .28 | .38 | .46 | .53 | .59 | .66 | .72 | .14 | .23 | .30 | .36 | .42 | .47 | .52 | .57 |
| | 2.0 | .07 | .12 | .17 | .22 | .27 | .32 | .37 | .41 | .07 | .11 | .15 | .18 | .21 | .24 | .27 | .30 |
| | 3.0 | .04 | .08 | .11 | .14 | .17 | .21 | .24 | .27 | .04 | .07 | .10 | .12 | .14 | .16 | .18 | .20 |
| | 4.0 | .03 | .06 | .09 | .11 | .13 | .15 | .18 | .20 | .03 | .05 | .08 | .09 | .11 | .13 | .14 | .15 |

**$R_w$ AT SERVICE LEVELS**

| In Service | $e_w/t$ | Flexure .01 | .02 | .03 | .04 | .05 | .06 | .07 | .08 | Uniform .01 | .02 | .03 | .04 | .05 | .06 | .07 | .08 |
|---|---|---|---|---|---|---|---|---|---|---|---|---|---|---|---|---|---|
| FOR ALL STEEL | .1 | .76 | .90 | 1.03 | 1.17 | 1.30 | 1.43 | 1.56 | 1.69 | .75 | .87 | .99 | 1.12 | 1.24 | 1.36 | 1.48 | 1.61 |
| | .2 | .56 | .67 | .78 | .88 | .99 | 1.09 | 1.20 | 1.30 | .55 | .64 | .74 | .83 | .92 | 1.02 | 1.11 | 1.20 |
| | .3 | .42 | .52 | .61 | .70 | .79 | .88 | .97 | 1.05 | .39 | .48 | .55 | .63 | .71 | .78 | .86 | .94 |
| | .4 | .31 | .40 | .48 | .56 | .63 | .71 | .78 | .86 | .28 | .36 | .43 | .49 | .56 | .62 | .68 | .75 |
| | .5 | .24 | .32 | .39 | .46 | .53 | .59 | .66 | .72 | .22 | .28 | .34 | .40 | .45 | .51 | .56 | .62 |
| | .6 | .19 | .27 | .33 | .39 | .45 | .51 | .56 | .62 | .18 | .24 | .29 | .34 | .39 | .43 | .48 | .53 |
| | .7 | .16 | .23 | .29 | .34 | .39 | .44 | .50 | .55 | .15 | .20 | .25 | .29 | .33 | .38 | .42 | .46 |
| | .8 | .14 | .20 | .25 | .30 | .35 | .40 | .44 | .49 | .13 | .18 | .22 | .26 | .30 | .33 | .37 | .41 |
| | .9 | .13 | .18 | .23 | .27 | .31 | .36 | .40 | .44 | .11 | .16 | .19 | .23 | .26 | .30 | .33 | .36 |
| | 1.0 | .11 | .16 | .21 | .25 | .29 | .32 | .36 | .40 | .10 | .14 | .18 | .21 | .24 | .27 | .30 | .33 |
| | 2.0 | .06 | .08 | .11 | .13 | .15 | .17 | .19 | .21 | .05 | .07 | .09 | .11 | .12 | .14 | .16 | .17 |
| | 3.0 | .04 | .06 | .07 | .09 | .10 | .12 | .13 | .15 | .03 | .05 | .06 | .07 | .08 | .09 | .11 | .12 |
| | 4.0 | .03 | .04 | .05 | .07 | .08 | .09 | .10 | .11 | .02 | .04 | .05 | .05 | .06 | .07 | .08 | .09 |

**Table A-9** Stress Ratios for Columns.

$f'_c = 4000$psi (27.6 N/sqmm)
$n = 8.11$; $g = .125$

$R_n = \dfrac{P_n/bt}{.85f'_c}$

$R_w = \dfrac{P_w/bt}{f_c}$

$e_n = M_n/P_n$

$e_w = M_w/P_w$

STEEL AT FLEXURE FACES ONLY (b; t; gt; ½ $A_s$ at each face)

STEEL UNIFORMLY DISTRIBUTED (b; t; gt; ¼ $A_s$ at each face)

**$R_n$ AT ULTIMATE LOAD**

|  |  | $e_n/t$ \ $A_g/bt$ | Flexure .01 | Flexure .02 | Flexure .03 | Flexure .04 | Flexure .05 | Flexure .06 | Flexure .07 | Flexure .08 | Uniform .01 | Uniform .02 | Uniform .03 | Uniform .04 | Uniform .05 | Uniform .06 | Uniform .07 | Uniform .08 |
|---|---|---|---|---|---|---|---|---|---|---|---|---|---|---|---|---|---|---|
| AT ULTIMATE LOADS | GRADE 40 STEEL | .1 | .88 | .97 | 1.05 | 1.14 | 1.22 | 1.31 | 1.39 | 1.48 | .87 | .96 | 1.04 | 1.13 | 1.21 | 1.29 | 1.38 | 1.46 |
| | | .2 | .69 | .77 | .85 | .92 | 1.00 | 1.07 | 1.14 | 1.21 | .67 | .75 | .82 | .89 | .96 | 1.02 | 1.09 | 1.16 |
| | | .3 | .54 | .62 | .70 | .76 | .83 | .89 | .96 | 1.02 | .51 | .58 | .65 | .71 | .77 | .83 | .89 | .95 |
| | | .4 | .41 | .52 | .58 | .65 | .70 | .76 | .82 | .87 | .37 | .46 | .53 | .59 | .64 | .69 | .74 | .79 |
| | | .5 | .30 | .42 | .50 | .56 | .61 | .66 | .71 | .76 | .27 | .36 | .43 | .49 | .55 | .59 | .64 | .68 |
| | | .6 | .22 | .33 | .42 | .49 | .54 | .59 | .63 | .68 | .20 | .29 | .36 | .41 | .46 | .51 | .56 | .60 |
| | | .7 | .16 | .27 | .36 | .43 | .48 | .53 | .57 | .61 | .15 | .24 | .30 | .35 | .40 | .44 | .49 | .53 |
| | | .8 | .13 | .22 | .30 | .37 | .43 | .48 | .51 | .55 | .13 | .20 | .25 | .30 | .35 | .39 | .43 | .46 |
| | | .9 | .11 | .19 | .26 | .32 | .38 | .43 | .47 | .50 | .11 | .17 | .22 | .27 | .31 | .34 | .38 | .41 |
| | | 1.0 | .09 | .16 | .23 | .28 | .34 | .39 | .43 | .47 | .09 | .15 | .19 | .24 | .27 | .31 | .34 | .37 |
| | | 2.0 | $k_n<g$ | .07 | .10 | .13 | .15 | .18 | .20 | .23 | .04 | .07 | .09 | .12 | .13 | .15 | .17 | .18 |
| | | 3.0 | $k_n<g$ | .04 | .06 | .08 | .10 | .12 | .13 | .15 | .02 | .04 | .06 | .08 | .09 | .10 | .12 | .13 |
| | | 4.0 | $k_n<g$ | .03 | .05 | .06 | .08 | .09 | .10 | .11 | .02 | .03 | .05 | .06 | .07 | .08 | .09 | .10 |
| | GRADE 50 STEEL | .1 | .90 | 1.01 | 1.12 | 1.23 | 1.34 | 1.45 | 1.55 | 1.66 | .89 | 1.00 | 1.10 | 1.21 | 1.31 | 1.42 | 1.52 | 1.62 |
| | | .2 | .71 | .81 | .91 | 1.00 | 1.10 | 1.19 | 1.28 | 1.37 | .69 | .78 | .87 | .95 | 1.04 | 1.12 | 1.21 | 1.29 |
| | | .3 | .56 | .66 | .75 | .83 | .91 | .99 | 1.07 | 1.15 | .52 | .61 | .69 | .76 | .83 | .91 | .98 | 1.05 |
| | | .4 | .44 | .54 | .63 | .70 | .78 | .85 | .92 | .99 | .39 | .49 | .56 | .63 | .69 | .75 | .81 | .87 |
| | | .5 | .33 | .46 | .54 | .61 | .67 | .74 | .80 | .86 | .29 | .39 | .47 | .53 | .59 | .64 | .70 | .75 |
| | | .6 | .25 | .38 | .47 | .53 | .59 | .65 | .71 | .76 | .22 | .32 | .39 | .46 | .51 | .56 | .61 | .65 |
| | | .7 | .19 | .31 | .41 | .48 | .53 | .58 | .64 | .69 | .17 | .27 | .33 | .39 | .45 | .49 | .54 | .58 |
| | | .8 | .15 | .26 | .35 | .43 | .48 | .53 | .58 | .62 | .14 | .22 | .29 | .34 | .39 | .44 | .48 | .52 |
| | | .9 | .13 | .22 | .31 | .38 | .44 | .48 | .53 | .57 | .12 | .19 | .25 | .30 | .35 | .39 | .43 | .47 |
| | | 1.0 | .11 | .19 | .27 | .34 | .40 | .45 | .49 | .53 | .11 | .17 | .22 | .27 | .31 | .35 | .39 | .43 |
| | | 2.0 | .05 | .08 | .12 | .15 | .18 | .21 | .25 | .28 | .05 | .08 | .11 | .13 | .15 | .17 | .19 | .21 |
| | | 3.0 | $k_n<g$ | .05 | .08 | .10 | .12 | .14 | .16 | .18 | .03 | .05 | .07 | .09 | .11 | .12 | .13 | .15 |
| | | 4.0 | $k_n<g$ | .04 | .06 | .07 | .09 | .11 | .12 | .14 | .02 | .04 | .05 | .07 | .08 | .09 | .10 | .11 |
| | GRADE 60 STEEL | .1 | .93 | 1.06 | 1.19 | 1.32 | 1.46 | 1.59 | 1.72 | 1.85 | .91 | 1.04 | 1.16 | 1.28 | 1.41 | 1.53 | 1.66 | 1.78 |
| | | .2 | .73 | .85 | .97 | 1.08 | 1.19 | 1.31 | 1.42 | 1.53 | .70 | .81 | .91 | 1.01 | 1.11 | 1.21 | 1.31 | 1.41 |
| | | .3 | .57 | .69 | .80 | .90 | .99 | 1.09 | 1.19 | 1.28 | .53 | .63 | .72 | .81 | .89 | .97 | 1.06 | 1.14 |
| | | .4 | .46 | .57 | .67 | .76 | .85 | .93 | 1.02 | 1.10 | .41 | .51 | .59 | .66 | .74 | .81 | .88 | .95 |
| | | .5 | .36 | .49 | .57 | .66 | .73 | .81 | .89 | .96 | .31 | .42 | .49 | .56 | .62 | .69 | .75 | .81 |
| | | .6 | .27 | .42 | .50 | .58 | .65 | .72 | .79 | .85 | .24 | .35 | .42 | .48 | .54 | .60 | .65 | .71 |
| | | .7 | .21 | .35 | .45 | .51 | .58 | .64 | .70 | .77 | .19 | .29 | .36 | .43 | .48 | .53 | .58 | .63 |
| | | .8 | .17 | .30 | .40 | .46 | .52 | .58 | .64 | .69 | .16 | .25 | .32 | .38 | .43 | .47 | .52 | .56 |
| | | .9 | .15 | .25 | .35 | .42 | .48 | .53 | .58 | .64 | .14 | .21 | .28 | .33 | .38 | .43 | .47 | .51 |
| | | 1.0 | .13 | .22 | .31 | .39 | .44 | .49 | .54 | .59 | .12 | .19 | .25 | .30 | .35 | .39 | .43 | .47 |
| | | 2.0 | .05 | .10 | .14 | .17 | .21 | .25 | .29 | .32 | .05 | .09 | .12 | .15 | .17 | .19 | .22 | .24 |
| | | 3.0 | .03 | .06 | .09 | .11 | .14 | .16 | .18 | .21 | .03 | .06 | .08 | .10 | .12 | .13 | .15 | .16 |
| | | 4.0 | .02 | .05 | .07 | .09 | .10 | .12 | .14 | .16 | .02 | .04 | .06 | .08 | .09 | .10 | .12 | .13 |

**$R_w$ AT SERVICE LEVELS**

|  |  | $e_w/t$ \ $A_g/bt$ | Flexure .01 | Flexure .02 | Flexure .03 | Flexure .04 | Flexure .05 | Flexure .06 | Flexure .07 | Flexure .08 | Uniform .01 | Uniform .02 | Uniform .03 | Uniform .04 | Uniform .05 | Uniform .06 | Uniform .07 | Uniform .08 |
|---|---|---|---|---|---|---|---|---|---|---|---|---|---|---|---|---|---|---|
| IN SERVICE | FOR ALL STEEL | .1 | .74 | .86 | .98 | 1.09 | 1.21 | 1.32 | 1.43 | 1.54 | .73 | .84 | .94 | 1.05 | 1.15 | 1.26 | 1.36 | 1.47 |
| | | .2 | .55 | .64 | .73 | .82 | .91 | 1.00 | 1.10 | 1.18 | .53 | .61 | .70 | .77 | .86 | .93 | 1.02 | 1.09 |
| | | .3 | .40 | .49 | .57 | .65 | .73 | .80 | .88 | .96 | .38 | .45 | .52 | .59 | .66 | .72 | .79 | .85 |
| | | .4 | .29 | .37 | .44 | .51 | .58 | .65 | .71 | .78 | .27 | .34 | .40 | .46 | .51 | .57 | .62 | .67 |
| | | .5 | .22 | .30 | .36 | .42 | .48 | .54 | .59 | .65 | .21 | .27 | .32 | .37 | .42 | .46 | .51 | .56 |
| | | .6 | .18 | .25 | .30 | .36 | .41 | .46 | .51 | .56 | .17 | .22 | .27 | .31 | .35 | .39 | .43 | .47 |
| | | .7 | .15 | .21 | .26 | .31 | .36 | .40 | .45 | .49 | .14 | .19 | .23 | .27 | .30 | .34 | .38 | .41 |
| | | .8 | .13 | .19 | .23 | .27 | .32 | .36 | .40 | .44 | .12 | .16 | .20 | .24 | .27 | .30 | .33 | .37 |
| | | .9 | .12 | .17 | .21 | .25 | .28 | .32 | .36 | .39 | .11 | .15 | .18 | .21 | .24 | .27 | .30 | .33 |
| | | 1.0 | .11 | .15 | .19 | .22 | .26 | .29 | .33 | .36 | .09 | .13 | .16 | .19 | .22 | .24 | .27 | .30 |
| | | 2.0 | .05 | .08 | .10 | .12 | .14 | .15 | .17 | .19 | .05 | .07 | .08 | .10 | .11 | .13 | .14 | .15 |
| | | 3.0 | .03 | .05 | .07 | .08 | .09 | .10 | .12 | .13 | .03 | .04 | .05 | .07 | .08 | .09 | .10 | .10 |
| | | 4.0 | .03 | .04 | .05 | .06 | .07 | .08 | .09 | .10 | .02 | .03 | .04 | .05 | .06 | .06 | .07 | .08 |

**Table A-10** Stress Ratios for Columns.

$f'_c$ = 5000psi (34.4 N/sqmm)
n = 7.25; g = .125

$R_n = \dfrac{P_n/bt}{.85f'_c}$

$R_w = \dfrac{P_w/bt}{f_c}$

$e_n = M_n/P_n$

$e_w = M_w/P_w$

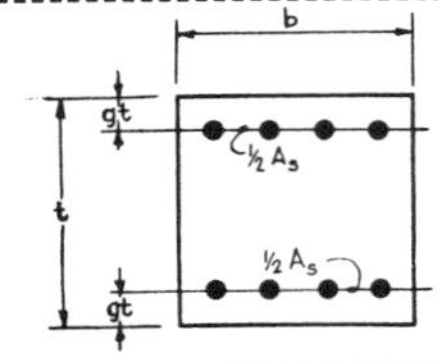

Diagram labels: b, t, gt, ½ $A_s$ (steel at flexure faces only); b, t, gt, ¼ $A_s$ (steel uniformly distributed)

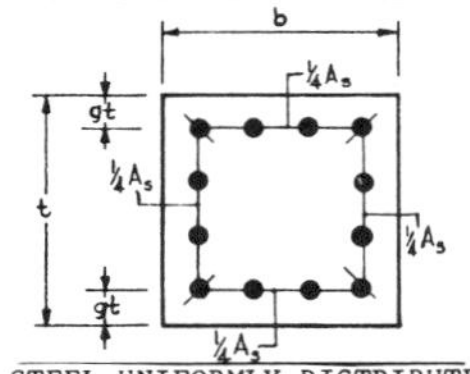

## $R_n$ AT ULTIMATE LOAD

AT ULTIMATE LOADS — GRADE 40 STEEL

| $e_n/t$ | Flexure faces .01 | .02 | .03 | .04 | .05 | .06 | .07 | .08 | Uniformly distributed .01 | .02 | .03 | .04 | .05 | .06 | .07 | .08 |
|---|---|---|---|---|---|---|---|---|---|---|---|---|---|---|---|---|
| .1 | .86 | .92 | .99 | 1.06 | 1.13 | 1.20 | 1.26 | 1.33 | .85 | .92 | .98 | 1.05 | 1.12 | 1.18 | 1.25 | 1.31 |
| .2 | .67 | .74 | .80 | .86 | .91 | .97 | 1.03 | 1.08 | .65 | .71 | .77 | .82 | .88 | .93 | .99 | 1.04 |
| .3 | .52 | .59 | .65 | .70 | .75 | .81 | .86 | .91 | .49 | .55 | .61 | .66 | .70 | .75 | .80 | .84 |
| .4 | .38 | .48 | .54 | .59 | .64 | .68 | .73 | .77 | .35 | .43 | .49 | .54 | .58 | .62 | .66 | .70 |
| .5 | .27 | .38 | .46 | .51 | .55 | .59 | .63 | .67 | .24 | .33 | .39 | .44 | .49 | .53 | .57 | .60 |
| .6 | .19 | .29 | .37 | .44 | .48 | .52 | .56 | .60 | .17 | .26 | .32 | .37 | .41 | .45 | .49 | .53 |
| .7 | .14 | .23 | .31 | .37 | .43 | .47 | .50 | .53 | .14 | .21 | .26 | .31 | .35 | .39 | .42 | .46 |
| .8 | .11 | .19 | .26 | .32 | .37 | .42 | .45 | .48 | .11 | .17 | .22 | .27 | .30 | .34 | .37 | .40 |
| .9 | .09 | .16 | .22 | .27 | .32 | .37 | .41 | .44 | .09 | .15 | .19 | .23 | .27 | .30 | .33 | .36 |
| 1.0 | .08 | .14 | .19 | .24 | .29 | .33 | .37 | .40 | .08 | .13 | .17 | .20 | .24 | .27 | .30 | .32 |
| 2.0 | $k_n<g$ | .06 | .08 | .11 | .13 | .15 | .17 | .19 | $k_n<g$ | .06 | .08 | .10 | .12 | .13 | .14 | .16 |
| 3.0 | $k_n<g$ | .04 | .05 | .07 | .08 | .10 | .11 | .13 | $k_n<g$ | .04 | .05 | .07 | .08 | .09 | .10 | .11 |
| 4.0 | $k_n<g$ | .03 | .04 | .05 | .06 | .07 | .08 | .10 | $k_n<g$ | .03 | .04 | .05 | .06 | .07 | .08 | .08 |

AT ULTIMATE LOADS — GRADE 50 STEEL

| $e_n/t$ | Flexure faces .01 | .02 | .03 | .04 | .05 | .06 | .07 | .08 | Uniformly distributed .01 | .02 | .03 | .04 | .05 | .06 | .07 | .08 |
|---|---|---|---|---|---|---|---|---|---|---|---|---|---|---|---|---|
| .1 | .88 | .96 | 1.05 | 1.14 | 1.22 | 1.31 | 1.39 | 1.48 | .87 | .95 | 1.03 | 1.12 | 1.20 | 1.28 | 1.36 | 1.44 |
| .2 | .69 | .77 | .85 | .92 | .99 | 1.07 | 1.14 | 1.21 | .67 | .74 | .81 | .88 | .94 | 1.01 | 1.08 | 1.14 |
| .3 | .53 | .61 | .69 | .76 | .82 | .89 | .95 | 1.01 | .50 | .57 | .64 | .70 | .75 | .81 | .87 | .92 |
| .4 | .41 | .50 | .57 | .63 | .69 | .75 | .81 | .86 | .37 | .46 | .51 | .57 | .62 | .67 | .72 | .77 |
| .5 | .30 | .42 | .49 | .55 | .60 | .65 | .70 | .75 | .27 | .36 | .43 | .48 | .53 | .57 | .61 | .66 |
| .6 | .21 | .33 | .42 | .48 | .53 | .57 | .62 | .67 | .20 | .29 | .35 | .41 | .45 | .49 | .53 | .57 |
| .7 | .16 | .27 | .36 | .43 | .47 | .51 | .56 | .60 | .15 | .23 | .29 | .35 | .39 | .43 | .47 | .51 |
| .8 | .13 | .22 | .30 | .37 | .42 | .46 | .50 | .54 | .13 | .20 | .25 | .30 | .34 | .38 | .42 | .45 |
| .9 | .11 | .19 | .26 | .32 | .38 | .42 | .46 | .49 | .11 | .17 | .22 | .26 | .30 | .34 | .37 | .41 |
| 1.0 | .09 | .16 | .23 | .28 | .34 | .39 | .42 | .46 | .09 | .15 | .19 | .23 | .27 | .30 | .34 | .37 |
| 2.0 | $k_n<g$ | .07 | .10 | .13 | .15 | .18 | .20 | .23 | .04 | .07 | .09 | .11 | .13 | .15 | .16 | .18 |
| 3.0 | $k_n<g$ | .04 | .06 | .08 | .10 | .12 | .13 | .15 | .02 | .04 | .06 | .08 | .09 | .10 | .11 | .13 |
| 4.0 | $k_n<g$ | .03 | .05 | .06 | .08 | .09 | .10 | .11 | .02 | .03 | .04 | .06 | .07 | .08 | .09 | .10 |

AT ULTIMATE LOADS — GRADE 60 STEEL

| $e_n/t$ | Flexure faces .01 | .02 | .03 | .04 | .05 | .06 | .07 | .08 | Uniformly distributed .01 | .02 | .03 | .04 | .05 | .06 | .07 | .08 |
|---|---|---|---|---|---|---|---|---|---|---|---|---|---|---|---|---|
| .1 | .90 | 1.00 | 1.11 | 1.21 | 1.32 | 1.42 | 1.52 | 1.63 | .89 | .98 | 1.08 | 1.18 | 1.28 | 1.37 | 1.47 | 1.57 |
| .2 | .70 | .80 | .89 | .98 | 1.07 | 1.16 | 1.25 | 1.33 | .68 | .76 | .84 | .92 | 1.00 | 1.08 | 1.16 | 1.24 |
| .3 | .54 | .64 | .73 | .81 | .89 | .96 | 1.04 | 1.12 | .51 | .59 | .66 | .73 | .80 | .86 | .93 | 1.00 |
| .4 | .43 | .53 | .61 | .68 | .75 | .82 | .89 | .95 | .38 | .47 | .53 | .60 | .66 | .71 | .77 | .83 |
| .5 | .32 | .44 | .52 | .58 | .65 | .71 | .77 | .83 | .28 | .38 | .45 | .50 | .55 | .61 | .66 | .70 |
| .6 | .24 | .37 | .45 | .51 | .57 | .63 | .68 | .74 | .21 | .31 | .38 | .43 | .48 | .52 | .57 | .61 |
| .7 | .18 | .30 | .40 | .46 | .51 | .56 | .61 | .66 | .17 | .26 | .32 | .38 | .42 | .46 | .50 | .54 |
| .8 | .15 | .25 | .34 | .41 | .46 | .51 | .55 | .60 | .14 | .21 | .28 | .33 | .37 | .41 | .45 | .49 |
| .9 | .12 | .21 | .30 | .37 | .42 | .46 | .50 | .55 | .12 | .18 | .24 | .29 | .33 | .37 | .41 | .44 |
| 1.0 | .11 | .18 | .26 | .33 | .38 | .42 | .46 | .50 | .10 | .16 | .21 | .26 | .30 | .34 | .37 | .40 |
| 2.0 | .04 | .08 | .11 | .14 | .17 | .20 | .24 | .27 | .04 | .08 | .10 | .13 | .15 | .16 | .18 | .20 |
| 3.0 | $k_n<g$ | .05 | .07 | .10 | .12 | .13 | .15 | .17 | .03 | .05 | .07 | .09 | .10 | .11 | .13 | .14 |
| 4.0 | $k_n<g$ | .04 | .06 | .07 | .09 | .10 | .12 | .13 | .02 | .04 | .05 | .06 | .08 | .09 | .10 | .11 |

## $R_w$ AT SERVICE LEVELS

IN SERVICE — FOR ALL STEEL

| $e_w/t$ | Flexure faces .01 | .02 | .03 | .04 | .05 | .06 | .07 | .08 | Uniformly distributed .01 | .02 | .03 | .04 | .05 | .06 | .07 | .08 |
|---|---|---|---|---|---|---|---|---|---|---|---|---|---|---|---|---|
| .1 | .73 | .83 | .94 | 1.04 | 1.14 | 1.24 | 1.34 | 1.44 | .72 | .81 | .91 | 1.00 | 1.09 | 1.19 | 1.28 | 1.37 |
| .2 | .54 | .62 | .70 | .78 | .86 | .94 | 1.02 | 1.10 | .52 | .60 | .67 | .74 | .81 | .88 | .95 | 1.02 |
| .3 | .39 | .47 | .54 | .61 | .68 | .75 | .82 | .89 | .37 | .44 | .50 | .56 | .62 | .68 | .73 | .79 |
| .4 | .28 | .36 | .42 | .48 | .54 | .60 | .66 | .72 | .26 | .33 | .38 | .43 | .48 | .53 | .58 | .63 |
| .5 | .22 | .28 | .34 | .40 | .45 | .50 | .55 | .60 | .20 | .26 | .30 | .35 | .39 | .43 | .47 | .52 |
| .6 | .17 | .23 | .29 | .33 | .38 | .43 | .47 | .51 | .16 | .21 | .25 | .29 | .33 | .37 | .40 | .44 |
| .7 | .15 | .20 | .25 | .29 | .33 | .37 | .41 | .45 | .13 | .18 | .22 | .25 | .28 | .32 | .35 | .38 |
| .8 | .13 | .18 | .22 | .26 | .29 | .33 | .37 | .40 | .11 | .15 | .19 | .22 | .25 | .28 | .31 | .34 |
| .9 | .11 | .16 | .19 | .23 | .26 | .30 | .33 | .36 | .10 | .14 | .17 | .20 | .22 | .25 | .28 | .30 |
| 1.0 | .10 | .14 | .18 | .21 | .24 | .27 | .30 | .33 | .09 | .12 | .15 | .18 | .20 | .23 | .25 | .27 |
| 2.0 | .05 | .07 | .09 | .11 | .12 | .14 | .16 | .17 | .04 | .06 | .08 | .09 | .10 | .12 | .13 | .14 |
| 3.0 | .03 | .05 | .06 | .07 | .08 | .10 | .11 | .12 | .03 | .04 | .05 | .06 | .07 | .08 | .09 | .10 |
| 4.0 | .02 | .04 | .05 | .06 | .06 | .07 | .08 | .09 | .02 | .03 | .04 | .05 | .05 | .06 | .07 | .07 |

**Table A-11** Development Length of Hooked Bars.

| DEVELOPMENT LENGTH TO BACK OF HOOK IN INCHES | | | | | | | | | | |
|---|---|---|---|---|---|---|---|---|---|---|
| CONCRETE STRENGTH | STEEL GRADE | BAR SIZES #3 | #4 | #5 | #6 | #7 | #8 | #9 | #10 | #11 |
| 3000 psi | 40 | 6 | 7 | 9 | 11 | 13 | 15 | 16 | 18 | 20 |
| | 50 | 7 | 9 | 11 | 14 | 16 | 18 | 21 | 23 | 25 |
| | 60 | 8 | 11 | 14 | 16 | 19 | 22 | 25 | 27 | 30 |
| 4000 psi | 40 | 6 | 6 | 8 | 9 | 11 | 13 | 14 | 16 | 17 |
| | 50 | 6 | 8 | 10 | 12 | 14 | 16 | 18 | 20 | 22 |
| | 60 | 7 | 9 | 12 | 14 | 17 | 19 | 21 | 24 | 26 |
| 5000 psi | 40 | 6 | 6 | 7 | 8 | 10 | 11 | 13 | 14 | 16 |
| | 50 | 6 | 7 | 9 | 11 | 12 | 14 | 16 | 18 | 19 |
| | 60 | 6 | 8 | 11 | 13 | 15 | 17 | 19 | 21 | 23 |

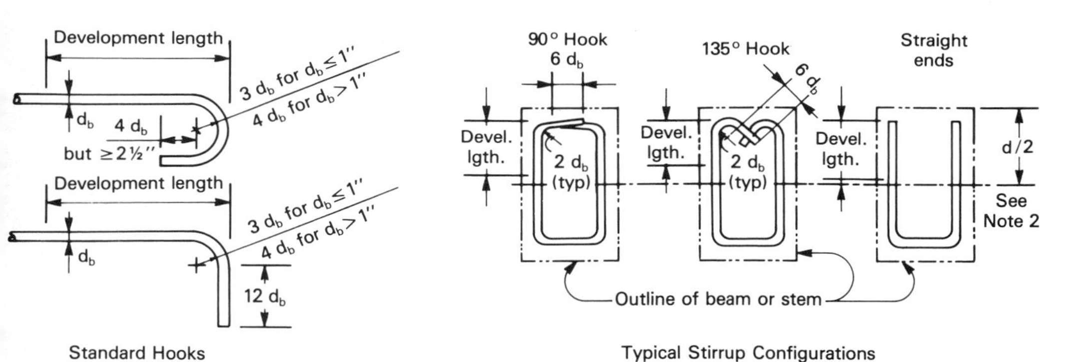


Standard Hooks

Typical Stirrup Configurations

**Table A-12** Development Length of Stirrups and Ties.

| LENGTH TO BACK OF HOOK OR TO END OF STRAIGHT BAR (inches) | | | | | | | | | | |
|---|---|---|---|---|---|---|---|---|---|---|
| | | END CONFIGURATION AND BAR SIZES | | | | | | | | |
| CONCRETE STRENGTH | STEEL GRADE | 90° HOOK #3 | #4 | #5 | 135° HOOK #3 | #4 | #5 | STRAIGHT #3 | #4 | #5 |
| 3000 psi | 40 | 7 | 8 | 8 | SEE NOTE 1 | | | 12 | 12 | 15 |
| OR | 50 | 7 | 8 | 8 | 5 | 6 | 6 | 12 | 12 | 15 |
| MORE | 60 | 7 | 8 | 9 | 5 | 6 | 7 | 12 | 12 | 15 |

Note 1) For 135° hook with Grade 40 steel, the strength is fully developed by the hook; no additional embedment is required.
Note 2) Full development length of stirrups and ties must fall within a distance of d/2 from the compression face of the beam.

**Table A-13** Development Length of Straight Deformed Reinforcing Bars.

| REQUIRED IMBEDMENT IN INCHES TO DEVELOP FULL STRENGTH | | | | | | | | | | | | |
|---|---|---|---|---|---|---|---|---|---|---|---|---|
| | | | DEVELOPMENT LENGTH | | | SPLICE LENGTHS IN TENSION | | | | | | SPLICE LGTH |
| ULTIMATE CONCRETE STRENGTH | BAR SIZE NO. | GRADE OF STEEL | TENSION | | COMPRESSION | CLASS A | | CLASS B | | CLASS C | | COMPRESSION |
| | | | TOP | OTHER | ALL BARS | TOP | OTHER | TOP | OTHER | TOP | OTHER | ALL BARS |
| 3000 psi | #3 | 40 | 12 | 12 | 8 | 12 | 12 | 12 | 12 | 14 | 12 | 12 |
| | | 60 | 13 | 12 | 8 | 13 | 12 | 16 | 12 | 21 | 15 | 12 |
| | #4 | 40 | 12 | 12 | 8 | 12 | 12 | 15 | 12 | 19 | 14 | 12 |
| | | 60 | 17 | 12 | 11 | 17 | 12 | 22 | 16 | 29 | 20 | 15 |
| | #5 | 40 | 14 | 12 | 9 | 14 | 12 | 18 | 13 | 24 | 17 | 13 |
| | | 60 | 21 | 15 | 14 | 21 | 15 | 27 | 20 | 36 | 26 | 19 |
| | #6 | 40 | 18 | 13 | 11 | 18 | 13 | 23 | 17 | 31 | 22 | 15 |
| | | 60 | 27 | 19 | 16 | 27 | 19 | 35 | 25 | 46 | 33 | 23 |
| | #7 | 40 | 25 | 18 | 13 | 25 | 18 | 32 | 23 | 42 | 30 | 18 |
| | | 60 | 37 | 26 | 19 | 37 | 26 | 48 | 34 | 63 | 45 | 26 |
| | #8 | 40 | 32 | 23 | 15 | 32 | 23 | 42 | 30 | 55 | 39 | 20 |
| | | 60 | 48 | 34 | 22 | 48 | 34 | 63 | 45 | 82 | 59 | 30 |
| | #9 | 40 | 41 | 29 | 16 | 41 | 29 | 53 | 38 | 69 | 50 | 23 |
| | | 60 | 61 | 44 | 25 | 61 | 44 | 80 | 57 | 104 | 74 | 34 |
| | #10 | 40 | 52 | 37 | 19 | 52 | 37 | 67 | 48 | 88 | 63 | 25 |
| | | 60 | 78 | 56 | 28 | 78 | 56 | 101 | 72 | 132 | 94 | 38 |
| | #11 | 40 | 64 | 46 | 21 | 64 | 46 | 83 | 59 | 109 | 78 | 28 |
| | | 60 | 96 | 68 | 31 | 96 | 68 | 125 | 89 | 163 | 116 | 42 |
| 4000 psi | #3 | 40 | 12 | 12 | 8 | 12 | 12 | 12 | 12 | 14 | 12 | 12 |
| | | 60 | 13 | 12 | 8 | 13 | 12 | 16 | 12 | 21 | 15 | 12 |
| | #4 | 40 | 12 | 12 | 8 | 12 | 12 | 15 | 12 | 19 | 14 | 12 |
| | | 60 | 17 | 12 | 9 | 17 | 12 | 22 | 16 | 29 | 20 | 15 |
| | #5 | 40 | 14 | 12 | 8 | 14 | 12 | 18 | 13 | 24 | 17 | 13 |
| | | 60 | 21 | 15 | 12 | 21 | 15 | 27 | 20 | 36 | 26 | 19 |
| | #6 | 40 | 17 | 12 | 9 | 17 | 12 | 22 | 16 | 29 | 20 | 15 |
| | | 60 | 25 | 18 | 14 | 25 | 18 | 33 | 23 | 43 | 31 | 23 |
| | #7 | 40 | 21 | 15 | 11 | 21 | 15 | 28 | 20 | 36 | 26 | 18 |
| | | 60 | 32 | 23 | 17 | 32 | 23 | 42 | 30 | 54 | 39 | 26 |
| | #8 | 40 | 28 | 20 | 13 | 28 | 20 | 36 | 26 | 47 | 34 | 20 |
| | | 60 | 42 | 30 | 19 | 42 | 30 | 54 | 39 | 71 | 51 | 30 |
| | #9 | 40 | 35 | 25 | 14 | 35 | 25 | 46 | 33 | 60 | 43 | 23 |
| | | 60 | 53 | 38 | 21 | 53 | 38 | 69 | 49 | 90 | 64 | 34 |
| | #10 | 40 | 45 | 32 | 16 | 45 | 32 | 58 | 42 | 76 | 54 | 25 |
| | | 60 | 67 | 48 | 24 | 67 | 48 | 87 | 62 | 114 | 82 | 38 |
| | #11 | 40 | 55 | 40 | 18 | 55 | 40 | 72 | 51 | 94 | 67 | 28 |
| | | 60 | 83 | 59 | 27 | 83 | 59 | 108 | 77 | 141 | 101 | 42 |
| 5000 psi | #3 | 40 | 12 | 12 | 8 | 12 | 12 | 12 | 12 | 14 | 12 | 12 |
| | | 60 | 13 | 12 | 8 | 13 | 12 | 16 | 12 | 21 | 15 | 12 |
| | #4 | 40 | 12 | 12 | 8 | 12 | 12 | 15 | 12 | 19 | 14 | 12 |
| | | 60 | 17 | 12 | 9 | 17 | 12 | 22 | 16 | 29 | 20 | 15 |
| | #5 | 40 | 14 | 12 | 8 | 14 | 12 | 18 | 13 | 24 | 17 | 13 |
| | | 60 | 21 | 15 | 11 | 21 | 15 | 27 | 20 | 36 | 26 | 19 |
| | #6 | 40 | 17 | 12 | 9 | 17 | 12 | 22 | 16 | 29 | 20 | 15 |
| | | 60 | 25 | 18 | 14 | 25 | 18 | 33 | 23 | 43 | 31 | 23 |
| | #7 | 40 | 20 | 14 | 11 | 20 | 14 | 25 | 18 | 33 | 24 | 18 |
| | | 60 | 29 | 21 | 16 | 29 | 21 | 38 | 27 | 50 | 36 | 26 |
| | #8 | 40 | 25 | 18 | 12 | 25 | 18 | 32 | 23 | 42 | 30 | 20 |
| | | 60 | 37 | 27 | 18 | 37 | 27 | 49 | 35 | 63 | 45 | 30 |
| | #9 | 40 | 32 | 23 | 14 | 32 | 23 | 41 | 29 | 54 | 38 | 23 |
| | | 60 | 47 | 34 | 20 | 47 | 34 | 62 | 44 | 81 | 58 | 34 |
| | #10 | 40 | 40 | 29 | 15 | 40 | 29 | 52 | 37 | 68 | 49 | 25 |
| | | 60 | 60 | 43 | 23 | 60 | 43 | 78 | 56 | 102 | 73 | 38 |
| | #11 | 40 | 49 | 35 | 17 | 49 | 35 | 64 | 46 | 84 | 60 | 28 |
| | | 60 | 74 | 53 | 25 | 74 | 53 | 96 | 69 | 126 | 90 | 42 |

# Index


## A

Accelerators, 11
Accuracy of computations, 21
ACI alternate design method. (*see* working stress method) 17
ACI Code, general description, reference to, 16, 19
ACI strength method. (*see* strength method) 17
Additives, 11
Aggregates, 1, 2
Air entrainment, 11
Allowable soil pressure, 220
Allowable stress, 18, 42, 253
Alternate method. (*see* working stress method) 17
Anchorage:
- flexural reinforcement, 134
- shear reinforcement, 144

Areas of reinforcing bars, 13, 34, 254, 255
Aspdin, Joseph, 2
ASTM bar designation, 33
Axial load, on:
- beams, tension or compression, 46, 69
- columns, 153, 157
- pedestals, 158

Axial stress, allowable, 253

## B

Balanced design in beams, 65
Bar cutoff points:
- in regions of negative moment, 130, 140–42
- in regions of positive moment, 134, 140–42

Bar dimensions and weights, 13, 34
Bar placement tolerances, 6, 32
Bars, reinforcing:
- areas by spacing, 86, 254
- areas of combinations of sizes, 255
- areas per bar, 34, 254
- ASTM designation, 33
- bends in, 96, 138, 155
- bond on, 14, 133, 147
- bundled, 147
- concrete cover over, 24, 139
- cutoff points, 134, 140–42
- deformed, 12, 24
- imperial sizes, 35
- lap splices, 137, 264
- longitudinal, 23
- metric sizes, 34
- selection of, 43
- smooth, 13
- stirrups, 96

stresses in, 253
UNESCO recommended sizes, 34
weights of, 34
Bar sizes, 13, 34
Beams:
bond development in, 133
with compressive reinforcement, 41
continuous, 183–89
critical moment locations, 192, 223, 236
critical shear locations, 98, 223, 224, 236
deflections of, 25–27
flexure in, 38
and girders, distinction between, 28
investigation of, 53, 79
shear in, 91
shear reinforcement, 94
sizes, guidelines and rules of thumb, 27
web, 97
Bearing capacity, 32, 224, 253
Bending, of longitudinal reinforcement, 96, 138, 155
Blocks, clay tile filler, 124
Bond, strength of concrete, 14, 133
Bond stress:
allowable values for, 150
deformed bars, 134
elastic analysis for, 150
smooth bars, 134
Building code, 16
Building frames, 183, 189–95

C

Cantilever walls, 28, 29
Capacity reduction factor $\phi$, 6, 32
Cement, Portland:
action of, 2
types of, 3
Cement-water-sand matrix, 2
Clay tile fillers, 124
Clear span with ACI coefficients, 192
Codes, 16, 19, 33
Coefficient of thermal expansion, 9
Columns:
ACI formula for, 158
axial loads, 153, 158
balanced condition, 163
buckling, 157, 159
capacity of minimum section, 155, 172
circular, 154
clear cover, 24
concentrically loaded, 158
design loads on, 193, 194
effective length, 160
elastic analysis, 174
Euler formula for, 156
failure mode, 156
flexure on, 158, 160
interior, unbalanced moment, 193, 195
lateral ties, 155
length effects, 157, 159
loads on, rules of thumb, 22, 154
long, 159
minimum dimensions, 22, 154
minimum eccentricity, 161
moments on, 157–66
practices in design, 154–56
rectangular, 154
shear on, 156
sizes of, 154–56
ties, 155
ultimate strength analysis, 160–66
unbraced length, 160
working loads on, 154, 195
Combined flexure and axial load:
on beams, 46–52
on columns, 158–66
on pedestals, 158
Comparison, strength method to working stress, 18, 83, 179–80
Compression reinforcement in beams, 41
Compression splices, 138
Compressive strength, standard cylinder, 3
Computations, accuracy of, 21
Concentric loads:
on columns, 158
on footings, 219
Concrete:
additives, 11
age-strength relations, 4
aggregates for, 1–2
compressive strength test for, 3
cover over reinforcement, 24
cracking in, 94, 246
creep in, 10
density of, 6
durability, 5–6
joists, 108, 123
lightweight, 6
matrix, 2
modulus of elasticity for, 9
normal weight, 6
paste, 2
quality control for, 5
related properties, 11
shrinkage in, 10

stress-strain relations in, 8
tensile strength in, 1, 91, 161
thermal coefficient for, 9
weight, 6
Continuity in beams and frames, 183–92
Continuous slabs, 27, 183, 245
Cracking:
control joints, 246
diagonal, 92–94
shear, 94
Creep in concrete:
effect of stress levels on, 10
effects on reinforcement, 26, 40, 174
equivalent strains, 10, 40
Critical sections:
on beams, 98, 134, 192
on footings, 223, 236
on punching shear, 224
Cylinder strengh test, 4

## D

Dead load:
approximate values, 22
factors to ultimate load, 33
Deflection:
ACI Code limits on, 25–26
analysis for, 18, 27
control of, 26
due to creep, shrinkage, cracking, 10, 95
long term, 10, 26, 177
short term, 10, 25
time-dependent, 26, 177, 179
Deformations, elastic, 25–27, 40
Deformations, inelastic, 10, 16, 62, 160
Derivations of design tables:
beams, 42, 65
bond development, 136, 138, 139, 146
columns, 166
Design:
ACI methods, 16
alternate method, 17
practices, 20, 154
strength method, 17
Design for axial load, 153
Design for flexure:
elastic range, 38
ultimate load, 62
Design for flexure plus axial load:
elastic range, 46
ultimate load, 69
Design loads, 33
Development length:
ACI Code requirements, 140
of bundled bars, 147
in compression, 136
of splices, 137
of standard hooks, 138
in tension, 136
Diagonal tension, 92
Diagrams, shear and moment, 184–89
Differential settlement, 31, 220
Dimensions, minimum, 22
Dowels, to footing, 228–29, 239

## E

Eccentricity on columns, 157, 160
Effective column length, 157, 160
Effective depth, 23
Effective flange width, 23, 109, 113
Elastic analysis, 17, 39, 147, 174
Elastic buckling in columns, 156
Elastic continuity analysis, 183, 189
Elastic deformations, 25–27
Elastic frames, 189
Elasticity in concrete, 1, 7
Ell beams, 109, 113
Embedment length, 133
End anchorage, 147–50
Envelopes, 183–89
Euler formula, 156

## F

Factored loads, 32
Flexural capacity, 17, 38, 62
Flexural stiffness, 195
Flexure:
in beams, 38, 62
in columns, 158
elastic, 39, 174
inelastic, 62, 160
Flexure formula:
elastic, 38
ultimate, 64–65
Flexure-shear cracking, 94
Floor slabs, 85
Floor systems:
concrete joist, 123
one-way slabs, 85
slab-beam-girder, 109
two-way joist system, 130
Footing rotations, elastic, 221

Footings:
grade beams, 239
spread, 222
strip, 235
types, 219
Foundation settlement, 220
Frame analysis, 189–93
Frames, braced, 172, 189
Frames, unbraced, 159

G

Girders and beams, definition, 28
Grade beams, 219, 239

H

Hollow columns, 154
Homogeneity, 1
Hooks, standard, 138
Hydration, of cement, 3

I

Inclined cracking, 93–94
Inertia, moment of, 42
Inflection points, 135, 208
Interaction diagrams, 179
Investigation of elastic flexure, 53
Investigation of ultimate flexure, 79

J

Joints:
in beams, 31
in columns, 155
construction, 31, 246
control, 246
temperature and shrinkage, 29, 246
Joists:
bridging, 124
dead load of, 127
one-way, 123
pan system, 125
shear in, 126
two-way, 130

L

Lateral ties, 155
Length effects, columns, 159
Live load factors, 33
Load factors, 33
Load for maximum moment, 189, 192
Load for maximum shear, 189, 192
Longitudinal reinforcement, 23, 155, 240
Long-term loading, 26

M

Mechanical connectors, 147
Methods of design, 17
Metric bar sizes, 34
Minimum dimensions, 22
Mixing, placing, curing, 6
Modulus of elasticity, 9
Moment coefficients, ACI, 192
Moment diagrams, maxima 190
Moment envelopes, 134, 184
Moment of inertia, 42
Moment magnification, 160
Moments:
coefficients, ACI, 192
column, 158
design for, 39
negative, 159
positive, 159
service, 177, 180
ultimate, 17
working, 17

N

Negative moments, 159
Neutral axis, 40, 62, 92, 94
Nodes in column buckling, 157
Nominal ultimate axial force, 153
Nominal ultimate moment, 64, 66
Nominal ultimate shear force 97
Nominal ultimate shear stress, 97–99

O

Overload, 32

P

Pans, joist system, 125
Pedestals, 158
Plastic flow, 10
Plasticizers, 12
Plastic section modulus, 64
Points of inflection, 135
Portland cement, 3
Positive moment, max. and min., 188
Positive moment reinforcement, 114

R

Rectangular sections:
  compressive reinforcement for, 41
  design of, 42
  investigation of, 53, 79
Reduction factor $\phi$, 32
Reinforcement:
  cover, minimum 24
  cutoff points for, 134, 140, 146
  deformed, 12–14
  development of 133
  grades of, 12, 33
  maximum ratio for, 22, 156
  minimum ratio for, 22, 155, 172
  modulus of elasticity, 13, 253
  spacing, 256
  standard hooks for, 138, 263
  temperature and shrinkage, 23, 24
  transverse, 23
  yield stress, 13, 33
Reinforcing bars:
  imperial sizes, 35
  metric sizes, 34
Retaining walls, 28
Retardants, 12
Roof slabs, 85
Rules of thumb, 27

S

Safety factor, 33
Secant modulus, 9
Service load, 18
Service stress, 17
Settlement, foundation, 31, 220
Shallow foundations, types, 218
Shear:
  in beams, 91
  factored, 32, 33
  inclined, 94
Shear and diagonal tension, 92
Shear diagrams, maxima, 189
Shear envelopes, 183–89
Shear reinforcement:
  bent-up bars, 96
  minimum area for, 99
  patterns of, 96
  stirrups, 96–99
Shear strength:
  strength method, 97
  strength reduction factor $\phi$, 32
Shear-tension cracking, 94
Short-term loading, 25
Shrinkage:
  prediction of, 10
  rules of thumb, 29
Sizes, minimum 22
Slab-beam-girder system, 109
Slab floors:
  cover over reinforcement, 24
  design of, 85
  rules of thumb for thickness, 28
  shear reinforcement in, 86
  thickness calculations, 87
Soil behavior, 220–22
Soil pressure, allowable, 220
Spacing of reinforcement, 256
Spread footings, 222
Steel:
  deformed and plain, 12
  properties of, 13, 33
  standard sizes, 34–35
  stress, allowable, 253
  types of, for reinforcement, 13
Stirrups:
  development of, 144
  effects of, 95
Strain, 7, 13, 40, 62, 161
Strength, of concrete, 3, 36
Strength method, 17, 62, 97, 160
Strength reduction factor $\phi$, 32
Strength-time curve, 4
Stress-strain curves:
  concrete, 8–9
  steel, 13
Super plasticizers, 12
Sustained load, 10, 26, 221

## T


Tables:
- allowable stresses, 253
- approximate service stresses in steel, 84
- areas of bars, 254–55
- bar sizes, UNESCO, 34
- bar sizes, U.S./Canadian, 34, 35
- coefficients for section constants, 257–59
- dead load of joist systems, 127
- design guidelines, 256
- development lengths of hooked bars, 263
- development lengths of stirrups and ties, 263
- development length of straight bars, 264
- properties of individual bars, 254
- steel areas for combinations of bars, 255
- stress ratios for columns, 260–62

Tee sections:
- approximate analysis, 121
- joists, 123
- negative moment on, 109
- neutral axis, location, 110, 112
- positive moment on, 110, 114
- shear in, 108, 116, 126

Temperature effects, 9, 29
Tensile strength, of concrete, 1
Tension lap splices, 137
Thermal expansion, coefficient of, 9
Ties, lateral 155
Time-dependent deformations, 26
Tolerances, 6, 32
Torsion, 21


## U


Ultimate compressive strain, 62, 161
Ultimate moment capacity, 62
Ultimate strength method, 17, 62, 97, 160
Undercapacity factor, 32
U-stirrups, multiple 96


## V


Vapor barrier, 246


## W


Waffle slabs, 130
Water, for concrete mix, 3
Water-cement ratio, 6
Water reducing agents, 11
Web reinforcement, 97
Weight, of concrete, 6
Working stress method:
- bond development in, 147
- column analysis in, 174
- flexural analysis in, 39
- shear analysis in, 104


## Y


Yield stress:
- concrete, idealized, 9
- steel, idealized, 13

1. Columns
2. Foundations
3. Beams/Girders/Tie's.